Notes for Introductory Statistics and Probability

K. M. Brown *(with contributions by C. P. Gregory and Y. Feinman)*

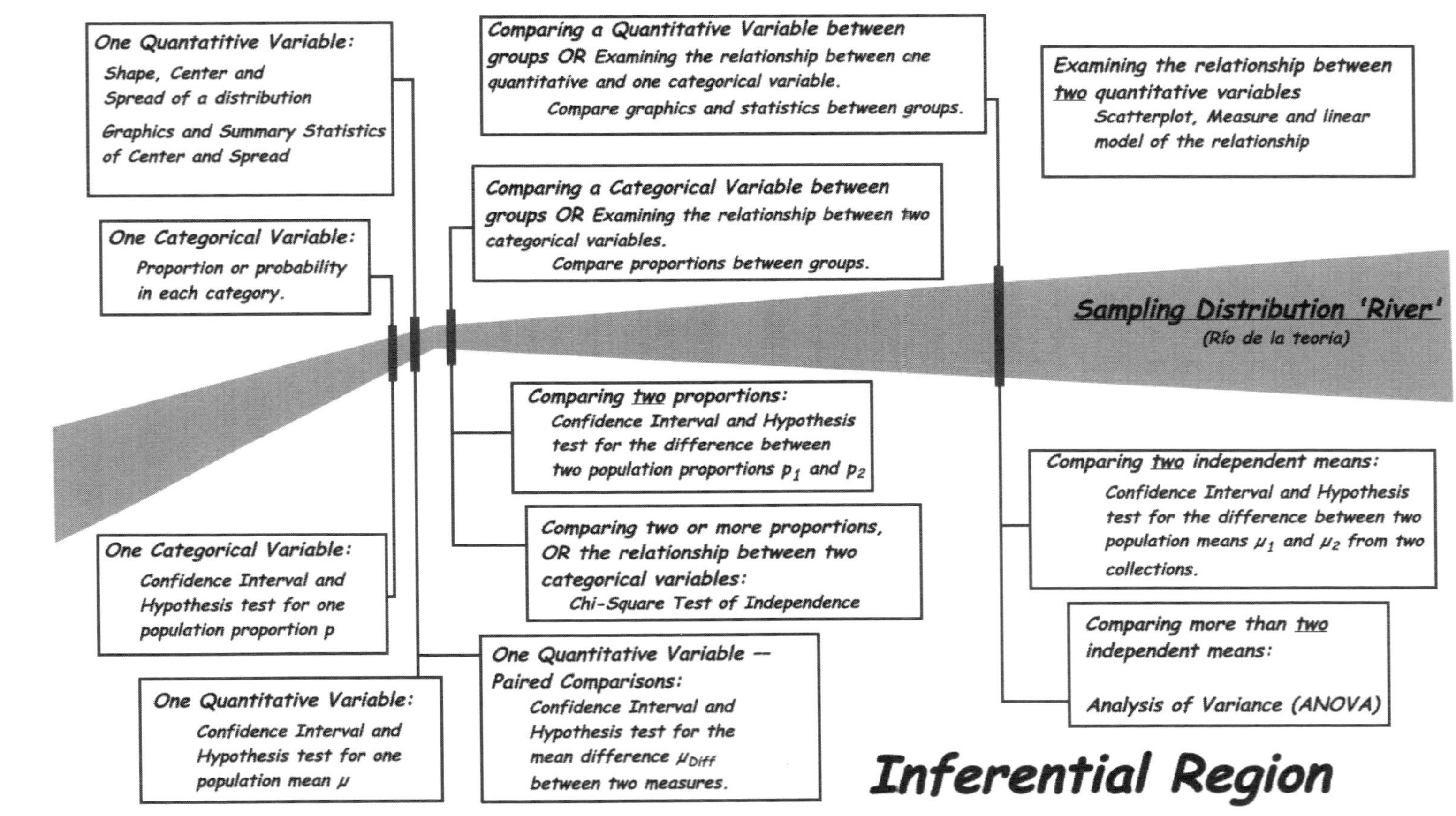
Descriptive Region
One Quantatitive Variable:
Shape, Center and
Spread of a distribution
Graphics and Summary Statistics
of Center and Spread
One Categorical Variable:
Proportion or probability
in each category.
Comparing a Quantitative Variable between groups OR Examining the relationship between one quantitative and one categorical variable.
Compare graphics and statistics between groups.
Comparing a Categorical Variable between groups OR Examining the relationship between two categorical variables.
Compare proportions between groups.
Examining the relationship between two quantitative variables
Scatterplot, Measure and linear model of the relationship
Sampling Distribution 'River'
(Río de la teoria)
Comparing two proportions:
Confidence Interval and Hypothesis test for the difference between two population proportions p_1 and p_2
Comparing two or more proportions, OR the relationship between two categorical variables:
Chi-Square Test of Independence
One Categorical Variable:
Confidence Interval and Hypothesis test for one population proportion p
One Quantitative Variable:
Confidence Interval and Hypothesis test for one population mean μ
One Quantitative Variable — Paired Comparisons:
Confidence Interval and Hypothesis test for the mean difference μ_{Diff} between two measures.
Comparing two independent means:
Confidence Interval and Hypothesis test for the difference between two population means μ_1 and μ_2 from two collections.
Comparing more than two independent means:
Analysis of Variance (ANOVA)
Inferential Region

Table of Contents

Unit 1: Describing distributions

Unit 2: Describing relationships between variables

Unit 3: Beyond description: trusting data

Unit 4: Inference for Categorical Variables

Unit 5: Inference for Quantitative Variables

Preface: Some Questions about the Course in Statistics

Why is there a sea star on the cover?

Well, it's a nice picture. But it is also because statistics is about everything – sea stars, muscles and anemones included, which is probably why your university system has statistics as a requirement for your program of study. Even without the sea star, think of all of the information (we call it data) that is collected from people when they use social media, shop on-line, even talk. Everywhere you turn, there are data --- masses of data. *The purpose of statistics is to make sense of these data, to see if the data answer questions, to see if the data tell a story.* Tons of data just sitting there do not do it, do not show what patterns or relationships there may be lurking there. Making pictures (a better term is graphics) is a good start. A good graphic may show patterns otherwise hidden, and we now have good software to make the graphics easily. But that is jumping ahead just a bit.

Is statistics hard?

A much better and more useful question is to ask what kinds of challenges you are likely to encounter (how the course is "hard"). Here is a list of them:

- There is a body of technical terminology that must be mastered – a new language, in a sense. Statisticians use what look like common words, but give those words a technical sense connected that deviates somewhat from the "normal" usage. An example is the word "random."
- Statistics uses mathematical symbolism, including some symbols that may be new to you. Having said that, being able to get the answers using formulas is *definitely not* the most important thing, as it may have been in previous mathematics courses. We have software to do calculation; we only need to know what the software is doing, which is harder. Still, the advice is: do not fear symbols; indeed, embrace symbolism!
- What *is* important, and harder is to be able to say what the numbers that are calculated and the graphics that are produced *mean* in the light of questions that we are asking. We call this *interpretation.* This text is focused on interpretation; it gives examples in the ***Notes*** section, and the ***Exercises*** give practice.
- Finally, expect some abstract ideas, and some logic that may take days – perhaps even weeks – to understand. These difficult ideas and logic come after some weeks into the course, and have to do with the thorny question of whether we can actually trust the data that are collected to represent reality faithfully.

What will we be doing?

The core of the course is in the ***Exercises***. They are designed to lead you through how statisticians analyze data (make sense of data) by actually analyzing data. The idea is that one learns by actually using the techniques that statistician use with data. The ***Notes*** provide backup with definitions of technical terms, and explanations. Probably a good strategy is to read through the ***Notes*** for a particular section before doing the exercises, without worrying too much if you do not understand it all at first reading; the questions in the ***Exercises*** will probably address some of the possible confusions.

Practical Requirements for the Course

- ***Online Resources*** It is likely that your college or university, as a part of its course management system, has a site devoted to the course that you are taking; that site should be your first recourse for information. Specifically, there may be online exercises and quizzes. Otherwise, the website for these materials is found at: http://www.cleonestats.blogspot.com/.
- ***Data Sets***: You will be analyzing data – lots of data. The data sets are ***not*** included with the software application. Rather, they will be found either at the course site at your college or university's course management system, or at http://www.cleonestats.blogspot.com/

For Instructors

§1.1 Starting Out

First Steps

Statistics is for asking and answering questions about ***collections.*** More specifically, statistics asks and answers questions about collections of ***data***. There can be collections of people, collections of places, such as towns or villages, collections of times, collections of animals, collections of good things to eat (apples, pears, apricots, etc.). Collections can be just about anything you can name.

Here is an example: here are data on a collection of students in four statistics classes. The data are probably similar to some data you may collect on the first day of class, and they are shown in spreadsheet format.

Combined ClassData Aut 09

	Gender	AgeYears	Height	Mother...	Poltical...	Langua...	Number...	Number...	Number...	Instruct...	Tattoo	Domina...
1	F	26	165.0	35	Liberal	2	6	7	2	60	Y	Right
2	M	19	166.0	27	Moderate	1	5	5	12	55	N	Right
3	F	19	154.0	32	Moderate	1	4	3	6	52	N	Right
4	F	19	165.0	24	Moderate	2	5	6	6	60	N	Right
5	M	22	175.3	32	Moderate	2	4	4	6	62	N	Right
6	M	19	187.0	35	Moderate	1	3	3	9	54	N	Right

Terminology: Cases, Variables (Quantitative and Categorical), Values

We need some terminology to navigate our way through what we see here. We refer to the objects in a collection as ***cases***. For this collection, the cases are students taking statistics. In the spreadsheet format, each row is a different case. Here, each row (a row runs horizontally) represents a different student. It is very important to be clear in your mind about the cases for any collection; otherwise, you will not know what you are talking about.

The numbers or the words that you see in the spreadsheet shown above are *data*; other words that could be used for this general term *data* are *measurements* or *observations*. The *data* for the very first case—or student—in the collection above shows that the student is female, twenty-six years old, 165 centimeters tall, whose mother was thirty-five when she was born, etc. More importantly, for the cases in a collection, ***data*** are collected on ***variables.*** A ***variable*** measures some aspect of each case, and that aspect potentially varies from case to case. The height of a student measured in centimeters (*Height* in the spreadsheet) is one example of a variable; gender is another example of a variable, and the number of languages a student speaks (*Languages* in the spreadsheet) is another. The *age* of the student is yet another variable. Notice that in spreadsheet format, the variables are the columns (a column runs vertically).

The ***values*** of a variable are the possibilities for that variable. It is obvious that the values for the variable *Gender* are male and female. For the variable *Age* the values can theoretically be any number, but for college students we do not expect to see any ages less than about fourteen, and we expect to see very few above fifty or so. That is, we can say that values of the variable *age* that are below fourteen and above fifty are not very *likely*. On the other hand, what *is likely* is that we will find students who have ages between nineteen and twenty-five. So, if we were asked to list the possible values for student ages, to be completely safe, we could write something like: "10, 11, 12...89, 90" with the understanding that the three dots (the ellipses) show that any number between twelve and eighty-nine can occur.

Or, we could write "any age greater than 10 and less than 90." If we want to list the *likely values,* we may write something like "16, 17, 18, 19, 20…48, 49, 50."

What about the variable *height*? If the cases are students, and we are measuring in metric units, we do not expect to see many cases with values less than about 120 centimeters (about three feet, eleven inches) or greater than 220 centimeters (about seven feet, two inches). These values for short and tall people are not *likely,* although these values are possible. For possible values, we could write "any number between 120 cm and 220 cm." If our collection were elementary school students rather than college students then we would expect to see a different range of heights.

Variables can be either ***quantitative*** so that the values are measured with numbers, or they can be ***categorical*** where the values are essentially not numerical. Obviously *height* and *age* are both quantitative variables, whereas *gender* and whether or not a student has a tattoo (the variable *Tattoo* in the spreadsheet) are categorical variables. Categorical variables may have more than two categories; the variable *PoliticalView* has the categories "liberal," "moderate," and "conservative." The distinction between quantitative and categorical variables is very important because what we do with quantitative variables is often different from what we do with categorical variables.

Types of Variables A ***quantitative variable*** has values that are *numerical* whereas a ***categorical variable*** has values that are essentially *non-numerical.*

Statistical Questions

We said that statistics is for answering questions about collections of data. But what kinds of questions do we typically ask? We will call the kinds of questions we ask ***statistical questions.*** Here are some examples of statistical questions, using the data above about statistics students.

1. Are male students or female students more likely to have a tattoo?
2. On average, are students in an evening section of a statistics course older than students in a day section of statistics? (We think of comparing two collections—the day students with the evening students.)
3. What percentage of students speaks just one language? What percentage speaks two languages? What percentage speaks three languages? Four? Five?
4. It makes sense that the more people there are in a household, the more cars there will be in that household. But is it true that for each additional person in a household there is an additional car? Probably not. So, can we say how many additional cars there will be in a household if another person is added?

We could go on with additional examples of statistical questions just for these data. One of the most important goals for this course is that you learn to use statistics to answer statistical questions and be able to express the answer to the questions in language that can be understood. At this point, however, notice two things about our four example statistical questions.

Some of the questions involve a comparison between parts of the collection—sub-collections—or a comparison between two collections. The first question compares the male and the female students (comparing sub-collections), and the second question compares the day students and the evening students (two collections.) Both situations involve the same thinking and procedures.

Secondly, all of these questions are about the *collection* and not about the *individual* students, who are the cases in this collection. We typically do *not* ask, "What is the age of the oldest stats student?" Or "Who got the highest test score, and what was that score?" Or "What is the biggest household that any student lives in?"

That we do not often ask questions about the oldest or the biggest—questions about extremes—may come as a surprise because questions involving the biggest or smallest or greatest or least are the first kinds of questions that occur to some people. The reason that the questions about extremes come to mind is probably because we are interested in who is best or who is greatest, especially in the sports or entertainment world ("Which player scored the most goals?" "Which movie is the most successful this summer?"). Likewise, in the consumer economy, we are concerned with getting the best deal. What you will encounter in this course largely ignores these questions about extremes or questions about single cases; rather, we will be much more interested in describing the collection as a whole or in comparing parts of a collection. Or we will be interested in looking at the relationship between variables in the collection. (The fourth statistical question in our list above is about the relationship between variables. If you cannot see how you would approach this question, do not worry; we will get to it.)

So, in using statistics, we are typically not very interested in the individual cases. In particular, we are typically not interested in the extreme cases. We are interested rather in describing the collection as a whole. You may in fact come to have philosophical objections to statistics' concern with collections at the expense of the individual. Fair enough. However, describing and generalizing about collections is what statistics is about.

How do we answer statistical questions? Here are some examples.

Answers to Statistical Questions: Example 1

Our first statistical question was: *Are male students or female students more likely to have a tattoo?*

How can we answer this question? We have to have some data. Each student was asked whether or not he or she had a tattoo. So, as a first step, we do some *calculation.* We count the number of male students who have tattoos and the number of female who have tattoos, and, in fact, in one of the exercises you will do that first for some data collected in 2008. Calculation is the first step, but we will see it is not the only step. We will illustrate this and the following steps using similar data from Pennsylvania State University.

At Pennsylvania State University, statistics students were also asked whether they had a tattoo or not. Here are the counts. The counts of students who have and did not have tattoos are shown in a Two-Way Table here. So eighteen female students had tattoos and thirteen male students had tattoos. Does that mean that that the females are more *likely* to have a tattoo? No!

PennState2

		Tattoo		Row Summary
		No	Yes	
Sex	Female	119	18	137
	Male	55	13	68
Column Summary		174	31	205

S1 = count ( )

Notice that in these statistics classes there were more females than males; there were 137 females and sixty-eight males. There are different numbers (or counts) of male and female students, but that does *not* stop us from answering our question. We can still answer our question even though the total counts of males and females are different. We use the idea of a ***proportion*** or ***percentage***.

A proportion looks at a count in a specific category *relative to* a total number. A proportion is a fraction. To get a proportion we divide our number of "successes" (here, females with tattoos) by the total number of females. If we want percentages, we will multiply these proportions by one hundred. Hence, for the females we calculate $\frac{18}{137} \cdot 100 \approx 13.14\%$. After ***calculation***, we ***interpret*** our calculations. We can say that 13.14% of the females in the Penn State collection have tattoos. Another way to talk about this is that out of one hundred female students, we expect about thirteen of them to have tattoos. Or we can say that the proportion of female students who have a tattoo is 0.1314. Or (as we shall learn later) we can say that the probability that a female has a tattoo is .1314 or 13.14%.

What about the males? For the males we calculate $\frac{13}{68} \cdot 100 \approx 19.12\%$. What can we say? We can say that 19.12% of the male students have tattoos. Or we can say that out of one hundred males, we expect about nineteen of them will have tattoos. Or we can say that the proportion of males that have tattoos is 0.1912. Or we can say that the probability that a male will have a tattoo at Penn State is 0.1912 or 19.12%

Proportion and Percentage

A ***proportion*** relates a count in a specific category to a relevant total count by dividing a specific count by the total count. Proportions must be between zero and one.

A ***percentage*** expresses a proportion as a quantity out of one hundred, so a proportion multiplied by one hundred gives the same fraction as a percentage. Percentages must be between zero and one hundred.

Our answer? From our calculation and our interpretation of the numbers, we can say that at Penn State, males are more likely than females to have a tattoo.

Answers to Statistical Questions: Calculation, Interpretation, and Truth

Notice that the answer to our statistical question involved two things. ***calculation*** and ***interpretation***. The *calculation* was to divide the number of females who had tattoos by the total number of females (and multiply by one hundred) and to divide the number of males with tattoos by the total number of males (and multiply by one hundred). But we do not stop with calculation.

The next step is ***interpretation***. *Interpretation* requires saying what the results of the calculation *mean* in the context of our original question. All of the sentences above that have "we can say" can be part of an interpretation. In the end, notice that it takes a number of sentences saying something about the numbers so that we can get to the answer to our question. Our conclusion is that at Penn State, males were more likely to have a tattoo than females, at least when these data were collected. To say this, we also have to say how the calculations back this up.

Answers to ***statistical questions*** always involve both some *calculation* (sometimes much, sometimes little) and some translating the calculations into language to answer the question. An answer to a statistical question is not complete without *interpretation*. You have been warned!

Two more issues. First, when we interpret, how far can we generalize? Will what we have found for the Penn State students necessarily apply to students in California, for example? This is a huge question and will be one of the major themes of the course, especially the part entitled ***inferential statistics.*** In that part we learn when and how we can generalize or, in statistical language, *infer*.

The second issue has to do with the truthfulness of the Penn State students. Suppose some of them are not telling the truth and are saying they have a tattoo when they do not have one or *vice versa.* If some of them are not telling the truth, we are in trouble. Our calculations and our interpretations are only as good as our data, and lying makes bad data. Lying is just one way that data may be made bad.

Answers to Statistical Questions: Example 2

Our third statistical question was:

> *What percentage of students speaks just one language? What percentage speaks two languages? What percentage speaks three languages? Four? Five?*

Once again, as an illustration, we will show a similar question with a similar collection and allow you to answer the third statistical question from some data that has been collected. As a part of the Census At School project, students in secondary schools in Australia were asked how many people lived "at home." These data show the number of people in the student's household, including the student. So, our question is:

> *What percentage of students lives in households of two people? What percentage lives in households with three people? What percentage lives in households of four?*

What should we do with the data? What calculations should we make? Households with zero people do not make sense (vacant houses make sense but not vacant households), but we can have households with values of 1, 2, 3, 4, 5, etc. people. Hence, a good way to proceed will be to count the number of cases where the variable *NumberHousehold* takes on the values 1, 2, 3, 4, 5, etc. It will be convenient to make a table to hold the results of this calculation. How far should we go? Looking over the data reveals that there is one student with thirty people in her household and two students who reported nine people in their home. Those are the biggest household sizes. Let us make a table with cells for what we think we see. Then we laboriously make a tally, going through the data and recording every single instance of 1, 2, 3, etc.

Number Household	**1**	**2**	**3**	**4**	**5**	**6**	**7**	**8**	**9**	**30**	**Total**
Number of cases	|	~~||||~~ ||	~~||||~~ |||	Etc.	Etc.				||	|	

Below are the numbers that we will see after quite a bit of work. (Software will do the tedious work.)

Number Household	**1**	**2**	**3**	**4**	**5**	**6**	**7**	**8**	**9**	**30**	**Total**
Counts	1	10	63	142	112	49	12	6	2	1	398

Once again, what we really want for comparison are proportions or percentages. If we add all the numbers of cases, we see the collection contains 398 cases. To get the percentage of students who live in households of one, two, three, etc. people, we divide each of the counts by the total count of 398 and multiply by one hundred. Here is what we get.

Number Household	1	2	3	4	5	6	7	8	9	30	Total
Counts:	1	10	63	142	112	49	12	6	2	1	398
Percent of cases:	0.25%	2.51%	15.83%	35.68%	28.14%	12.31%	3.02%	1.51%	0.50%	0.25%	100.00%

If you add the percentages, we should get 100% since our tally includes *all* the students in the collection. (Sometimes the sum will be very, very close to 100% but not exactly 100% because we round to, say, two decimal places; for this particular calculation, we do in fact get 100%.)

That is the *calculation;* but what about the *interpretation?* What do the calculations mean?

Perhaps the first thing that draws your attention is the student whose household has thirty people. We call such data ***outliers*** because the data point lies outside the main body of the data. The figure of 30 could be a mistake (it could be "3") or perhaps this student lives in some kind of commune or perhaps she was not taking the questionnaire seriously. Statisticians are typically not so much interested in the extremes as in the main body of data. Statisticians do pay attention to outliers, but they are not usually the main focus of attention.

What can we say? One thing is that the great majority of secondary students live in households that have between three and six people (about 92%), and a second thing is that over 60% live in households that have either four or five people. A relatively smaller proportion of Australian students live in households of more than six people; the number is about 18%.

Notice that these (60%, 18%, etc.) figures do not really have meaning until we compare them with something else. Would we get the same kinds of numbers for secondary level students in North America rather than in Australia? (My guess is yes, but we would have to see.) For students in Western Europe? For students in West Africa? Southeast Asia? The Indian subcontinent? Suppose we could compare these numbers with the sizes of households two centuries ago? Lesson: *interpretation* involves comparison, either between collections or within a collection (are there differences for different parts of Australia?), or to a standard. Interpretation is based upon comparison.

Embracing Symbols—and the Meaning of "n"

Why do we have symbols? For fast recognition—at least that is what the commercial world thinks. Students often see symbols as a burden, something extra to learn. No! Learn symbols—indeed, embrace symbols–as a way to make a mental map of statistics. Embrace symbols as another connector relating what are otherwise entirely disconnected things. Here is your first symbol, n, and its definition.

Notation for the number of cases in a collection

n stands for the number of cases in a collection.

For the Penn State collection of students, $n = 205$, and for the Australian students, $n = 398$.

Summary to Starting Out: Terminology and Perspective

The main idea of this section is to introduce essential and basic *terminology* and essential *perspective* in order to be able to continue. Here is a summary of the terminology introduced above and some advice: good students seek to understand how these terms relate to each other.

Terminology:

- ***Cases*** The objects (people, places, things, or times) about which we have *data* in a *collection*
- ***Data*** The measurements or observations on *variables* for the *cases* in a *collection*
- ***Collection*** The sum total of the data for all the cases together. Collections may be divided into sub-collections, where the data for the sub-collections are compared (e.g., comparing males and females).
- ***Variable*** An aspect or characteristic of *cases* in a *collection* whose *value* potentially varies from case to case and which can be measured numerically (a *quantitative variable*) or by categories (a *categorical variable*)
    - ***Attribute*** Another name for variable
- ***Value of a variable*** The numbers or categories that are logically possible for a *variable*
- ***Quantitative variable*** A variable whose *values* are expressed as numbers or numerically
- ***Categorical variable*** A variable whose *values* are expressed non-numerically
- ***Statistical Question*** A question about a *collection* of *cases* that can be answered by doing *calculations* on *data* and then *interpreting* the results of the calculation to answer the question
- ***Notation*** Symbols that stand for quantities or operations that are part of statistical calculations

Perspective:

- The goal of statistics is to answer questions about data—that is, about measurements on variables for cases in a collection or in several collections.
- The focus of statistical analysis is to say something about collections and not necessarily about specific members or elements of the collection.
- The language of *probability* pervades statistical thinking and analysis, and this language is discussed in the next section. Statistical analysis deals in *likelihoods* ("How likely is it that a student from Australia lives in a household of five or more people?") rather than in certainties.
- Statistics always involves both ***calculation*** and ***interpretation***.

§1.2 The Language of Statistics: Probability

Smuggled In

Our first statistical question (introduced in the last section) was:

Are male students or female students more likely to have a tattoo?

We answered this question in Example 1 for the Penn State students by calculating and interpreting the proportions of male and female students who had tattoos. For the females we calculated that $\frac{18}{137} \bullet 100 \approx 13.14\%$ had tattoos, whereas for the males we calculated that $\frac{13}{68} \bullet 100 \approx 0.1912 = 19.12\%$ had tattoos. We concluded that *for the Penn State students,* males are more likely to have a tattoo than females.

PennState2

		Tattoo		Row Summary
		No	Yes	
Sex	Female	119	18	137
	Male	55	13	68
Column Summary		174	31	205

S1 = count( )

In our statistical question we actually smuggled in the idea of probability by using the word "likely." We probably have some notion of how the word "likely" is used. One meaning that we will give to the word "probability" (or to the word "likely") is this: "If we chose a student completely at random from all the females on the Penn State campus, the probability that she would have a tattoo is 0.1314 or 13.14%." Or if we chose a student completely at random from all the male students at Penn State, the probability that he would have a tattoo is 0.1912 = 19.12%. Now what does "completely at random" mean? If students walked around the Penn State campus completely randomly then the chance that you—in your own random wandering—would meet a student (either male or female) with a tattoo would be $\frac{31}{205} \approx 0.1512$, or about 15.12%, since altogether there were thirty-one students out of 205 that had a tattoo. But students do not walk the campus completely at random; perhaps the ones with tattoos tend to be together and those without tattoos together. Our picture is flawed because no one actually walks around a campus completely randomly. But our definition of randomness demands just that.

Here is a better picture of randomness. If we were able to put all the Penn State students in a giant bin (like the lottery bins) and draw one out randomly then the probability that the student chosen would have a tattoo is about 15.12%. The ideas that we have from everyday life about probability, chance, and likelihood can carry us a certain distance, but we need specific terminology and notation to use probability without getting into trouble.

Language and Notation

We will begin with the example of students having (or not having) a tattoo. The first bit of terminology that we use is the idea of an ***event.*** An event can be nearly anything (not just something for which you buy tickets), as long as we can define it well. In our first example, we can define two events: "having a tattoo" and "not having a tattoo." In accord with what we see in the table, we will assign:

Y to be defined as the event: "Student has a tattoo"

N to be defined as the event: "Student does not have a tattoo"

Then, to express the idea of probability, we will use the notation $P(Y)=\frac{31}{205}=0.1512$. This is read: "the probability of Y is 0.1512." Or, in more expanded form, we say: "The probability that a student has a tattoo is 0.1512." Notice that this *P(Y)* is similar to the function notation you learned in algebra: *f(x)* is read "*f* of *x*". Notice also that the calculation does not belong inside the parentheses. What goes inside the parentheses is the name of the event, often designated by a letter. Remember: embrace notation!

Events Can Also Refer to Numbers

In the last section we asked these statistical questions about Australian students:

What percentage of students lives in households of two people? What percentage lives in households with three people? What percentage lives in households of four? (The number of people in the household includes the student.)

We can now express these questions in probability terms as using the language of events:

What is the probability that an Australian student lives in a household of two people?

What is the probability that an Australian student lives in a household of three people?

What is the probability that an Australian student lives in a household of four people?

How do we use the probability notation to express the answers? Since the questions refer to different numbers, what we will usually do is to let a capital letter stand for the number, in this case the number of people in a household. So, in this example, we would have:

***X* = 2** to be defined as the event: "The Australian student lives in a household of two people."

***X* = 3** to be defined as the event: "The Australian student lives in a household of three people."

Then, if we have the data that were shown in Section 1.1, we would write: $P(X=3)=\frac{63}{398}=0.1583$ and read: "The probability that an Australian student (in this collection) lives in a household of three people is 0.1583." (Recall that the total number of students was n = 398.)

Number Household	1	2	3	4	5	6	7	8	9	30	Total
Counts:	1	10	63	142	112	49	12	6	2	1	398
Percent of cases:	0.25%	2.51%	15.83%	35.68%	28.14%	12.31%	3.02%	1.51%	0.50%	0.25%	100.00%

Notice that in the notation $P(X=3)=\frac{63}{398}=0.1583$ there is an "=" sign *inside* the parentheses and also *outside* the parentheses in this notation. The equals sign inside the parentheses refers to the event "household of three people," and the second equals sign is actually the verb "is" in our interpretation: "The probability that an Australian student (in this collection) lives in a household of three people *is* 63/398 which *is* 0.1583."

Example of Interpretation: The answer to our statistical question "*What is the probability that an Australian student in our collection lives in a household of four people?*" would be calculated (as before) as the proportion 142/398 and expressed as $P(X=4)=\frac{142}{398}=0.3568$, which is read, "The probability that an Australian student (in this collection) lives in a household of four people *is* 0.3568, or 35.68%"

Probabilities are proportions. You may well be more comfortable thinking in terms of percentages rather than in proportions. Get used to thinking in proportions; in essence, they are the same as percentages since percentages are just proportions multiplied by one hundred. We can see that probabilities are proportions in the way we calculate probabilities; look at our examples:

Tattoo example: $$P(Y) = \frac{\textit{Number of students who have a tattoo}}{\textit{Total number of students}} = \frac{31}{205} = 0.1512$$

Students in households of size four:

$$P(X = 4) = \frac{\textit{Number of students who live in households of size 4}}{\textit{Total number of students}} = \frac{142}{398} = 0.3568$$

That probabilities are proportions means that the smallest a probability can be is zero, and the largest a probability can be is 1, and these two ends of the interval of possible probabilities can be given meanings. If $P(E) = 0$ then the event E did not happen, and if $P(E) = 1$ then the event E was certain to happen. If we choose a student completely at random and we knew that $P(Y) = 0$ then we would know that no student had a tattoo, and if $P(Y) = 1$ we would know that *every* student had a tattoo. The box just below summarizes the terminology and notation about probability.

Definition and Notation for the Probability of an Event E:

We express the *probability* that event E happens using the notation and formula:

$$P(E) = \frac{\text{Number of Cases where Event } E \text{ is true}}{\text{Total number of Cases}}$$

$$0 \le P(E) \le 1$$

Sample Space: It is useful (but sometimes hard) to think of *all* the possibilities in a specific situation where we are applying probability. When we do this, we speak of the ***sample space S*** of all possible events. We often use braces $\{\cdots\}$ to indicate the elements in a sample space. In our tattoo example the student can either have a tattoo or not have a tattoo, so there were just two possibilities. This would be written in symbols as $S = \{N, Y\}$. For the example of the households for the Australian students, the sample space can be written $S = \{1, 2, 3, \ldots\}$ where the dots indicate that the numbers continue. Before we look at the data, we consider all the events that logically, and from our experience with similar data, could happen, even if we think the likelihood is very small. It is extremely unlikely that an Australian student lives in a household of 265 people, but it is possible, so we include this number in our sample space. It is not possible that the household size is a negative number, so we do not include negative numbers or numbers such as 3.658. In the context, only positive integers make sense.

The elements that we include in the sample space must also be what we call ***mutually exclusive.*** By mutually exclusive we mean that no two events in the list of events can happen at the same time. A student either has a tattoo or a student does not have a tattoo, so having and not having a tattoo *are* mutually exclusive. A student (at any particular point in time) lives in a particular size household; so the events $X = 4$ and $X = 5$ should not happen at the same time for the same student.

The various sizes of households are mutually exclusive (we may have to be very careful how we ask the question about size of household).

One of our statistical questions was about whether males or female are more likely to have a tattoo. It would be quite *wrong* to list the sample space for this question as $S = \{N, Y, M, F\}$ because obviously Y (has a tattoo) can happen at the same time as F (female), and so the events Y and F are *not* mutually exclusive. We shall see how we handle the situation of *non*-mutually exclusive events in the next section.

The elements of a sample space must also be ***exhaustive,*** which simply means that the sample space includes *all* the possibilities for our application of probability. It is for that reason that we listed the sample space for the number of people in a student's household as $S = \{1, 2, 3, \ldots\}$. We have listed everything that is logically possible.

> ***Sample space***
>
> A ***sample space*** is a complete listing (therefore, an ***exhaustive*** listing) of all the ***mutually exclusive*** events (therefore, events that cannot occur together) that are possible when applying probability language. The notation commonly used for sample space is $S = \{\ \cdots\ \}$ where the events are listed in the braces.

Applying Probability to Our Questions. The application of our probability notation and calculation to the questions about the sizes of households for students is straightforward. Our statistical questions were:

What is the probability that an Australian student lives in a household of two people?

What is the probability that an Australian student lives in a household of three people?

What is the probability that an Australian student lives in a household of four people?

The calculated answers, in probability notation, and the written interpretations are:

$P(X = 2) = \dfrac{10}{398} \approx 0.0251$ — The probability that we find an Australian high school student in a two-person household *is* (or equals) 2.51%.

$P(X = 3) = \dfrac{63}{398} \approx 0.1583$ — The probability that we find an Australian high school student in a three-person household *is* (or equals) 15.83%.

$P(X = 4) = \dfrac{142}{398} \approx 0.3568$ — The probability that we find an Australian high school student in a four-person household *is* (or equals) 35.68%.

However, our first statistical question about gender and tattoos is a bit more complicated because it involves two variables for each case (each student): the student's gender and whether or not the student has a tattoo—and we have already seen that the events associated with gender and the events associated with whether a student has a tattoo are not mutually exclusive. The questions about the size of households really involved just one variable measured for each student—namely, the size of the student's household, even though there were many different possible *values* for the variable. So, we need ways of using probabilities with events that are not mutually exclusive.

Let us return to our first statistical question and the table of data that goes with it, shown below.

Are male students or female students more likely to have a tattoo?

We have answered this question before by calculating the proportion for the females, $\frac{18}{137} \approx 0.1314$ and comparing this proportion with the proportion we calculated for the males, $\frac{13}{68} \approx 0.1912$. How do we apply the probability notation to this question?

Since there are two variables for each student, we must notice that there are four possible events in this case, the same number as the number of interior cells in the table. These four possible events are:

"*F* and *N*" 119 students Female and does *not* have a tattoo.
"*F* and *Y*" 18 students Female and *does* have a tattoo.
"*M* and *N*" 55 students Male and does *not* have a tattoo.
"*M* and *Y*" 13 students Male and *does* have a tattoo.

PennState2

		Tattoo		Row Summary
		No	Yes	
Sex	Female	119	18	137
	Male	55	13	68
Column Summary		174	31	205

S1 = count( )

These four events are represented by the four cells of the table, and if we made a sample space for this application it would be $S = \{F \text{ and } N, F \text{ and } Y, M \text{ and } N, M \text{ and } Y\}$. Notice that every student can only be in one of these four interior cells (so these four combinations are *mutually exclusive*), and these four cells exhaust the possibilities for this application (so the four combinations are *exhaustive*). So how are we to handle the situation where we have two variables and the events are combinations of what were simple events?

"And," "Or," "If," and "Not" Probability Calculations and Interpretations

There are four different probability calculations that we can use, and for the impatient reader, we will reveal now that the one we want for our statistical question ("Are male students or female students more likely to have a tattoo?") is the "***if***" calculation. However, the other calculations are also very important.

And (Intersection) In some ways, the simplest calculation is the one that calculates the probability that a student "occupies" one of the four interior cells of the table. For example, the probability that a student is *both* female *and does* have a tattoo is $P(F \text{ and } Y) = \frac{18}{205} \approx 0.088 = 8.8\%$, and the probability that a student is both a male and *does* have a tattoo is $P(M \text{ and } Y) = \frac{13}{205} \approx 0.063 = 6.3\%$.

If we compare these two probabilities, we see that the one for the females is larger, and this tells us that, for this collection, the likelihood of finding (or choosing at random) a tattooed female is greater than the likelihood of finding a tattooed male. However, this "and" calculation does *not* answer our question of whether the probability of having a tattoo is greater for males or females. In fact, this calculation confuses the issue because there are actually two probabilities at play. One probability is the likelihood that a student is a female rather than a male in this collection (remember we did not have equal numbers of males and females). The second probability is the likelihood of having a tattoo, which may be different for males and females. Useful as the calculation of "cell" probabilities is (the "and" calculation), the calculation does not answer our statistical question about the relative likelihood of a tattoo among male and female students.

An "and" probability is commonly referred to as the ***intersection*** of the two events, and you can see from the table relating gender and tattoos that this word makes some sense because the cell containing the eighteen tattooed females is the intersection of the "female" row and the "tattoo yes" column. The calculation of the probability of an ***intersection*** ("and") is given in the box below.

Intersection is also a good word to remember to guard against a common error: the first thing that comes to mind for some people with the word "and" is the word "addition." Banish the thought! "And" or intersection refers to the number where *both* event A *and* event B have happened.

There is a situation where $P(A \text{ and } B)$ *must* be zero, and that is if the events A and B are mutually exclusive events. It is impossible for a single student to both have and not have a tattoo at a single point in time, so $P(Y \text{ and } N)=0$. Similarly, we know that $P(X=4 \text{ and } X=5)=0$ because the events of household sizes of four and five are mutually exclusive. Our term *mutually exclusive* is reserved for events where it is impossible for the events to occur at the same time. However, if we found that $P(F \text{ and } Y)=0$ for some collection of data, we would not conclude that the events F and Y were mutually exclusive; we would just know that there were no tattooed females in that collection.

Definition and Notation for the Probability of the Intersection of Two Events A and B:

$$P(A \text{ and } B)=\frac{\text{Number of cases where event } A \text{ and event } B \text{ are both true}}{\text{Total number of cases}}$$

Interpretation: $P(A \text{ and } B)$ is read: "The probability that both A and B are true is…"

Or (Union) A second kind of probability calculation that we could make with more than one event is designated by the word "or" and is commonly known as ***union.*** The notation that we will use is *P(A or B)*; as an example, we will calculate the probability *P(F or Y)*. Your first inclination may be to think of "either *F* or *Y*". Not so! In probability language and calculations, the word "or" does not have an exclusive meaning; rather, it will always mean "or including and." Hence the event "*F* or *Y*" in our example will include all of the females who also have tattoos (there are eighteen of them) as well as all the females who do not have tattoos (because they are females: there are 119 of them) and all of the males who have tattoos (because they have tattoos: there are thirteen of them). If you are not reluctant to write in your ***Notes,*** it may be a good idea to circle or shade in these numbers in the table. With these numbers, we can do the

PennState2

		Tattoo		Row Summary
		No	Yes	
Sex	Female	119	18	137
	Male	55	13	68
Column Summary		174	31	205

S1 = count( )

calculation: $P(F \text{ or } Y)=\frac{18+119+13}{205}=\frac{150}{205}\approx 0.732=73.2\%$. If we think of our "randomly choosing" interpretation for probability, we can read this as saying that "the probability of randomly choosing a Penn State student who is either female or has a tattoo or is both female and has a tattoo."

There is another way to calculate the probability of the union of two events that is sometimes useful. Notice that if we add 18 + 119 we get the total number of females, which is 137. And if we add 18 + 13 we get the 31 tattooed students. Now we can calculate the probability of choosing a female as $P(F)=\frac{137}{205}$, and we can also calculate the probability of having a tattoo, $P(Y)=\frac{31}{205}\approx 0.1512$.

Now these two added together is $P(F)+P(Y)=\frac{18+119}{205}+\frac{18+13}{205}=\frac{18+119+18+13}{205}$, almost what we have in $P(F\ or\ Y)=\frac{18+119+13}{205}=\frac{150}{205}$ *except* that we in our calculations have added the eighteen tattooed females twice. But these eighteen tattooed females are the number in the intersection *F and Y*. So, we can do the calculation:

$$\begin{aligned}P(F\ or\ Y)&=P(F)+P(Y)-P(F\ and\ Y)\\&=\frac{137}{205}+\frac{31}{205}-\frac{18}{205}\\&=\frac{168}{205}-\frac{18}{205}\\&=\frac{150}{205}\end{aligned}$$

Recall the warning at the end of the last section about thinking that "and" must mean addition, when in fact it does not mean addition. The rule that we have shown above for calculating the probability of a union ("or") of two events is commonly called the ***addition rule*** since it involves a sum. The box below gives the rule in general. Whether you use this rule or simply determine what should be included in the union of two events A and *B* depends on the data that you have at hand.

The probability of a union of events is *not* what we want to answer with our question about the relative likelihood of tattoos between males and females. However, this "or" (union) is useful for other calculations. Think of our household collection of data again.

Number Household	1	2	3	4	5	6	7	8	9	30	Total
Counts:	1	10	63	142	112	49	12	6	2	1	398
Percent of cases:	0.25%	2.51%	15.83%	35.68%	28.14%	12.31%	3.02%	1.51%	0.50%	0.25%	100.00%

We may want to calculate the probability that an Australian student lives in a household of either size four *or* size five (household sizes that we think are fairly common), and we could write $P(X=4$ *or* $X=5)$. We know that $P(X=4)=\frac{142}{398}\approx 0.3568$, and $P(X=5)=\frac{112}{398}\approx 0.2814$, and we know that $P(X=4\ and\ X=5)=0$ because these two events (household size of four and household size of five) are mutually exclusive. We can use the formula we used above and get:

$$\begin{aligned}P(X=4\ or\ X=5)&=P(X=4)+P(X=5)-P(X=4\ and\ X=5)\\&=\frac{142}{398}+\frac{112}{398}-\frac{0}{398}\\&=0.3568+0.2814-0\\&=0.6382\end{aligned}$$

Notice that if we know that two events are *mutually exclusive* then we know that the probability of their intersection must be zero, and we can ignore the intersection probability of the formula. We can extend this idea. Suppose we want to calculate the probability that an Australian student lives in a household that is smaller than four people. We can write this as $P(X<4)$.

With our data, this means that we want to calculate $P(X<4)=P(X=1\ or\ X=2\ or\ X=3)$; household sizes of one or two or three are all that are possible that are less than 4. These events are all *mutually exclusive*, and so we can calculate:

$$\begin{aligned}P(X<4)&=P(X=1\ or\ X=2\ or\ X=3)\\&=P(X=1)+P(X=2)+P(X=3)\\&\approx 0.0025+0.0251+0.1583\\&=0.1859\end{aligned}$$

(We have used the calculations shown in the table above.)

Definition and notation for the probability of the union of two events—addition rule:

$$P(A\ or\ B)=\frac{\text{Number of cases where either event } A \text{ or event } B \text{ or both are true}}{\text{Total Number of cases}}=P(A)+P(B)-P(A\ and\ B)$$

Addition Rule for the Probability of the Union of Two Mutually Exclusive Events:

$$P(A\ or\ B)=P(A)+P(B)$$

Interpretation: $P(A\ or\ B)$ is read: "The probability that either A or b or both A and B are true is..."

If (Conditional) To answer our question about whether males or females are more likely to have a tattoo, we have to use a third (and extremely useful!) type of probability calculation called ***conditional*** probability. Once again, it will be helpful to refer to the table. What we really want to do is to *compare* the probabilities for females and for males, so it makes sense to consider *just* the females and then *just* the males. We have already shown this, but the notation showing that we are doing this is

PennState2

		Tattoo		Row Summary
		No	Yes	
Sex	Female	119	18	137
	Male	55	13	68
Column Summary		174	31	205

S1 = count ( )

$P(Y \mid F)=\frac{18}{137}\approx 0.1314$. The vertical line in the notation indicates that we are just considering the females, and the probability can be interpreted in a number of ways. One interpretation is: "The probability that a student has a tattoo *given that* the student is female is 13.14%." Alternately, we could say, "The probability that a student has a tattoo *if* the student is female is 13.14%." There are other valid expressions in English, but languages are generally not as precise as the mathematical notation (which is one reason we have the notation!). We want to compare $P(Y \mid F)$ with $P(Y \mid M)$.

To calculate $P(Y \mid M)$, we *restrict* our denominator to the males and calculate

$P(Y \mid M)=\frac{13}{68}\approx 0.1912$, to get the probability that a student has a tattoo *if* (or *given that*) the student is a male. We can conclude (as we have done before) that for the Penn State data, males are more *likely* to have a tattoo than females.

Notice that for the calculation $P(Y \mid F)=\frac{18}{137}\approx 0.1314$, the numerator has just the eighteen tattooed females; that is, it has those students who are in the intersection of events F and Y. And in the denominator, we have just those students where event F is true.

We can use these observations to make a rule for calculating conditional probability (see the box below), and we can use these to make a rule using other probabilities. Notice that $P(F \text{ and } Y)=\frac{18}{205}$ and also that $P(F)=\frac{137}{205}$, and so: $P(Y \mid F)=\frac{P(F \text{ and } Y)}{P(F)}=\frac{\frac{18}{205}}{\frac{137}{205}}=\frac{18}{205}\times\frac{205}{137}=\frac{18}{137}\approx 0.1314$. Notice how the fractions work out so that the end of the calculation gives the number of "tattooed females" divided by "the total number of females." This kind of calculation can always be done; it is a rule, although if you have the data, it is often easier just to calculate the conditional probability directly as we did earlier.

Definition and notation for conditional probability:

$$P(A \mid B)=\frac{\text{Number of Cases where event } A \text{ and event } B \text{ are both true}}{\text{Number of cases where event } B \text{ is true}}=\frac{P(A \text{ and } B)}{P(B)}$$

Interpretation: $P(A \mid B)$ is read: "The probability that A is true *given* that B is true is..."

Not Our probability toolkit has one more rule—one that is very useful. As an example, notice that from our table of male and female students who have tattoos, we could calculate the probability that a student does not have a tattoo in two ways. We could calculate $P(N)=\frac{174}{205}\approx 0.8488$, or we could get the same thing by calculating: $P(N)=P(\text{not } Y)=1-P(Y)=1-\frac{31}{205}\approx 1-0.1512=0.8488$. In this example, since there are just two events, "*not Y*" is just "*N*" but the same principle works if we have more than two events. In the last section we worked out $P(x<4)=0.1859$, the probability that an Australian lives in a household with fewer than four people. Now it is easy to get the probability that an Australian student lives in a household with four or more persons. The calculation of this probability is just $P(X\geq 4)=1-P(X<4)\approx 1-0.1859=0.8141$. (Interpretation: the probability that an Australian student has the experience of living with three or more other people is over .80.)

Definition and notation for the probability of "Not Event A"

$$P(\text{not } A)=1-P(A)$$

Interpretation: $P(\text{not } A)$ is read: "The probability that the event '*not A*' is true is..."

This rule works with conditional probabilities, with intersections, and with unions, although of course you have to work out the meaning of the *not* in these situations. Here is an example. We could calculate the probability that a student does not have a tattoo *if* that student is a female by calculating:

$$P(N \mid F)=P(\text{not } Y \mid F)=1-P(Y \mid F)=1-\frac{18}{137}\approx 1-0.1314=0.8686$$

Independence and Conditional Probability

We answered the statistical question "Are male students or female students more likely to have a tattoo?" by calculating and comparing the conditional probabilities $P(Y \mid F) = \frac{18}{137} \approx 0.1314$ and $P(Y \mid M) = \frac{13}{68} \approx 0.1912$, and concluded that for the students in the Penn State collection, men are more likely to have a tattoo. There is another similar and very important use of conditional probability, in which we compare a conditional probability with the probability *without* the condition. In our example, we would compare the conditional probability of having a tattoo *given* that the student is female $P(Y \mid F) = \frac{18}{137} \approx 0.1314$ with the probability that a student (any student, not just the females) has a tattoo, $P(Y) = \frac{31}{205} \approx 0.1512$. Since $0.1314 \neq 0.1512$, we say that the events Y (having a tattoo) and F (being a female student) are ***not independent***. If these two probabilities had come out to be equal, we would say that the two events having a tattoo and being a female student were ***independent.*** The formal definition is:

PennState2

		Tattoo		Row Summary
		No	Yes	
Sex	Female	119	18	137
	Male	55	13	68
Column Summary		174	31	205

S1 = count( )

Independent Events

Two events A and B are ***independent*** if $P(A \mid B) = P(A)$.

Two events A and B are ***not independent*** if $P(A \mid B) \neq P(A)$.

We will see that the idea of independence of events, and, by extension, independence of variables, is important in the second part of the course. Here are a number of comments about the idea.

First, if it is true that $P(A \mid B) = P(A)$ then it will also be true that $P(B \mid A) = P(B)$; the two events can be shown to be independent (or not independent) by using either event as the condition. Often, however, only one of the conditional probabilities actually makes sense.

Secondly, notice that our example used actual data and that we found the events Y and F to be *not independent*. With actual data it is very rare to find strict independence between events; that is, it is very rare to find that $P(A \mid B) = P(A)$. One of the most important uses of the idea of *independence* is in building a model of what the data in a collection would be if two variables (notice the extension beyond events) were independent. Then having built the model, we can compare the data we have to the "ideal" model of independent variables. However, we will wait until later sections to develop this.

The third comment is a kind of warning. It is common to confuse the notions of *mutually exclusive* and *independent*. If two events are *mutually exclusive* then $P(A \text{ and } B) = 0$. If two events are *independent* then $P(A \mid B) = P(A)$. These notions are quite different, and you can show mathematically that if two events are mutually exclusive, the events cannot be independent.

CAS Australia 08 B

		PtsDog		Row Summary
		No	Yes	
Hand	Ambidextrous	13	40	53
	Left handed	16	35	51
	Right handed	171	295	466
Column Summary		200	370	570

S1 = count( )

The fourth comment is also a warning. It is tempting to turn the idea of independence around and argue that *since* we cannot see any connection between events or variables *then* the events or variables *must* be independent. At this point, the independence of events for any given collection of data can only be shown by using the definition $P(A \mid B) = P(A)$ and not by any pre-conceived notions about the events involved. Here is an example.

Our statistical question is: "Are ambidextrous Australian students more likely to have a pet dog?" This sounds like a silly question; that is exactly why it was chosen. It *is* silly; we cannot think of any good reason why ambidextrous Australians would be more likely to have a pet dog. If the two events are "being ambidextrous" (= A) and "having a pet dog" (= D), then we would expect $P(D \mid A) = P(D)$.

Here are some data, and for these data $P(D) = \frac{370}{570} \approx 0.6491$ but $P(D \mid A) = \frac{40}{53} \approx 0.7547$, so the two events are *not* independent. When we get to §4.4, we shall develop a technique to determine whether this calculation really indicates that ambidextrous students are more likely to have a pet than, say, right handed students. Until that point, all we can say is that for our data the events A (being ambidextrous) and D (having a pet dog) are apparently not independent. We have deliberately chosen a silly example; there are other examples where the connection between the events (or variables) may or may not exist. A summary of the section is given on the next page.

Summary: Introduction to the Language of Probability

Probability is the language of statistics. You may regard this section as a kind of first introduction—even a kind of phrase book—to that language.

- ***Probability*** expresses the idea of *likelihood* of an *event* E using a number between 0 (completely unlikely) to 1 (certain), so that $0 \le P(E) \le 1$.
- $P(E)$ is read "the probability of event E..."
- With actual data, probabilities are calculated as fractions and expressed using the notation:

$$P(E) = \frac{\text{Number of Cases where Event } E \text{ is true}}{\text{Total number of Cases}}$$

- ***Sample space*** A complete listing (therefore, an ***exhaustive*** listing) of all the ***mutually exclusive*** events (therefore, events that cannot occur together) that are possible when applying probability language.
- ***Probability of the Intersection of Two Events A and B ("And")*** uses the notation and calculation:

$$P(A \text{ and } B) = \frac{\text{Number of cases where event } A \text{ and event } B \text{ are both true}}{\text{Total number of cases}}$$

- ***Interpretation of an "And" probability***

  $P(A \text{ and } B)$ is read: "The probability that both A and B are true is..."

- ***Probability of the Union of Two Events A and B ("Or")—Addition Rule***

$$P(A \text{ or } B) = \frac{\text{Number of cases where either event } A \text{ or event } B \text{ or both are true}}{\text{Total Number of cases}} = P(A) + P(B) - P(A \text{ and } B)$$

- ***Interpretation of an "Or" probability***

  $P(A \text{ or } B)$ is read: "The probability that either A or b or both A and B are true is..."

- ***Conditional Probability of Two Events A and B ("If")***

$$P(A \mid B) = \frac{\text{Number of Cases where event } A \text{ and event } B \text{ are both true}}{\text{Number of cases where event } B \text{ is true}} = \frac{P(A \text{ and } B)}{P(B)}$$

- ***Interpretation of a conditional probability***

  $P(A \mid B)$ is read: "The probability that A is true *given* that B is true is..."

- ***Probability of "Not Event A"*** $P(\text{not } A) = 1 - P(A)$
- ***Interpretation of a "Not" probability***

  $P(\text{not } A)$ is read: "The probability that the event '*not A*' is true is..."

- ***Independence of two events A and B***

  Two events A and B are ***independent*** if $P(A \mid B) = P(A)$.

  Two events A and B are ***not independent*** if $P(A \mid B) \ne P(A)$.

§1.3 Distributions and their pictures

The Important Idea of a Distribution

In the first section we looked at the basic goal of statistics, namely, to ask and answer questions about collections of data. To answer questions, we need some terminology, so we also looked at the ideas of ***collections, cases, variables,*** and **values.** To answer statistical questions, we saw that we must do both *calculation* and *interpretation,* where interpretation is saying what the calculations mean in terms of the question.

Now we come to what at first glance looks like more terminology; in fact, it is more than terminology. It is an idea that is "deep in the heart of statistics." This very important idea is the idea of a ***distribution.*** Here is a definition:

> ***Definition of a distribution of a variable in a collection***
> A ***distribution*** consists of the number of cases at each value of a variable in a collection.

If your first reaction to this definition is puzzlement, you should know that you are probably more familiar with distributions than you think. If you have been a student for some time, you have probably heard the term "grade distribution." Here is a grade distribution for some statistics classes taught by a professor in Costa Rica.

Grades Stats		
LetterGrade	A	6
	B	47
	C	85
	D	67
	F	28
	Column Summary	233
S1 = count ( )		

What we see fits the definition of a ***distribution*** because it shows the values of a variable ("A," "B," etc.) of the categorical variable *LetterGrade* and the number of cases (students) who earned each letter grade. Compare this sentence with the definition in the box above. We can (and will!) often show the proportions (or percentages) of cases at each value of a variable instead of the counts. When we use proportions we still have a distribution. We have a distribution because we have brought two things together: the values of the variable and the number or the proportion or the percentage at each value.

Combined Class Data Aut 08		
Languages	1	60
	2	54
	3	11
	4	1
	Column Summary	126
S1 = count ( )		

You may see why we call this idea a distribution: we can think of the cases (the students) as being "distributed" into the different grade categories, much like mail is put into pigeonholes. (Think about this carefully, especially if you usually think that it is the grades that are distributed to the students!) Here the students are distributed to the grades.

We use the idea of a distribution for quantitative variables as well as categorical variables. In the exercises for §1.1 you calculated a distribution for a quantitative variable, and shown here is what you should have found. What this shows is that there were sixty students who spoke just one language, fifty-four students who speak two languages, etc. What you got fits the definition of a ***distribution*** because you have the counts of the cases (the number of students) at each value of the variable *Languages.*

Graphics for Distributions: Dot Plots

Here is still another example of a distribution of a quantitative variable. However, this time we have arranged the data so that we can compare two distributions for the variable *NumberHousehold* (= "Number of People in the Household"). One row shows the distribution for a collection of California Statistics students and the other row shows the distribution for the same variable for secondary students in Western Australia. The Summary Table also shows proportions for each household size for the students in each place.

CA and WA Comparison

		NumberHousehold									Row Summary
		1	2	3	4	5	6	7	8	14	
Place	California	4 0.0625	9 0.140625	11 0.171875	21 0.328125	13 0.203125	2 0.03125	3 0.046875	0 0	1 0.015625	64 1
	Western Australia	0 0	2 0.031746	12 0.190476	27 0.428571	16 0.253968	3 0.047619	2 0.031746	1 0.015873	0 0	63 1
	Column Summary	4 0.0314961	11 0.0866142	23 0.181102	48 0.377953	29 0.228346	5 0.0393701	5 0.0393701	1 0.00787402	1 0.00787402	127 1

S1 = count( )
S2 = rowProportion

We are ready (that is, we have done the calculations) to answer this statistical question: *Are there differences in the distributions of household size when we compare students from California with students from Western Australia?*

It appears that a greater percentage of the California college students live in very small households (households of size one or two) compared with the Western Australian secondary students. Perhaps this greater percentage is because the California students are college students and are older so that some of them have set up their own households; they are not all living with their parents as the Australian secondary students are. Using the probability notation of §1.2, we can see that

$P(X<3 \mid C)=\dfrac{13}{64}\approx 0.2031$ is bigger than $P(X<3 \mid WA)=\dfrac{2}{63}\approx 0.0317$.

We can compare the numbers, and we should. However, oftentimes it is helpful to depict a distribution with a graphic. The simplest graphic we can make is called a ***dot plot***. Here is an example for the California students' data.

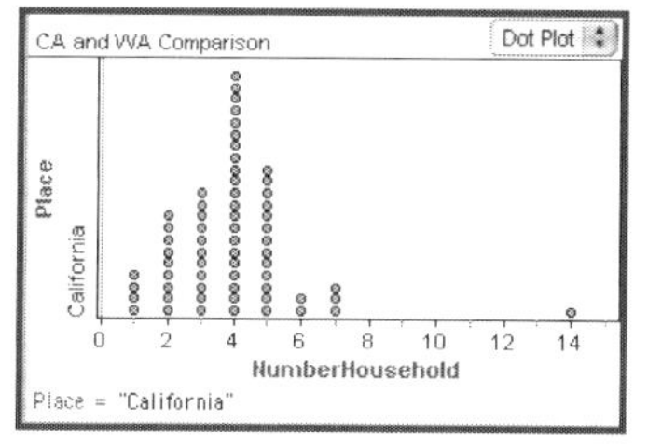

You can make a dot plot (by hand) by marking the values of the variable at equally spaced intervals on the horizontal axis (the x axis) and then stacking the dots (representing the cases) vertically for each value of the variable. Notice that a dot plot nicely shows our definition of a distribution in that a dot plot shows the number of cases at each value if you count the dots carefully.

Since we are comparing, it is often helpful to make a double (or triple, or more!) dot plot to show the comparison between two or more distributions. Here is a comparison for the household size data for the California and Western Australian students. It is fairly easy to see from the graphic that the Western Australia distribution for household sizes is just a bit to the right of the distribution for the California students.

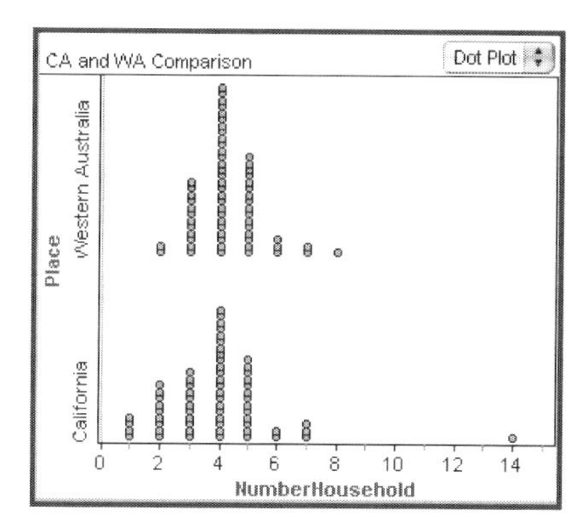

A dot plot is just one kind of graphic for a distribution. We will look at more graphics for distributions, and we will let software do the construction, usually.

Back of the Envelope Graphics: Stemplots

The idea of the ***stemplot*** is to organize—usually by hand—a collection of data so that it is easy to depict a distribution of a variable. Here is an example. We want to look at the distribution of the variable *Age* for just the twenty students who had tattoos in the ***CombinedClassDataSpr08.*** Here (on the right) are the data.

CombinedClassdDataSpr08

	Gender	Age
1	F	19.0
2	M	18.0
3	M	19.0
4	F	20.0
5	F	20.0
6	M	20.0
7	M	19.0
8	F	21.0
9	F	20.0
10	M	45.0
11	F	30.0
12	F	20.0
13	M	19.0
14	F	20.0
15	F	21.0
16	F	20.0
17	M	19.0
18	M	21.0
19	F	34.0
20	F	17.0

Tattoo = "y"

Age	17	18	19	20	21	22	Etc..	33	34	Etc	Total
Number of cases	|	|	卌	卌||	Etc.	Etc..	Etc..		|	Etc.	

We could do this, but there is a better way if we are working by hand. For each take each age, say 34, we consider the number in two parts: $3 \mid 4$.

Mathematically, the "3"is the tens digit, and the "4" is the units digit. The tens digit we will call the *stem* and the units digit we will call the *leaf.* Then we make a vertical list of what we think are all the possible *stems*: in this example, we have students in their teens, in their twenties, in their thirties, etc., perhaps even one or two who are as old as fifty-something. We would make a list of stems (shown in the first step) with a bar on the right of the numbers.

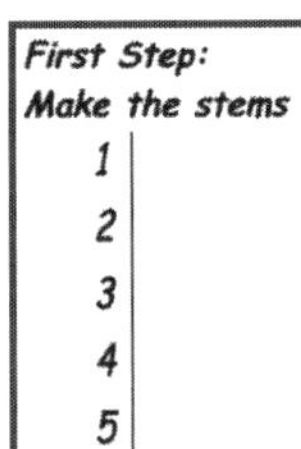

The second step (shown below) is to enter the *leaves* to the right of the bar but in the correct place. For the first six of the ages of the students with tattoos, the second step is shown below. Then the third step is to put the leaves in order. This can be done by re-arranging the leaves as they have been entered. Some people prefer to order the data first or to rank order the data as they are making the stemplot.

In any case, we will always want stemplots to be *rank ordered*—that is, with the numbers ranked from lowest to highest. The completed ordered stemplot for these data is shown here also.

Second Step:
Enter the leaves

```
1|9 8 9
2|0 0 0
3|
4|
5|
```

Third Step:
Order the leaves

```
1|7 8 9 9 9 9 9
2|0 0 0 0 0 0 0 1 1 1
3|0 4
4|5
5|
```

Notice that the plot we end up with shows the number of cases at each *group* of values. The stemplot is a graphic that shows a distribution with some of the values grouped together. Nonetheless, it is still a distribution, even with the values of the variable grouped together.

Split Stem Stemplots

Stem	Leaves
1	
1	9
2	00113334444
2	555666667788
3	00112222334
3	555667777
4	
4	5
5	
5	6

Male Students' Mothers' Ages

Sometimes it is helpful to split the stems of a stemplot. We have not shown it here, but for the data on *Age*, we could make one line for all those who are twenty to twenty-four (i.e., "20, 21, 22, 23, or 24") and another line for those twenty-five to twenty-nine. We would do the same kind of thing for the ages ten to nineteen, thirty to thirty-nine, forty to forty-nine, etc.

Notice that each of the stem-lines has five possible values assigned to it: we will call these ***stems of five.*** Here is an example of the use of split stems.

One of the variables measured for the ***CombinedClassDataSpr08*** was the age of the student's mother when the student was born. Here is a split stem stemplot for the variable *MothersAge* for just the male students from that collection.

Stem	Leaves
1	9
2	0011
2	333
2	4444555
2	6666677
2	88
3	0011
3	222233
3	4555
3	667777
3	
4	
4	
4	5
4	
4	
5	
5	
5	
5	6
5	

Male Students' Mothers' Ages

Notice also that the stem-lines for 40–44 and for 50–54 are left blank, but the stem-lines are still shown in the stemplot. This shows clearly that there were no mothers aged 40–44 or 50–54, although there was one mother aged forty-five and another fifty-six. The stemplot, like the dot plot, shows gaps in the data.

We could also split the stems in another way; we could have five stem-lines with just two possible values for the variable for each stem-line. We will call these ***stems of two.*** The same distribution is shown here with the *stems of two*. There are no strict rules to decide between the stems with fives values and the stems with two values. It is up to the maker of the stemplot, and both too few and too many stems can conceal important features. What can we do with an ordered stemplot? We can answer questions about distributions. For example:

What percentage of the male students had mothers less than thirty years of age when the student was born?

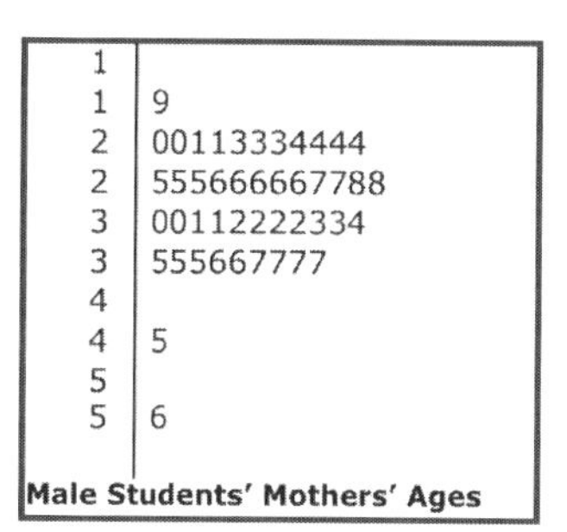

Stem	Leaves
1	
1	9
2	00113334444
2	555666667788
3	00112222334
3	555667777
4	
4	5
5	
5	6

Male Students' Mothers' Ages

Since the data are ordered, we can simply count the number of students whose mothers were under thirty, divide by the total number of students (here there are forty-six students), and multiply by one hundred. This comes out to $\frac{24}{46} \times 100 \approx 52.17\%$. Or, putting this calculation into the probability notation introduced in §1.2, we would have $P(X < 30 \mid M) = \frac{24}{46} \approx 0.5217$ where, in this case, X stands for values of the variable *MothersAge.*

More Bars in More Places: Histograms

A bar is a very common graphic device to show a quantity. Often the bars are vertical; sometimes they are horizontal. But bars are plentiful, graphically and otherwise. When the lengths of bars are used to show the frequencies of a value or a group of values in a distribution, then we have a ***histogram.*** See the graphic on the next page.

To show the idea of a histogram, we have copied the stemplot for the male students' mothers' ages. Now imagine a bar made around each stem-line and imagine the entire graphic rotated ninety degrees, so that the variable is shown on the *horizontal* axis instead of on the vertical axis. Here is what we have, shown just to the right and below:

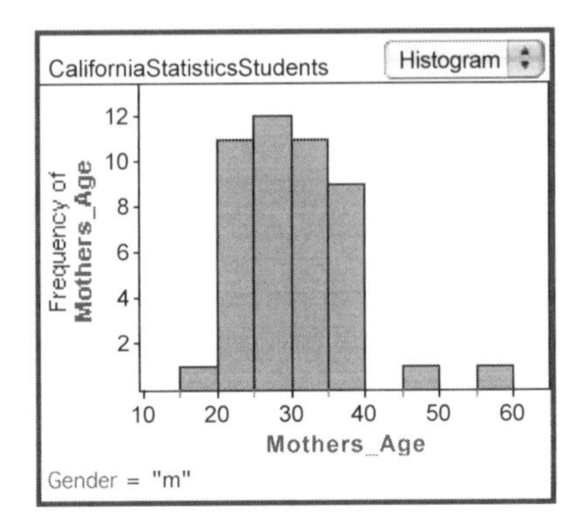

Then, on the *vertical* axis for this histogram, we have put a scale showing the ***frequency*** or ***count*** for each five-year grouping of the mothers' ages. If we look at the first bar we can see that there was just one student whose mother's age was in the interval $15 \leq MothersAge < 20$.
The second bar shows that there were eleven students whose mothers' ages were in the interval $20 \leq MothersAge < 25$.

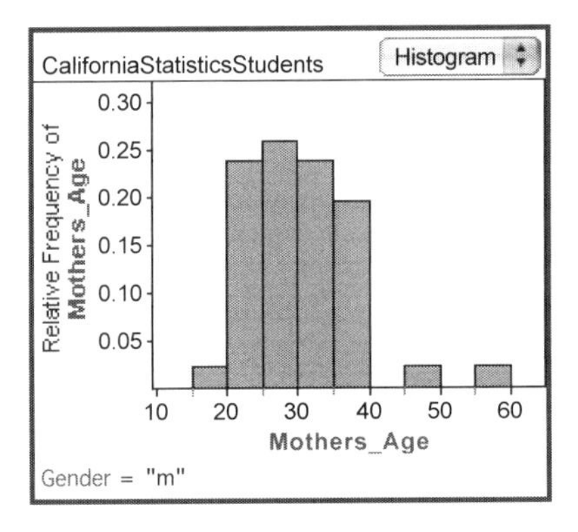

The interval "$20 \leq MothersAge < 25$" means that the only mothers' ages in that bar are the ages that are greater than or equal to twenty but less than twenty-five, and hence not including age twenty-five. To determine that there was one student in the interval $15 \leq MothersAge < 20$ and that there were eleven students in the next interval of the variable $20 \leq MothersAge < 25$, it is helpful to draw a horizontal line along the top of the bar to the scale on the left hand side. With these data (where the n is small), we can be pretty accurate reading from the graphic and getting the numbers; with a histogram from a much larger collection, you may not be able to read the count exactly.

If we want the *percentage* or the *proportion* of students whose mothers were younger than twenty-five, we would first estimate the number of students (as we have done just above) and then divide by the total number of students. Here the total number of male students is $n = 46$, so we get

$\frac{12}{46} \cdot 100 \approx 26.09\%$, in proportions 0.2609, or in probability notation $P(X < 25 \mid M) = \frac{12}{46} \approx 0.2609$.

Most often, histograms are made using software, and it is easy to change the vertical scale from frequencies to ***relative frequencies***, which are the frequencies divided by the total. In other words, the relative frequencies are the proportions. When this is done, you can read off the proportions for each bar.

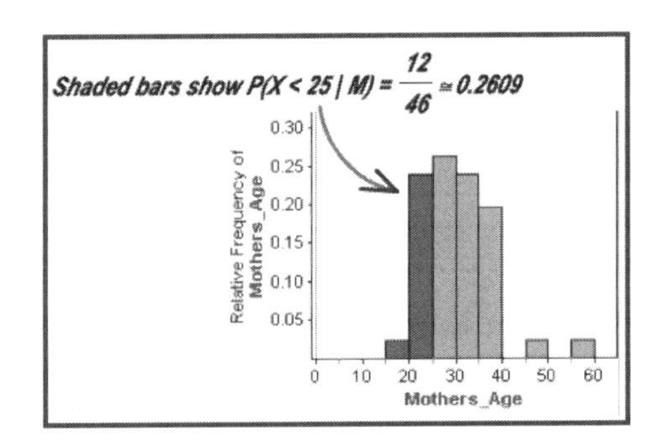

Showing Proportions with Histograms The proportion of students whose mothers were younger than twenty-five (for just the male students) that we calculated, $P(X < 25 \mid M) = \frac{12}{46} \approx 0.2609$ can be displayed on a histogram by shading in the bars for mothers ages less than twenty-five. This is shown in the graphic here. Notice that the proportion 0.2609 is neither on the x-axis nor on the y-axis but rather refers to the *area* in the shaded bars.

Histogram Bin Width The width of the interval $15 \leq MothersAge < 20$ as well as the interval $20 \leq MothersAge < 25$ is five years, as are the widths of all of the other bars in the histograms shown. This width—the width of the bars in a histogram—is known as ***bin width,*** and it can be changed.

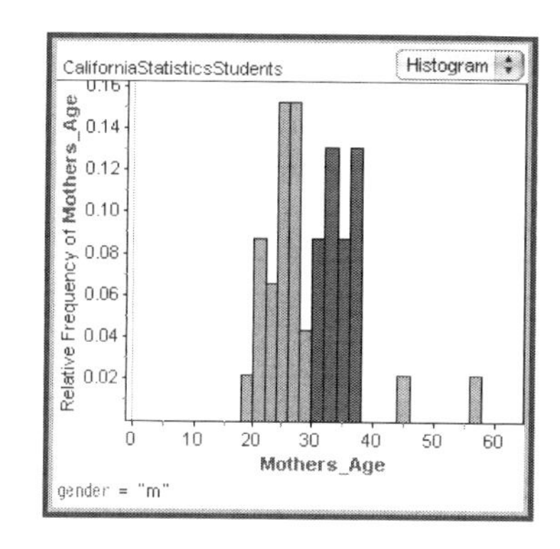

Here is the same distribution of *MothersAge* for the male students shown with bin width of width two. With a bin width of two, the age twenty-five will be in the middle of one of the bars if the bars start at age eighteen, and so we cannot show the proportion of mothers' ages less than twenty-five. What we can show (with this bin width) is the proportion of mothers of students in the interval $30 \leq MothersAge < 38$. Who decides on the *bin width*? If you are the researcher, you get to decide. Choosing too many bars or too few bars may not reveal the features of a distribution well. There is no one right answer as to the correct number of bars.

More Bars in More Places: Other Bar Graphics

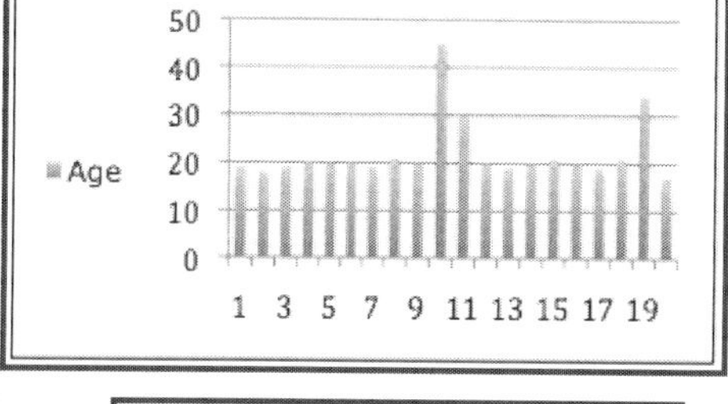

Not every graphic that has bars is a *histogram*. A histogram is a graphic used to show a *distribution*; that is, it shows the values of the variable and the number (or proportion) of cases at each value or group of values. The graphic here (which was made with Excel®) shows the ages for just the $n = 20$ students who have tattoos. This is a bar graph (it has bars), but it is *not* a histogram. It does not show the number or proportion of cases at each value, but rather it shows the *value* for each case. What this graphic is telling us is that the tenth tattooed student has an age of forty-five. You may also see that the bars are separated rather than against one another, as in our examples of histogram. For *quantitative* variables, the bars are drawn (by convention) without spaces between them.

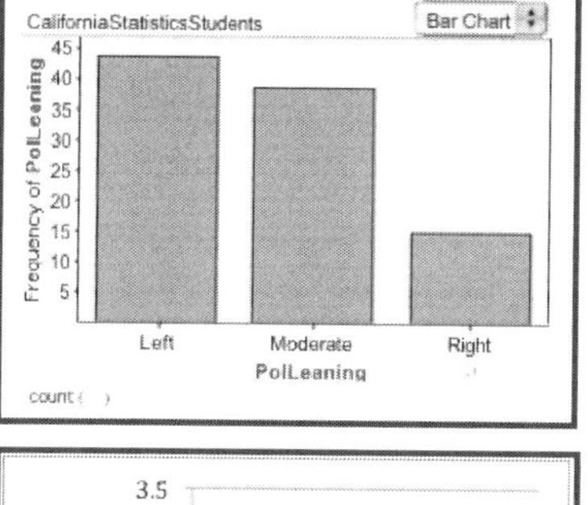

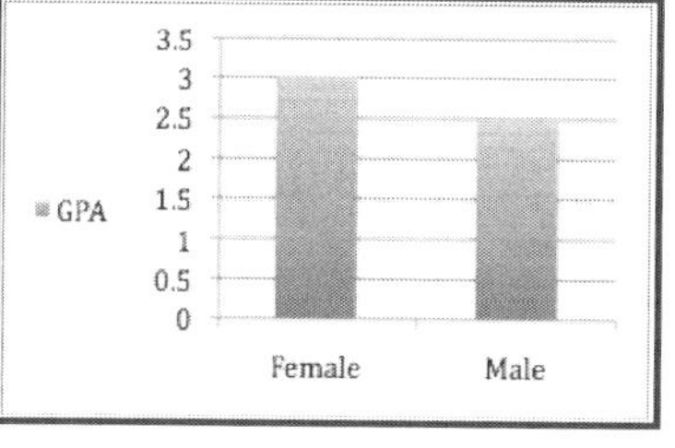

For categorical variables, a bar graph may also show a distribution, and when it does, the bars conventionally *do* have spaces between them. Here is an example that shows the number of students whose political leaning was "left," "moderate," or on the "right."

However, here is a bar graph used with categorical data where the graphic does *not* show a distribution; the variable on the vertical axis is GPA, and the chart shows the average GPA for males and female students. This kind of ***bar chart*** is common, is legitimate, and gives a good graphic presentation of the difference in average GPA between male and female students. However, what it shows is *not* a distribution either of the variable GPA or of the variable gender.

Example: Remember When You Were Thirteen—Interpreting Histograms

In their early teenage years, girls mature more rapidly than boys. You may recall a time in your school years when it seemed that many of the girls were as tall as the boys or taller. Perhaps this happened around age thirteen. Here is a statistical question that comes from this observation. *How do the proportions of students who are between 160 and 180 centimeters tall compare for male and female secondary students who are thirteen years old? Are the proportions similar?*

We can answer this question using the histograms for the male and female students from the Australian secondary students collection, but usually we need to determine from the graphic what the *bin width* is. By inspecting the bars, we see that the bin width is five centimeters, and that allows us to determine the boundaries of the interval 160 cm ≤ Height < 180. Here we have shaded in the interval. Notice that our histogram has relative frequencies so that we can read off the proportions from the vertical scale on the left. For the female students, we read off: 0.28 + 0.23 + 0.19 + 0.02 = 0.72 or in percentages, 72%, or $P(160 \le Height < 180 \mid F) = 0.72$

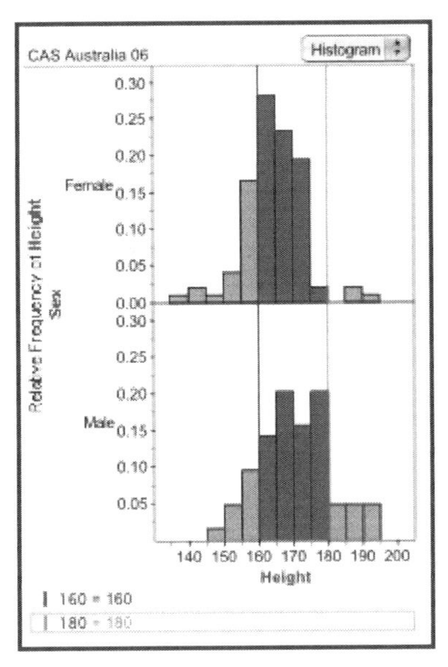

You can confirm that, for the male students, reading off the heights of the bars sums up to about 0.69. We could also work with the actual numbers from the data and find that 75 out of 104 thirteen-year-old girls have heights in the interval 160 cm ≤ Height < 180, and that works out to the proportion $\frac{75}{104} \approx 0.721$. There are 45 out of the 65 thirteen-year-old boys with heights in the interval 160 cm ≤ Height < 180, or $\frac{45}{65} \approx 0.692$. The proportion of thirteen-year-old boys in the "mid-heights" (in the interval 160 cm ≤ Height < 180) is similar to the proportion of thirteen-year-old girls in the "mid-heights," and the histogram exhibits this feature. The sum of the heights of the bars for the heights of the thirteen-year-old boys and girls is similar. It is not the whole story, however, and you will get the chance to add to it in the exercises.

As you know from §1.2, we can interpret these calculations as probabilities: suppose we choose a thirteen-year-old Australian secondary female student completely at random from our collection. What is the *probability* that she is between 160 and 180 centimeters tall? Our answer is: $P(160 \le Height < 180 \mid F) = 0.72$. For the boys, the *chance* or *likelihood* is about the same: we would calculate $P(160 \le Height < 180 \mid M) = 0.69$ and say the probability that a thirteen-year-old Australian male student is at least 160 centimeters but not taller than 180 centimeters is about 69%.

Summary: Distributions and Their Pictures

- ***Distribution of a variable*** The number of cases at each value of a variable in a collection. We may also speak of the proportion of cases at each value or group of values of a variable, and this is also a distribution.
- ***Graphics for Distributions*** Distributions may be displayed by: ***dot plots, stemplots, histograms.***
- ***Dot plots and histograms*** typically have the values of the variable on their horizontal axis, but the interval of values used for the bars of a *histogram* (the **bin width**) may vary.
- ***Dot plots*** typically do not have a scale on their vertical axis, but
- ***Histograms*** may have frequencies (counts), or relative frequencies, expressed as proportions or percentages on their vertical axis.
- Not all ***bar graphs*** are ***histograms.***
- ***Stemplots*** should always display the data in rank order, but the choice of whether the stems are ***stems of one, two, five,*** or ***ten*** is left to the researcher.

§1.4 Shape, Center, Spread of Quantitative Variables

A collection of data on colleges and universities is the subject of this section and our question is: how do colleges and universities differ? To answer this general question, we will look at three features of distributions of *quantitative* variables: ***shape, center,*** and ***spread***.

In the last section, we looked at the important idea of a *distribution* and then we looked at various kinds of graphics to display distributions. You should be familiar with *dot plots, stemplots,* and *histograms,* and we will be using these, as well as the idea of probability covered in §1.2. But in this section, we have data on colleges and universities—that is, the cases are colleges and universities. Many of the variables that we look at, such as the tuition fees for the different colleges, are *quantitative* rather than categorical. For quantitative variables, we can analyze the ***shape,*** the ***center,*** and the ***spread*** of distributions. As we will see, these terms have specific, technical meanings within statistics, and the meanings are not necessarily what you would think they would be.

Shapes of Distributions: Right and Left Skewness

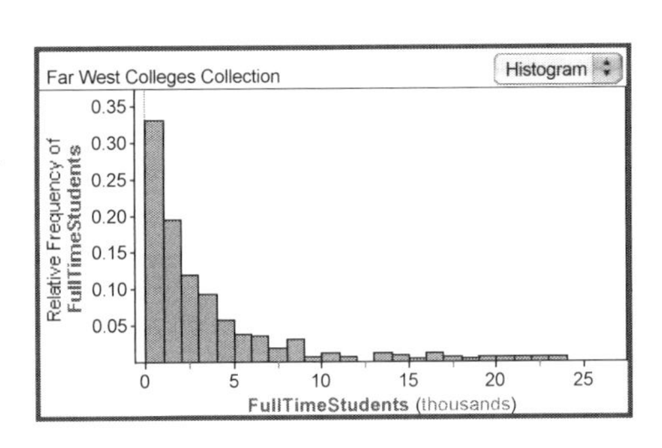

Every distribution has a form that you can see in a graphic; the technical term we use for this form is ***shape.*** We will use three terms to describe the shape of a distribution: ***right-skewed, left-skewed,*** and ***symmetric.*** Here is an example of a ***right-skewed*** distribution. It is called *right-skewed* because there is a "tail" on the right or high side of the values of the variable.

The cases in our collection are colleges and universities, and the histogram tells us that whereas most of the colleges have fewer than 5000 full-time students, there are some institutions—not many but some—where the number of full-time students is much greater, perhaps even over 20,000.

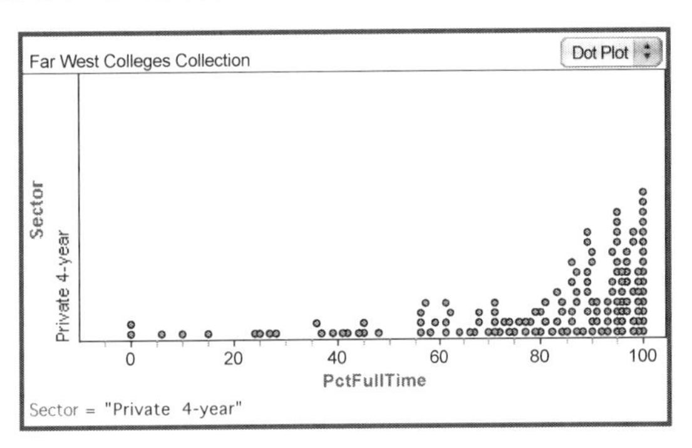

A ***left-skewed*** distribution has the "tail" on the left or low side of the variable. Here is an example of the variable "Percent Full-Time Students" for the private four-year colleges and universities. The graphic shows that most of the private colleges and universities have a relatively high percentage of their students as full-time students, but there are also some (but not many) private colleges or universities that have a much lower percentage of full-time students. (In one example we have used a *histogram* and in another a *dot plot*; either graphic can be used to show the shape of a distribution.)

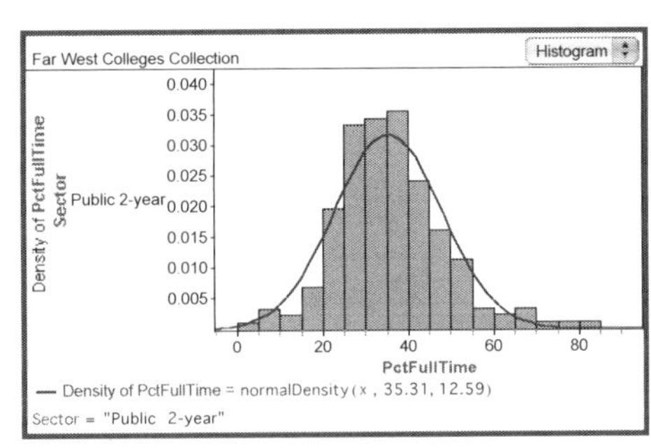

If a distribution is neither *right-* nor *left- skewed* then we may say that the distribution is ***symmetric.*** Distributions of real data seldom are exactly symmetrical, but they may be close to symmetrical. Here is an example of a distribution that is nearly symmetrical.

About half of the two-year colleges have less than 35% of their students full-time, and about half of the two-year colleges have more than 35% of their students full-time. There is no appreciable skew either to the right or to the left for this distribution, although the tail on the right-hand side of the distribution is a bit longer.

The curved line that you see in the graphic is an example of a ***density curve***. For now, think of a *density curve* as a smooth version of the *shape* of a distribution. A density curve can be thought of as a kind of idealization or as a ***model*** for the actual data. Like the histogram itself, the density curve's area includes 100% of the entire distribution; that is, if you calculate its area, the sum of the relative frequencies should be one, or 100%. The density curve that we have drawn for the distribution of *PctFullTime* is one that is perfectly symmetrical. This density curve is the one that we will see in §1.7 and is called the ***Normal Distribution.***

Here (in the box) are three density curves that show left-skewed, right-skewed, and symmetric shapes of distributions and definitions for each of the shapes of the curves.

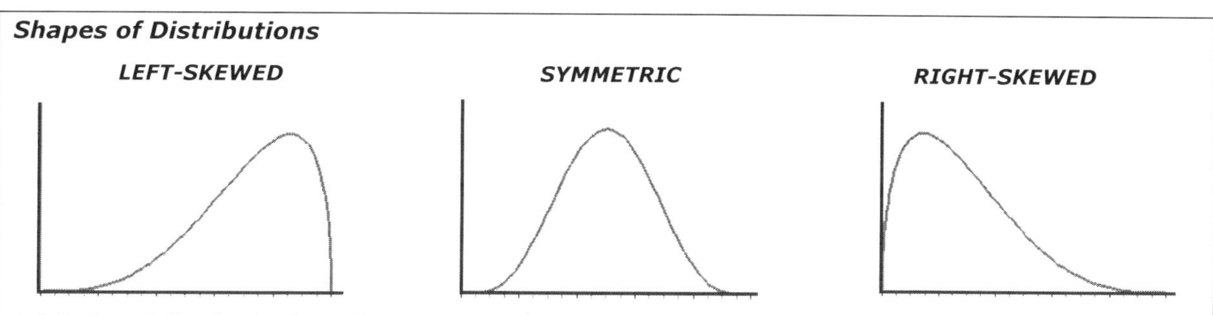

A *left-skewed distribution* has a longer *tail* on the *left* side of the distribution.
A *right-skewed distribution* has a longer *tail* on the *right* side of the distribution.
A *symmetric* distribution has roughly *equal* length *tails* on the right and left of the distribution.

All of these shapes are for distributions that have a single "peak." Sometimes the shapes of distributions do not fall into one of these neat categories. Sometimes we have too little data to say anything definitive about shape, and sometimes the shape is a mix. Quite often we encounter distributions that appear to have more than one peak. If a distribution appears to have just two peaks then that pattern is called ***bimodal*** (from two modes) and refers to a shape where there are two "humps."

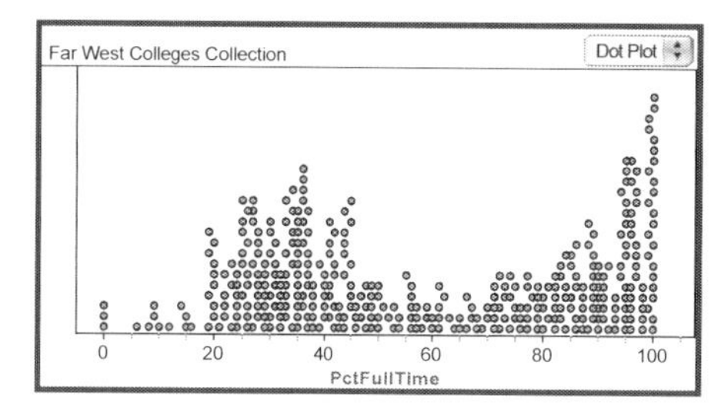

From the colleges and universities data, the variable *PctFullTime* provides an example of a *bimodal* distribution. A bimodal pattern may sometimes indicate that there are actually two groups in the data; by making plots of each of the groups, the bimodality can be explained. Of course, you have to have some idea what the groups are likely to be; for now, be assured that this knowledge is less of a problem than what you may think. Here, if we distinguish among public four-year, private four-year, and public two-year colleges then perhaps we can see what is happening with the *bimodality*.

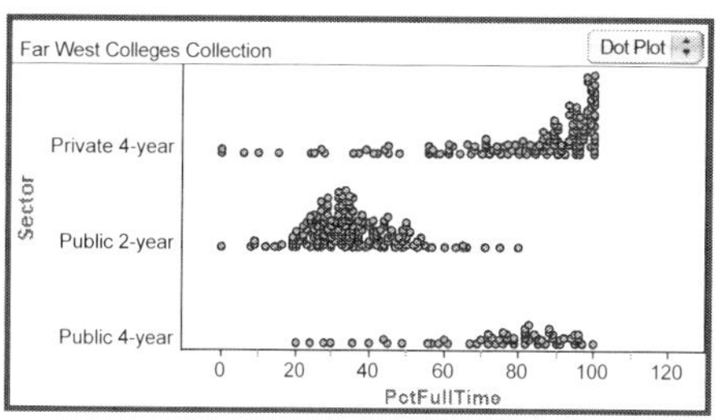

The right-hand "mode" comes from the left-skewed distribution of *PctFullTime* for the private four-year colleges and universities, while the left "mode" comes from the nearly symmetrical distribution for *PctFullTime* for the public two-year colleges.

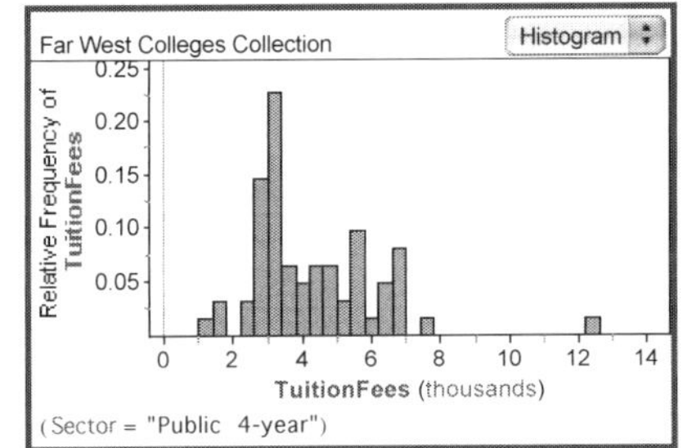

Center or Location of a Distribution: "On Average"

The idea of center When we say: "*On average*, tuition fees are about four thousand dollars per year for public four-year universities," we are speaking of what statisticians call the ***center*** of the distribution. We think of a value that characterizes the ***location*** of the distribution; we are answering the question: "Where–on the scale of the variable—*is* the distribution?" We look at center especially when we are comparing distributions and recall that comparison is important in statistics. Once again, the principle of comparison is important: "On average...four thousand dollars . . . $4000" really becomes important when "on average" the tuition fees for private schools are about $12,000.

Example: Let us look again at the distributions of the percent of full-time students (*PctFullTime*) for the three types of educational institutions; this time we will look at histograms, but we are looking at exactly the same three distributions as we did above just before the start of this section. The top histogram is for the private four-year colleges and universities; for this distribution, most of the colleges are toward the right end of the variable *PctFullTime*. That is, for most of these private four-year colleges, a high percentage of their students are full time. There are private four-year colleges that have a lower percentage of full-time students, but there are not many of these colleges. Therefore, we say that for the private four-year colleges, the distribution is mostly "centered" or "clustered" in the 90%+ area.

Now look at the histogram for the public two-year colleges (this is the second histogram). For them, most of the distribution is "centered" or "clustered" around 30% to 40%; the center is lower. Most of the two-year schools have only about 30% or 40% of their students full time rather than 90% or more full time.

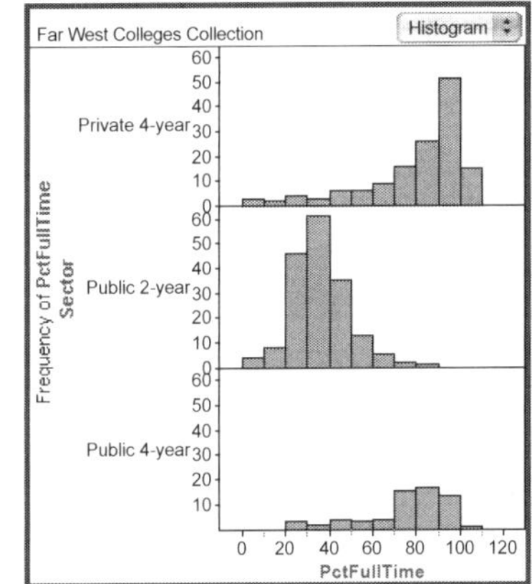

- Our question is: "*Where* on the scale of the variable is the distribution?"
- Our answer is: the distribution for the public two-year colleges is centered about 35%. We say that the center or location of the two-year college distribution is lower on the variable *PctFullTime* than the distribution for the private four-year colleges.

The distribution for the public four-year resembles the distribution for the private four-year colleges; the distribution is "centered" or "clustered" or "located" at around 80% full-time students. The comparison shows us that two-year colleges have a lower percentage of full-time students than do four-year educational institutions.

Higher, Lower, Bigger, Smaller, Greater, Lesser: Think Horizontally When we use these words to refer to the center of a distribution, and the values of the variable are on a horizontal axis, then a *higher, greater,* or *bigger* center means that the distribution is farther to the *right*—corresponding to larger numbers mathematically. A *lower, smaller,* and *lesser* center refers to a distribution primarily to the left. Think horizontally.

The height of the bars (or dots) does not mean the center is high. Look again at the histogram above for the *PctFullTime* for the public four-year colleges and compare it to the histogram for the two-year colleges. The histogram for the two-year colleges has the highest bars, but it is the type of college with the *lowest* percent full time, on average. The histogram for the public four-year schools is to the right on our variable, and so is higher. A high bar in a histogram does not necessarily tell us that the center is "high," and a high stack of dots in a dot plot does *not* tell us that the center of a distribution is high. The stack of dots just tells us that there happen to be many cases at that value.

The value with the greatest number of cases is known as the ***mode,*** but generally the *mode* is not a reliable indicator of the location of a distribution. A distribution may have several places with the same height, or mode. To know where the location or the center of a distribution is, you need to look at *where* the distribution as a whole resides on the variable.

The Spread of a Distribution and Its Meaning

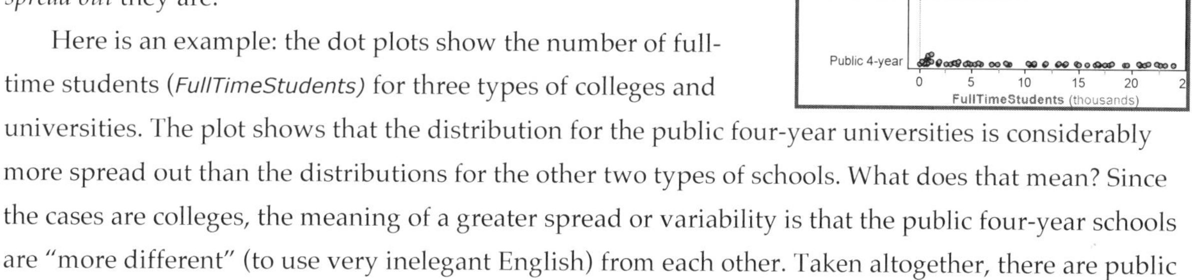

The third thing we look at is the ***spread*** of the distribution. Distributions differ in their location—the center—but also in how *spread out* they are.

Here is an example: the dot plots show the number of full-time students (*FullTimeStudents*) for three types of colleges and universities. The plot shows that the distribution for the public four-year universities is considerably more spread out than the distributions for the other two types of schools. What does that mean? Since the cases are colleges, the meaning of a greater spread or variability is that the public four-year schools are "more different" (to use very inelegant English) from each other. Taken altogether, there are public four-year colleges and universities that range from extremely small to extremely large. The sizes of private four-year colleges and universities and of two-year colleges are not as variable.

A start at understanding spread can be made by looking at the ***range*** of values for the variable. (The ***range*** is the highest number minus the lowest number.) We can see that the number of full-time equivalent students for the private schools is mostly from less than 100 full time students to about 8000 students (with one exceptional school), but for the public four-year colleges and universities, the numbers range from under 100 students to as many as 25,000 students.

In summary, there is a great *variety* or *diversity* in *FullTimeStudents* (the number of full-time students) for the public four-year schools and much more variety or "differentness" than with the private schools.

Outliers and Putting It Together

For the private four-year colleges there is one college that has a value for *FullTimeStudents* of about 16,000 and that the number of full-time students for this university is much more than most of the other private four-year institutions. (This happens to be the University of Southern California.) Cases that are far from the majority of the cases in a distribution are called ***outliers.*** There can be *outliers* on the high end of a distribution or the low end of a distribution, and, at this point, we will "detect" outliers just by looking; we will see a more exact way of detecting outliers later in §1.7.

Example and a warning What can we say about the *shape* and *center* of the three *FullTimeStudents* distributions? *Shape?* The distribution for the private four-year schools appears to be right-skewed, as does the distribution for the public two-year colleges. It is difficult from the graphic to say whether the distribution for the public four-year institutions is right-skewed, left-skewed, or symmetrical. *Center?* It looks as though the center or location of *FullTimeStudents* for the private four-year colleges is the least. We could say: "On average, the number of *FullTimeStudents* is the smallest for the private four-year colleges," even though we have not calculated the averages (we will do that in the next section). We could say that the center or location for *FullTimeStudents* for the two-year schools is somewhat higher because that distribution is generally somewhat to the *right* of the private schools distribution.

The ideas of *shape, center,* and *spread* are really aids to interpretation. They allow us to say something about the data. We will soon calculate numbers that indicate the centers and spreads of distributions, but, at this point, make certain to grasp the ideas and differences between *shape, center,* and *spread*. Here is the warning: work on making certain that you understand the differences between *shape, center, spread,* and the idea of a *distribution*.

Extended Example: Shape, Center, and Spread of Percent over Age Twenty-Four

Our overall statistical question is:

What differences are there in the distributions of the variable "percent of students over age twenty-four" (*PctOverAge24*) *between the private four-year colleges, public two-year colleges, and public four-year colleges, and what do the differences mean?* However, we will break this down into a number of smaller questions and show answers to these smaller questions (in "handwriting" font).

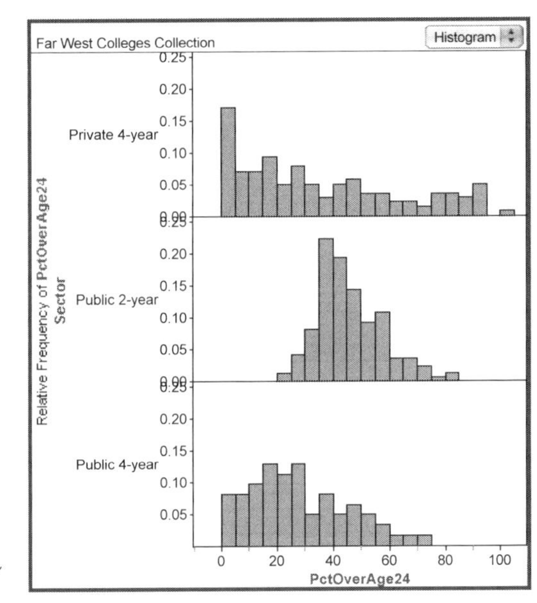

a. What are the cases for these data?

   The cases are colleges and universities.

b. What can you say about the shapes of the distributions?

   All three distributions appear to be somewhat right-skewed, since all three have a tail on the high side of the variable.

c. What can you say about the centers of the distributions?

   The distribution of the percentage over age twenty-four for the two-year schools is to the right of the distribution for the public four-year institutions. Since the spread of the distribution for the private colleges is so great, it is more difficult to say just from the graphic where the "center" is, but it is probably lower than for the two-year colleges. (We need some numbers!)

d. What can you say about the spreads of the distributions?

We have already noted that the spread of the distribution for the private colleges and universities on the variable "Percent Over Age 24" is big. This means that there is great variety or diversity in the private educational institutions in the "percent of students over age twenty-four"; some private institutions have almost none of their students over twenty-four, whereas other schools have a high percentage over age twenty-four. The least spread appears to be for the two-year colleges, so that with respect to the percentage over twenty-four, these colleges are quite similar. The spread for the variable is greater for the public four-year schools but not as great as for the private institutions

e. What does this all mean in the context of the data?

It makes sense that the percentage of students over age twenty-four in the two-year schools should be higher, on average, because some of the functions of two-year colleges are for retraining and also to give students who did not have the opportunity to go to college after high school a second chance when they are a bit older. That there is such a great variety (a large spread) in the private colleges suggests that some of these private institutions are vocational or professional in nature, whereas others in the private category are "traditional" liberal arts colleges that take students from high school.

Evenly Distributed and Warned

The exercises are designed to fend off common errors, but here are two or three warnings about common errors concerning shape, center, and spread.

First, make certain that you have the three features of distributions straight. Do not confuse spread with shape. Shape refers to the form of a distribution, whereas spread refers to how spread out over the variable the cases are.

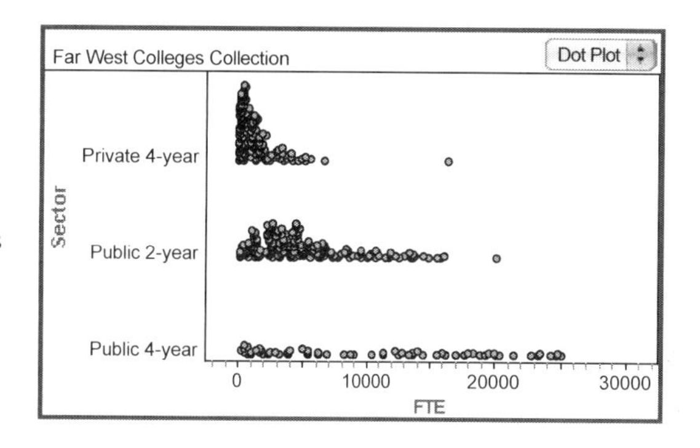

Secondly, learn to use statistical language properly. Sometimes, people who are beginning to study statistics will say, "The data are widely distributed" when what they mean to say (and which they *will* say when they learn the language) is that "the distribution has a wide spread."

Third warning—about "evenly distributed": people who have learned statistics will almost never say, "The data are evenly distributed," but this is a very common statement among beginners. The distribution of *FullTimeStudents* for the public two-year colleges is not at all "evenly distributed"; there are many colleges where the number of students is about three thousand to six thousand but *not* many colleges at all where the number of students is as high as fifteen thousand.

The term "evenly distributed" may be more apt for something like the *FullTimeStudents* distribution for the public four-year colleges and universities. However, statisticians would say that those data appear to ***uniformly*** distributed and not "evenly" distributed. A distribution where the probability of each of the values (or ranges of values) is the same is called a ***uniform*** distribution and not an "even" distribution. A good example of a distribution that we expect to be at least fairly uniform is the month of the year in which people are born.

Summary: Shape, Center, and Spread of Distributions of Quantitative Variables

We use the ideas of *shape, center,* and *spread* to characterize the distributions of quantitative variables. Generally we use these ideas to compare distributions. We have concentrated on looking at pictures of distributions to get an idea of the concepts; in the next sections, we shall learn about numerical summaries of these characteristics.

- ***Shape*** Whether a distribution is left-skewed, right-skewed, symmetric (if single-peaked), or bi-modal

  A ***left-skewed*** single-peaked distribution has a longer *tail* on the *left* side of the distribution.

  A ***right-skewed*** single-peaked distribution has a longer *tail* on the *right* side of the distribution.

  A ***symmetric*** single-peaked distribution has roughly *equal* length *tails* on the right and left.

  A ***bi-modal*** distribution is one with two distinct peaks.

- ***Center or Location*** The location of a distribution of cases on the values of the cases
  - When the values of a distribution are graphed on a horizontal axis, greater means greater in the mathematical sense.
  - A distribution with a greater (or higher, or bigger) center is one to the *right* on the variable depicted horizontally.
  - The ***mode*** of a distribution is a value or groups of values with the highest count. The mode does not necessarily indicate that a distribution has a high center compared to other distributions.
  - A distribution may have several modes—that is, values or groups of values with the same count.
- ***Spread*** The amount of variability, diversity, or "differentness" in a distribution compared with other distributions or to a standard
- ***Outliers*** Cases that stand outside the general mass of the data

§1.5 Numerical Summaries of Center/Location

The Idea of Center or Location: Where Is the Distribution?

We have already seen that the idea of the "center" of a distribution refers to its location on the scale of possible values. Suppose you were asked: "Comparing community colleges, four-year public universities, and four-year private universities, in which kind of institution are tuition fees the most and in which the least?" You would probably have a ready answer; you might even say, "On average, community colleges have the lowest tuition fees, private universities have the highest, and public four-year schools are between these." Here are three histograms on showing the distributions of tuition fees in colleges and universities in the Far West that backs up your idea.

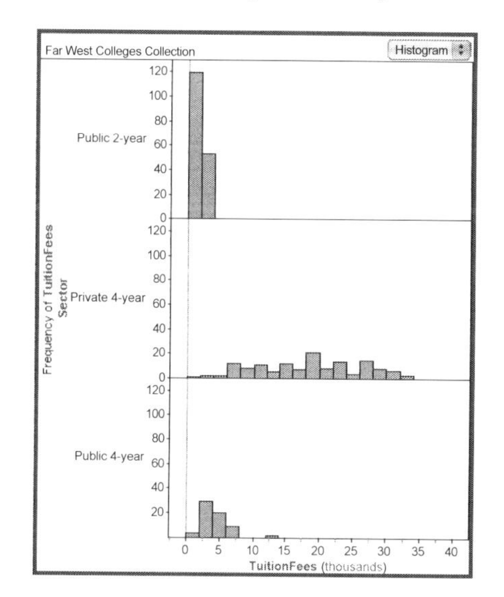

The graphic backs up your idea because the distribution for the public four-year college tuition fees is generally to the right of the publically funded schools, and the distribution for the public two-year colleges tuition fees is to the left of the distribution for fees for public four-year institutions. (You may also notice that there are big differences in the *spreads* of the distributions of tuition fees; the privately funded schools have a huge amount of variability compared with the publically funded schools.)

The fact that you may have answered "on average" may hint at the idea that we should be able to summarize the location (or center) of these distributions with a single number, and it may hint at the number you have in mind: the "average." This section is about averages, both the one you already know and another, about which you know less. The purpose of these averages is to summarize, with one number, the location of a distribution. These two averages, or measures of center, are called the ***mean*** and the ***median.***

To illustrate the calculation of these averages, we will look at a second example with a smaller collection. The cases are again colleges and universities, but this time we are considering just the twenty-five publically funded undergraduate institutions in Oregon, and we are interested in two variables; one is the categorical variable that distinguishes the two-year (community colleges) from the four-year institutions, and the other is (for each institution) the percentage of students that are over 24 years old.

We expect that the *percentage over twenty-four years* will be higher on average for the two-year schools. The dot plot (below) appears to show this, since the distribution for the "Percent over 24" is generally to the right of the distribution for the four-year schools.

Mean, x-bar, and sigma

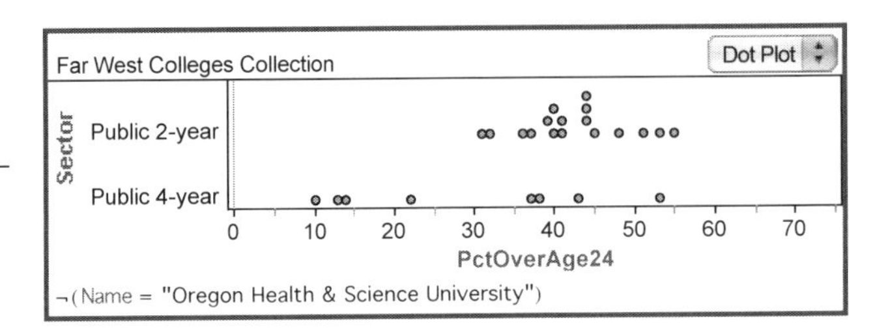

The average that you know about—the one that you learned in elementary school—is "the sum of all the things divided by the number of things." This measure is called the ***arithmetic average*** or the ***mean*** when we use it in statistics. The symbol that we use for the *mean* is $\bar{x}$, which is pronounced "*x-bar.*" The mean for the variable *PctOverAge24* for the eight four-year schools can be calculated as

$$\bar{x} = \frac{43+37+13+53+38+22+10+14}{8} = \frac{230}{8} = 28.75 .$$

Num	Name	Pct Over Age24	Two-year or Four-Year
1	Eastern Oregon University	43	Four
2	Oregon Institute of Technology	37	Four
3	Oregon State University	13	Four
4	Oregon State University, Cascades Campu	53	Four
5	Portland State University	38	Four
6	Southern Oregon University	22	Four
7	University of Oregon	10	Four
8	Western Oregon University	14	Four
9	Blue Mountain Community College	40	Two
10	Central Oregon Community College	36	Two
11	Chemeketa Community College	45	Two
12	Clackamas Community College	44	Two
13	Clatsop Community College	51	Two
14	Columbia Gorge Community College	55	Two
15	Klamath Community College	31	Two
16	Lane Community College	44	Two
17	Linn-Benton Community College	37	Two
18	Mt Hood Community College	32	Two
19	Oregon Coast Community College	44	Two
20	Portland Community College	53	Two
21	Rogue Community College	48	Two
22	Southwestern Oregon Community College	41	Two
23	Tillamook Bay Community College	39	Two
24	Treasure Valley Community College	40	Two
25	Umpqua Community College	41	Two

For the four-year schools, on average, 28.75% of the students are over age twenty-four. In symbolic form (so that we can apply this formula to any data set), we write:

$$\bar{x} = \frac{x_1 + x_2 + \cdots + x_n}{n} = \frac{1}{n}\sum_{i=1}^{n} x_i$$

Using "x's" with the subscripts reflects our general way of writing the value for each of the cases in turn. In our example, $x_1 = 43\%$ and is the value for *percentage over age twenty-four* (*PctOverAge24)* for Eastern Oregon University, $x_2 = 37\%$, the value for Oregon Institute of Technology, etc. For this example, the sample size is $n = 8$; that is, there are eight cases in the collection.

The Greek letter $\sum$, pronounced "sigma," is used to show a long sum where we do not want to write out all of the x_i—especially where we are summing many of them. The definition is quite simple: $\sum_{i=1}^{n} x_i = x_1 + x_2 + \cdots + x_n$, where the little numbers at the foot and head of the $\sum$ tell the reader where the sum begins and ends. You may or may not have been introduced to this symbol in your previous mathematics studies, but you need to become familiar with it because we will use this symbol often in statistics. There is an exercise in this section that takes you through how this symbol is used.

Mean value for a distribution of a quantitative variable

$$\bar{x} = \frac{x_1 + x_2 + \cdots + x_n}{n} = \frac{\sum_{i=1}^{n} x_i}{n}$$

You should be able to confirm that the mean for the variable *Percentage over age 24* is $\bar{x} \approx 42.42$;the mean percentage of students who are over age twenty-four for the two-year colleges is higher than the mean percentage of students over age twenty-four for the four-year institutions

Median: the value of the middle case

The definition of the median value for the distribution of a quantitative variable is given in the box.

> **_Median value for a distribution of a quantitative variable_**
>
> The ***median*** is the value that divides a distribution so that half the values are less and half the values are more.
>
> To find the median M:
>
> 1. Rank order the data from lowest to highest.
> 2. If the number of cases n is odd then the median M is the value of the middle number, whose location from the smallest value will be found by the formula $\frac{n+1}{2}$.
> 3. If the number of cases n is even then the median M is the mean of the values of the two middle numbers, and the location from the smallest value will (still) be found by the formula $\frac{n+1}{2}$.

Here is the calculation of the median value for the percentage over age twenty-four for the $n = 17$ Oregon two-year institutions. The first step is to rank order the data. If you make an ordered stem plot of the data, you will be able to put the data in order easily. Then calculate the *location* of the median using $\frac{n+1}{2} = \frac{17+1}{2} = \frac{18}{2} = 9$, so the median M is to be found at the ninth case from the start, since the total number of cases $n = 17$ was an odd number. That case is shown on the stem plot in bold italics. The median value is $M = 41$ percent over age twenty-four. Confirm from the stem plot that this median does indeed divide the distribution into two equal halves; there are eight cases that are less and eight cases that are more.

Stem	Leaves
1	
1	
2	
2	
3	12
3	679
4	001***1***444
4	58
5	13
5	5

There was an even number of four-year Oregon institutions. To find the median, again rank order the data from lowest to highest. A stem plot does this easily. Now calculate the location: $\frac{n+1}{2} = \frac{8+1}{2} = \frac{9}{2} = 4.5$. That the location is 4.5 shows that the median is between the fourth case and the fifth case since 4.5 is halfway between 4 and 5. The median is the mean of the values of these two. These two cases are shown on the stem plot in bold italics. The calculation of the median is thus $M = \frac{22+37}{2} = \frac{59}{2} = 29.5$ percent over age twenty-four. Usually, we will let software calculate means and medians, especially where the number of cases is large; it is good to know how the calculations are done.

Stem	Leaves
1	034
1	
2	***2***
2	
3	
3	***7***8
4	3
4	
5	3
5	

Here is table of summary statistics showing the mean (S1, the first number listed for each type of college), the median (S2, the second number listed for each type of college), and the count (S3, the third number listed for each type of college) of the *Percentage over age 24* distributions. For the public two-year schools, the mean is 42.41%, the median is 41%, and the number of colleges is $n = 17$, whereas for the public four-year schools, the mean is 28.75%, the median is 29.5%, and the number of colleges is $n = 8$.

Far West Colleges Collection

		PctOverAge24
Sector	Public 2-year	42.4118 41 17
	Public 4-year	28.75 29.5 8
Column Summary		38.04 40 25

S1 = mean ( )
S2 = median ( )
S3 = count ( )

¬(Name = "Oregon Health & Science Univer

Both the mean and the median tell us that the average percentage of students over age twenty-four is higher in the two-year institutions than it is in the four-year institutions.

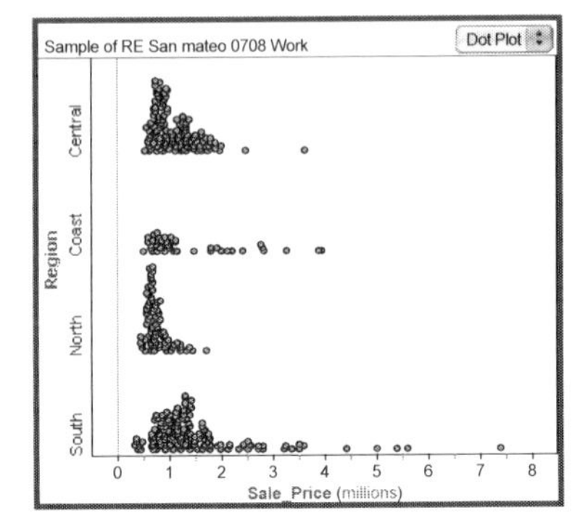

The Median Is Resistant but the Mean Is Not Resistant

Resistant to what? Answer: the median is a measure that is *resistant* to (or not affected by) to the influence of outliers and skewness, whereas the mean *is* sensitive to (or affected by) skewness and outliers. Here is why: the calculation of the mean involves a sum, whereas the calculation of the median does not. The sum in the calculation of the mean will be "pulled" in the direction of skewness: if a distribution is right-skewed, the sum in the mean will contain some very large numbers; if the distribution is left-skewed, the sum will contain relatively small numbers.

Here is an example showing the implications of the resistant nature of the median and the sensitive nature of the mean. The dot plots (above) show the sale prices of a sample of houses sold in San Mateo County in 2007–2008, categorized by regions within the county. All of the distributions are right-skewed but the Coast and the South especially so. Our question: on average, did houses sell for more in the Central region or the Coast region? It would be hard to make a judgment based on the dot plots; we need some numbers.

Sample of RE San mateo 0708 ...

		Sale_Price
Region	Central	1109303 969500 128
	Coast	1382200 966000 42
	North	758247 680000 78
	South	1489789 1264440 152
Column Summary		1214086 979500 400

S1 = mean ( )
S2 = median ( )
S3 = count ( )

Here are the means and medians for the sale prices of houses in the four regions. Notice how if we compare the *mean* sale prices, we would get the impression that houses sold for more on the Coast compared with the Central region. However, that mean is being "pulled" by the skewness of the Coast distribution. A more appropriate measure for the location of the distributions is the *median* sale price, and that measure shows that the average prices are about the same.

The Shape of a Distribution, the Mean, the Median, and the Balance Point

Since the median is *resistant* to skewness and the mean is not *resistant*, we can work backwards and detect the shape of a distribution by looking at the mean and the median of a distribution.

> ***Shape of a Distribution and Measures of Center***
>
> When a distribution is *right-skewed* then the mean > median.
>
> When a distribution is *left-skewed* then the mean < median.
>
> If a distribution is exactly *symmetrical* then the mean = median.

When working with actual data, we will very seldom encounter the situation where the mean and the median are *exactly* equal. However, if the mean and the median are close in value then we can say that the distribution is nearly symmetrical.

Compare the means and the medians of the sale prices of the houses in the San Mateo real estate sample shown above in the summary table and you will see that the distribution that appears to be least *right-skewed* (the one for the North region) is also the one that that has the smallest difference between the mean and the median.

Here is the plot of the sale prices for the South region showing the mean and the median.

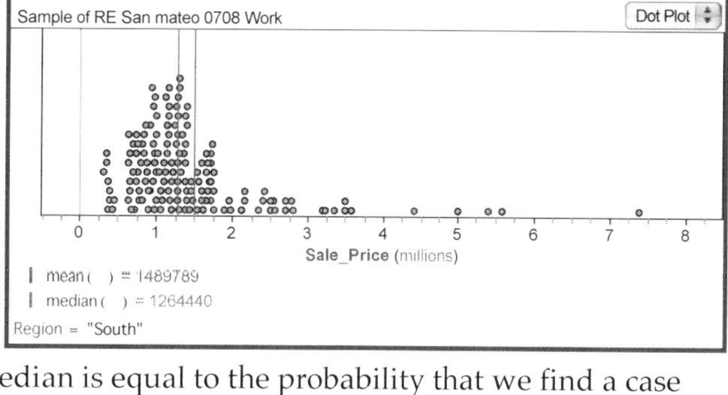

The median is the value that divides the distribution into two equal parts, so that there is the same number of cases above as below the median. In our probability notation $P(X < Median) = P(X > Median) = 0.50$, the probability that we find a case greater than the median is equal to the probability that we find a case less than the median. But what about the mean? What the mean shows for any distribution is the "center of mass" or (in other words) the balance point. If you had a huge heavy right-skewed distribution (like our friend Hiroyuki here), you would be able to carry it by balancing the distribution on your shoulder at the mean. The tail on the right side would just balance the clump on the left. Of course, if the distribution is symmetric then the balance point is at the center, since the tail on the left equals the tail on the right. For skewed distributions, the tails determine where the balance point is.

Extending the Summary: The Five-Number Summary

The median defines the location of a distribution by showing the value that divides the distribution into two equal halves. The median for the percent over age twenty-four for the two-year colleges is M = 41 and for the four-year schools it is M = 29.5. We can extend this idea by looking at the value that defines the point in the distribution so that 25% of the distribution is less than that point, and hence 75% is more than that point. We can call this ***Quartile 1*** or ***Q1*** because it is the point that defines the first quarter of the data. In a similar way, we define ***Quartile 3*** or ***Q3*** as the point where 75% of the distribution is less than that point and 25% of the distribution is greater than that point. What we are doing is breaking the data into four equal parts.

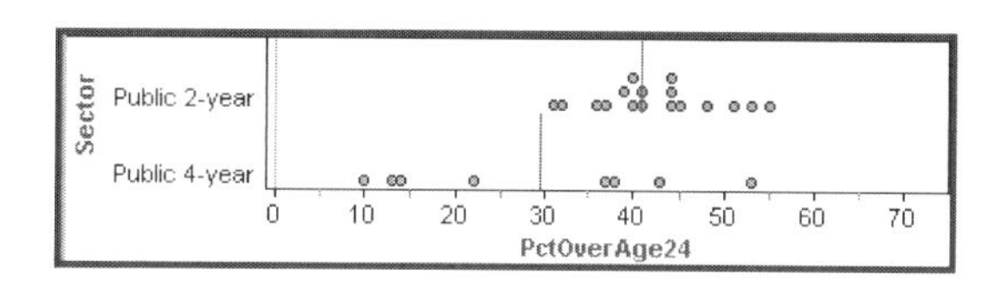

1	034
1	
2	2
2	___***M = 29.5***
3	
3	78
4	3
4	
5	3
5	

The way we will do this when we are working with small collections of data by hand is to recognize that one quarter is just one half of one half; we will work with the two halves of the data as divided by the median. We will find the median (one half) of the lower half of the data to get the ***Q1***. To get the ***Q3*** we will get the median of the upper half of the data. Here is an example using the Oregon colleges and universities data.

To get the ***Q1*** for *Percentage over age 24* for the two-year colleges, we apply the rules for finding the median in the box above to the eight cases that are below the median, leaving out the median itself, $M = 41$.

1	
1	
2	
2	
3	12
3	679
4	001**1**444
4	58
5	13
5	5

Since there are n =8, we calculate the location $\frac{n+1}{2}=\frac{8+1}{2}=4.5$ and this tells us that the $Q_1=\frac{37+39}{2}=\frac{76}{2}=38$. For ***Q3*** we make a similar calculation for the eight cases *above* the median of $M=41$. The location calculation again gives 4.5, which puts us between 45 and 48, so we have: $Q_3=\frac{45+48}{2}=\frac{93}{2}=46.5$. The calculation for the quartiles for the variable *Percentage over age 24* for the four-year schools follows the same pattern; we get the median of the lower half of the cases to get the Q_1 and then get the median of the upper half of the data to get the Q_3.

Definition of Quartiles

Q1 is the value that divides a distribution so that 25% of the values are less, and

Q3 is the value that divides a distribution so that 75% of the values are less.

Hand calculation of ***Quartiles***: find the median of the lower or upper half of the data.

Warning: Calculators and software may use a more complicated formula than this hand formula, so your hand calculations may not match your software calculations; however, the differences will be extremely small.

When we put together the minimum, the Q_1, the median, the Q_3, and the maximum for a distribution, we have what is called the ***five-number summary.*** The five-number summary for the distribution of the variable *Percentage over age 24* for the two-year colleges is: Min = 31, Q_1 = 38, M = 41, Q_3 = 46.5, Max = 55. For the four-year colleges and universities, the five-number summary is Min = 10, Q_1 = 13.5, M = 29.5, Q_3 = 40.5, Max = 53.

Five-Number Summary for the Distribution of a Quantitative Variable

Minimum, Q_1, Median, Q_3, Maximum

Had you reflected on the matter, you might have concluded that using just one number—the mean or the median—to characterize an entire distribution can be misleading. Using just one number may be "over-simplifying"; we may be able to have a better summary using five numbers.

A Graphic Based upon the Five-Number Summary: the Box Plot

One thing that we can do with these five numbers is to make a "schematic" graphic to show a distribution. That graphic is called a ***box plot*** or sometimes a ***box and whisker plot.*** Here are details.

Elements of a box plot (PctOverAge24 for the Colleges and Universities data)

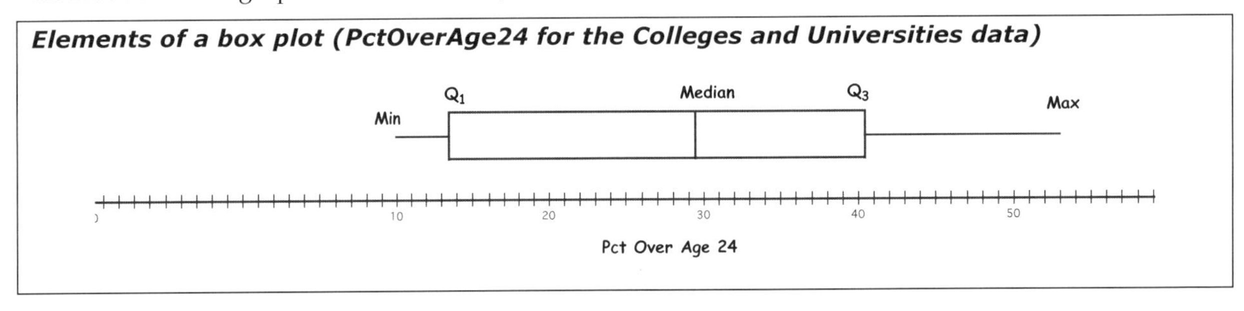

The middle part of the box plot is called (quite naturally) the box, and the lines connecting the quartiles to the extremes are sometimes called "whiskers" (think cats). Box plots are especially useful for comparing two or more distributions, since you can see quickly the locations of the distributions on the variable and also get an idea of how much spread or variability there is in each of the two (or more) distributions. Here are box plots comparing the *Percentage over age 24* distributions for the two-year and four-year colleges.

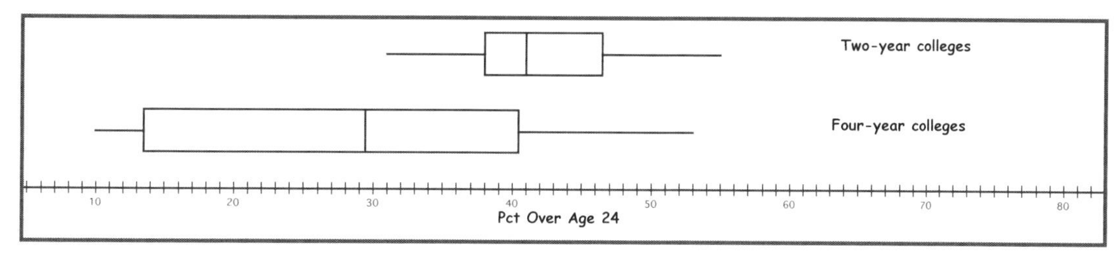

We can see what we have already concluded—that, on average, the percentage of students over age twenty-four is higher in two-year colleges than it is in four-year colleges. We can also see that there seems to be greater variability among the four-year institutions. Variability is very important topic and is where we are going next.

Summary: Numerical Measures of Center

The idea is to summarize the location of distributions by a just a few numbers. The measures that we use are the *mean (or arithmetic average)*, the *median*, and the *five-number summary*. The five-number summary also has the advantage of showing other features of distributions and can be turned into a graphic.

- ***Mean*** The "balance point" or center of mass of a distribution, calculated by:

$$\bar{x} = \frac{x_1 + x_2 + \cdots + x_n}{n} = \frac{\sum_{i=1}^{n} x_i}{n}$$

- ***Median*** The value that divides a distribution so that half the values are less and half the values are more. To find the median *M*:
  - Rank order the data from lowest to highest.
  - If the number of cases *n* is odd then the median *M* is the value of the middle number, whose location from the smallest value will be found by the formula $\frac{n+1}{2}$.
  - If the number of cases *n* is even then the median *M* is the mean of the values of the two middle numbers, and the location from the smallest value will (still) be found by the formula $\frac{n+1}{2}$.
- ***Resistant measures*** are measures *not* influenced by skewness or outliers in a distribution.
  - The *median* is a *resistant* measure of center and not influenced by skewness or outliers.
  - The *mean* is *not resistant* to skewness and outliers and it works most accurately for symmetric distributions.
- ***Shape of a Distribution and Measures of Center***
  - When a distribution is *right-skewed* then the mean > median.
  - When a distribution is *left-skewed* then the mean < median.
  - If a distribution is exactly *symmetrical* then the mean = median.
- ***Quartiles***
  - Q_1 is the value that divides the distribution so that 25% of the values are less.
  - Q_3 is the value that divides the distribution so that 75% of the values are less.
- ***Five-Number Summary for the Distribution of a Quantitative Variable***

  Minimum, Q_1, Median, Q_3, Maximum, where
- ***Box Plot*** is a graphic of a made from the five-number summary, as shown here:

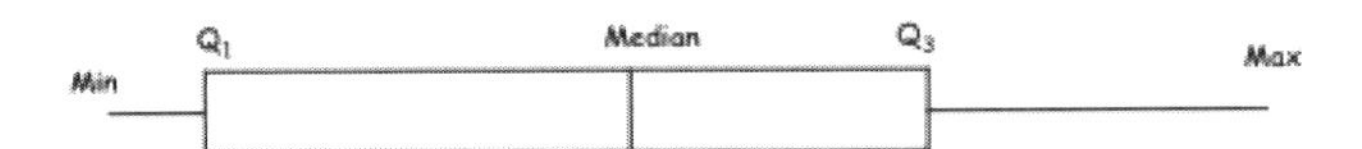

§1.6 Variation: Measurement and Interpretation

Variability (or Spread) and its challenges

Look again at the distributions of tuition fees for the three types of colleges and universities. It is clear that on average the tuition fees at the private colleges are higher than the tuition fees at the four-year public institutions, and the four-year public schools fees are on average somewhat higher than the tuition fees at community colleges. However, there is another thing to notice about the tuition fees at the private colleges. The tuition fees are much more spread or variable or diverse among the private colleges than they are for the public institutions. In other words (in extremely clumsy and poor English), the tuition fees are "more different" among the colleges.

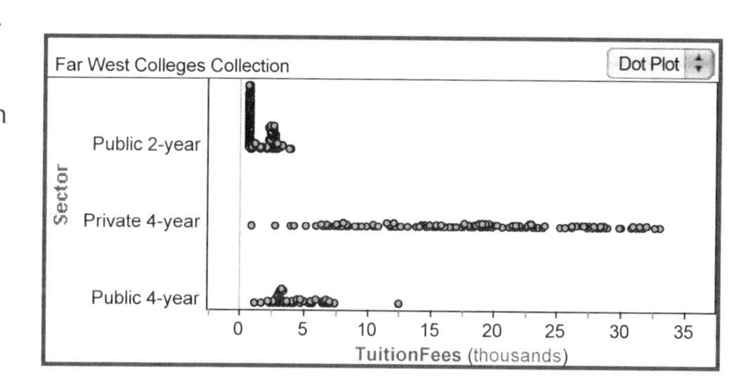

Try Exercise 1: Here is the question asked in Exercise 1:

"Imagine that you are explaining to a friend (perhaps by e-mail) what the graphic [the one above] tells you about tuition fees in public and private colleges. You can attach the graphic to your e-mail. You need to tell your friend what the dot plot shows about the center (or location) of the distributions but *also* about the differences in the *spreads* of the distributions between the private and the public institutions."

We will get to how you might have answered the question in Exercise 1 shortly. Let us say that there are good reasons for the greater variability in fees for the private schools. The state schools most often have their fees set or regulated by government in one way or another. The left "spike" of tuition fees you see in the two-year college distribution are the fees charged by California community colleges, and these fees are almost the same for every two-year college in the entire state. Private institutions, on the other hand, are more variable in many ways, including their sources of funding; private colleges range all the way from small church-related schools with just a few hundred students to major research universities such as Stanford.

So how did you explain to your friend what the graphic shows you? Was it like this?

"Fred [your friend's name], it is really hard to give you a figure for typical cost of a private college because the costs are all over the place, although on average they cost more than the state schools. The range of costs is enormous! There are some colleges that charge as little as $2500 per year for tuition, and some that cost over $30,000 a year—and everything in between. For state schools, the range is much smaller; you can expect to pay between about $1000 to $8000 per year for a state four-year school in the West."

The challenge of explaining variability. We think that you should have found the questions in Exercise 1 challenging and difficult. People are not as comfortable describing *variability* as they are describing the *location* of distributions.

To say "on average" comes fairly naturally; talking about variability does not. So here is a warning. Of the things to learn in this section—terminology, calculation, and interpretation—it will be interpretation, or "explaining what the numbers mean" in the context of the data that you will find the hardest. However, we will try to give some guidance; as starters, know that you will find words such as "variability" and "diverse" and "different" to be useful.

A good *start* to interpretation is to use the idea of a range, as in: "the range of fees for the state schools is between about $1000 and $8000 for tuition." That is: the range is $7,000 = $8,000 – $1,000 for the state schools, but for the private colleges, the range is $27,500 = $30,000 – $2,500. The idea of range is a good start; we will go on to use more accurate measures of variability.

Measures of Spread: Range and Inter-Quartile Range

We will measure the spread or variation in a distribution in three different ways.

The simplest way to measure spread is to use the range, which has been introduced in the example.

Definition of Range

Range = Maximum value – Minimum value

Here is a **Summary Table** showing the means and the five-number summaries for the *TuitionFees* for the three types of colleges and universities.

Far West Colleges Collection

	Sector			Row Summary
	Private 4-year	Public 2-year	Public 4-year	
TuitionFees	18501.6	1340.67	4285.26	8225.21
	885	624	1214	624
	11820	672	3030	799
	19083	788	3656	3035
	25244	2383	5496	14775
	33012	3882	12506	33012
	139	172	62	373

S1 = mean( )
S2 = min( )
S3 = Q1 ( )
S4 = median( )
S5 = Q3 ( )
S6 = max ( )
S7 = count ( )

From the definition of range as *Maximum value – Minimum value* you should easily be able to get the ranges from the five number summaries for the three categories of colleges and confirm that the private colleges have the greatest range, at $32,127.

Range, however, has a major disadvantage: range is not *resistant*; outliers affect its calculation. Our casual explanation above from the dot plot said that the range or the public four-year colleges was about seven thousand dollars, and yet for this example our definition of the range gives:

Range = $12,506 – $1,214 = $11,292

There is one state school whose fees are considerably greater, and this makes the range as a measure of spread greater than what we would casually see from the dot plot. Range is a bad and misleading measure of variability in this instance.

Warning about range

Range is a bad measure of variability because it is not resistant to outliers.

A second (and better) measure of spread that avoids the outlier problem also uses the five-number summary. It is called the ***Inter-Quartile Range*** or ***IQR.*** It is nearly as easy to interpret as range.

Definition of Inter-Quartile Range (IQR)

$$IQR = Q_3 - Q_1$$

Interpretation of the IQR

The IQR is the Range of the Middle 50% of a Distribution.

As you can confirm from the five-number summary, the ***IQR*** for the four-year state schools is $IQR = 5496 - 3030 = 2466$, for the two-year public schools it is $IQR = 2383 - 672 = 1711$, and for the private colleges it is $IQR = 25244 - 11820 = 13424$. What does this give us?

The meaning of the *IQR* can be seen better by looking at box plots. The box in a box plot incorporates the middle 50% of the distribution, since 25% of the distribution is to the left of Q_1 and another 25% of the distribution is to the right of Q_3. In our example of the tuition fees, notice how clearly the bigger spread of the private colleges is seen in the graphic: the box for the private four-year colleges (showing its IQR) for tuition fees is much wider than the IQRs for the other two types of institutions.

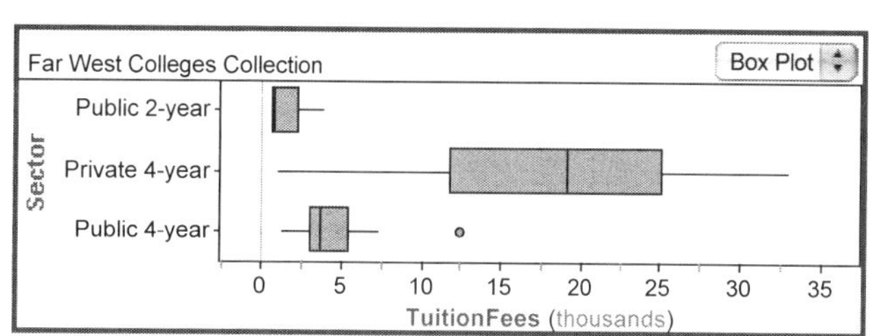

Measures Based on Distance from the Mean: Variance and Standard Deviation

The most common measures of variation are based on getting "something like" an average distance of the data in a collection from the mean of the data. We will use the data on the Oregon two-year and four-year schools and the percentage of students over age twenty-four. Here are dot plots of the data, showing the location of the means for the two groups of colleges. The names of the formulas for measuring spread from the mean are ***variance*** s^2 and ***standard deviation s.***

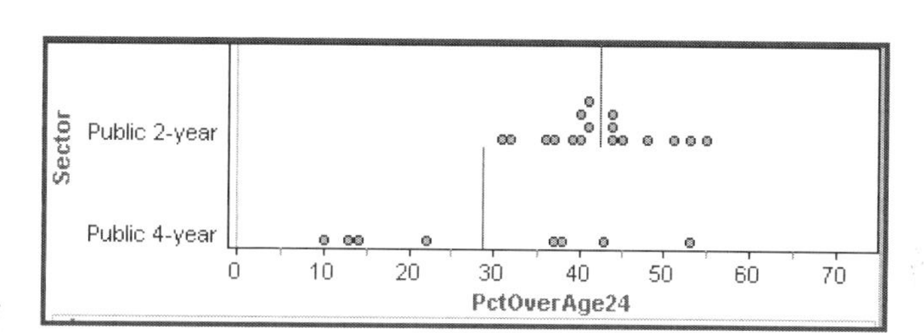

Definition of variance

$$s^2 = \frac{1}{n-1}\sum_{i=1}^{n}(x_i - \bar{x})^2$$

Definition of standard deviation

$$s = \sqrt{\frac{1}{n-1}\sum_{i=1}^{n}(x_i - \bar{x})^2}$$

Example. As you can see, there is basically just one formula since the standard deviation s is just the square root of the variance s^2. The way the formula works is shown (below) for the computation of the variance of *percentage over age 24*. The calculation starts by getting the difference between the value of each case x_i and the mean for the data: $x_i - \bar{x}$. For these data $\bar{x} = 28.75$. Since in this example $n = 8$, there will be eight of these differences. Then each of these differences is squared and then these squares are summed up and the sum divided by $n - 1$; here it will be 7.

Num	Name	Pct Over Age24	Two-year or Four-Year
1	Eastern Oregon University	43	Four
2	Oregon Institute of Technology	37	Four
3	Oregon State University	13	Four
4	Oregon State University, Cascades Campu	53	Four
5	Portland State University	38	Four
6	Southern Oregon University	22	Four
7	University of Oregon	10	Four
8	Western Oregon University	14	Four

What we get from this calculation is the variance s^2; the standard deviation s is found by getting the square root of this number. (In the second line of the calculation on the next page we have not shown two terms of the sum so that the sum will actually fit on the page.)

$$s^2 = \frac{1}{n-1}\sum_{i=1}^{n}(x_i - \bar{x})^2$$
$$= \frac{1}{7}\left[(43-28.75)^2 + (37-28.75)^2 + (13-28.75)^2 + \cdots + (22-28.75)^2 + (10-28.75)^2 + (14-28.75)^2\right]$$
$$= \frac{1}{7}\left[(14.25)^2 + (8.25)^2 + (-15.75)^2 + (24.25)^2 + (9.25)^2 + (-6.75)^2 + (-18.75)^2 + (-14.75)^2\right]$$
$$= \frac{1}{7}[203.0625 + 68.0625 + 248.0625 + 588.0625 + 85.5625 + 45.5625 + 351.5625 + 217.5625]$$
$$= \frac{1807}{7} \approx 258.2143$$

The variance for the *percentage over age 24* for the four-year schools is $s^2 = 258.2143$, and the standard deviation is $s = \sqrt{s^2} = \sqrt{258.2143} \approx 16.07$.

Looking at the variance formula closely and thinking graphically

The standard deviation is close to being the average distance of the data from the mean, and can be thought of as: "the average amount that a case value differs from the mean. "

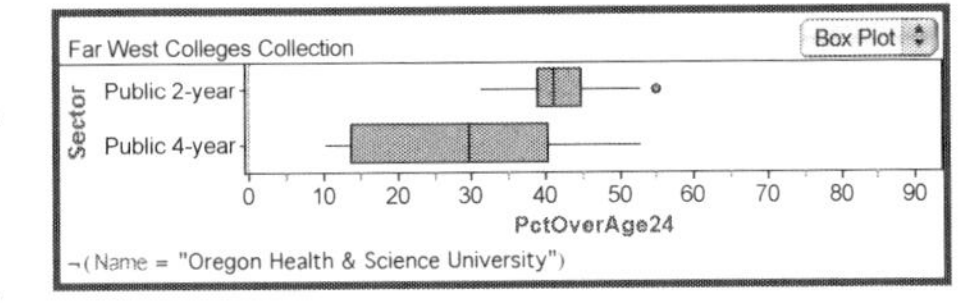

We will look at an example to develop this idea: the distances (either negative or positive) for our Oregon two- and four-year schools example are shown in the graphic below. (The boxplots are shown here.)

We have said that the spread or variation for the two-year colleges is smaller than the spread for the four-year schools for the variable *percentage over age 24.* Look at the distances (the "lines") in the plot below; the "lines" for the four-year schools look longer--on average—than the "lines" for the two-year colleges. The two-year colleges have a few "deviations from the mean" that are big, but even these are not as big as the ones for the four-year schools, and the two-year schools have many very short deviations. Each of these lines—the deviations or distances—is one of the $x_i - \bar{x}$ in the formula.

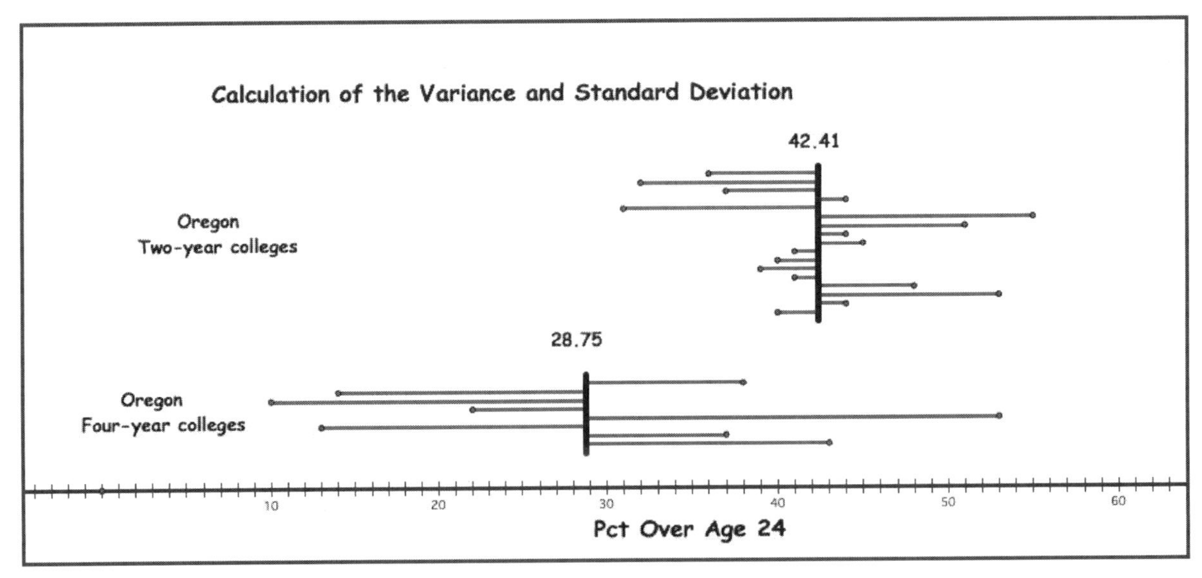

Why $\sum_{i=1}^{n}(x_i - \bar{x})$ ***is zero and why deviations are squared*** If we try to get the average of these distances, we will find that $\sum_{i=1}^{n}(x_i - \bar{x})$ will always be equal to zero, and so getting the average deviation $\frac{\sum_{i=1}^{n}(x_i - \bar{x})}{n}$ will always be zero. One way of seeing that the sum will always be zero is to see that the mean is the balance point of the data. That is, the sum of the lengths of the deviations on the left of the mean will just equal the sum of the lengths on the right-hand side of the mean. Another way is to work out: $\sum_{i=1}^{n}(x_i - \bar{x}) = \sum_{i=1}^{n} x_i - \sum_{i=1}^{n} \bar{x} = n\bar{x} - n\bar{x} = 0$

The solution to our dilemma is to square each of the deviations, so we are getting the average of the squared deviations. By taking the square root of the variance s^2 we get the standard deviation s so that we get a measure that has the same units as our original data.

Why the sum is divided by n − 1. In the formula $s = \sqrt{\frac{1}{n-1}\sum_{i=1}^{n}(x_i - \bar{x})^2}$ we divide by $n - 1$, and not n. Now, if the number in our collection is big, dividing by $n - 1$ will come out to be almost the same as dividing by n, so numerically there is not much difference. However, the question remains as to *why* we divide by one less than n. The complete answer can better be appreciated at the end of the course, or perhaps not even until into the next statistics course. For now, we can only give a hint of the reasoning. Later in the course, we will be concerned with not only what we actually see from our data but also with "what could possibly be." If our data have a mean $\bar{x}$ and this mean is calculated by the formula $\bar{x} = \frac{1}{n}\sum_{i=1}^{n} x_i$ then for that particular mean $\bar{x}$ and for n cases, the sum $\sum_{i=1}^{n} x_i$ is fixed. If the sum is fixed then only $n - 1$ cases in the data can vary freely.

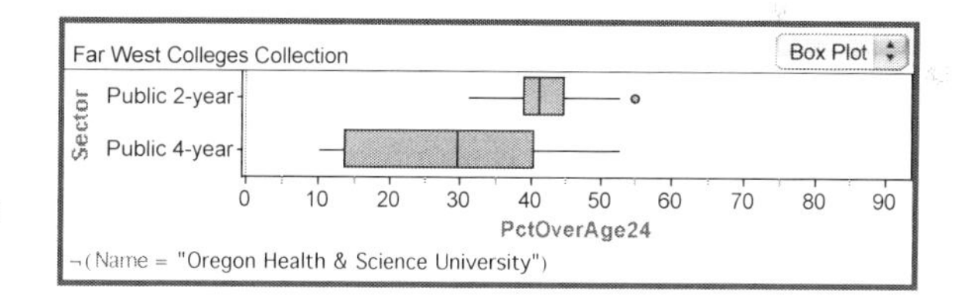

Interpretation: the important and hard part

Should the variance and standard deviation for the *percentage over age 24* for the *two-year schools* be bigger or smaller? Looking at the box plots it seems that the amount of spread or variation in *percentage over age 24* is less, so the s^2 and the s should be smaller. A smaller number for the variance and for the standard deviation shows a smaller spread—a larger number shows a larger spread or variation. Notice that the *IQR* for the two-year schools is also smaller than the *IQR* for the four-year schools. Notice also that although the measures of center/location for the two-year schools are big (these colleges have on average a greater percentage of students over twenty-four), the measures of spread are smaller.

Far West Colleges Collection

	Sector		Row Summary
	Public 2-year	Public 4-year	
PctOverAge24	42.4118	28.75	38.04
	41	29.5	40
	6.73664	16.069	12.1603
	6	27	8
	17	8	25

S1 = mean( )
S2 = median( )
S3 = s( )
S4 = iqr( )
S5 = count( )

¬(Name = "Oregon Health & Science University")

The two-year schools have on average a higher percentage over age twenty-four, but there is much less variability among these schools. The four-year schools have a lower percentage on average, but there is much more variability amongst the schools in *PctOverAge24*.

Far West Colleges Collection

	Sector			Row Summary
	Private 4-year	Public 2-year	Public 4-year	
PctOverAge24	34.539	45.4686	27.0323	38.3677
	26	43	24.5	39
	28.9486	11.343	17.3091	21.6583
	43	16	24	28
	141	175	62	378

S1 = mean ( )
S2 = median ( )
S3 = s ( )
S4 = iqr ( )
S5 = count ( )

Even without graphics to give us a visual picture, we should be able to use the numbers to compare the location and the spread of several distributions. Here are the numbers not just for the Oregon colleges and universities but also for all the Far West institutions on the same variable *PctOverAge24*.

From these numbers we can say that the percentage of students over age twenty-four is higher on average for the two-year colleges, but the amount of variability among these colleges is smaller than it is for the four-year institutions. The mean and the median (S1 and S2 in the Fathom summary table) for the two-year colleges are both higher than for the four-year schools, but the standard deviation (S3) and the *IQR* (S4)—the measures of variation—are smaller than those measures for the four-year schools. The spread of the percentage over age twenty-four is especially great for the private colleges, with an inter-quartile range of 43%, whereas the *IQR*s for the other two types of institutions are considerably smaller. In general, the larger the value of our measures of spread, *IQR*, s (the standard deviation) and s^2 (the variance), the more variability there is in the data and the more spread out the data are.

Interpretation of Measures of Variation

The *bigger* the value of the standard deviation or inter-quartile range, the *more variability* the data have.

Resistance Again

Recall that the mean $\bar{x}$ is *not resistant*; the value of the mean can be affected by skewness and outliers. The variance and standard deviation are also sensitive to outliers and skewness; part of the reason that this is so is that the variance and the standard deviation use the mean in their calculation.

However, notice that in the formula for the standard deviation $s = \sqrt{\frac{1}{n-1}\sum_{i=1}^{n}(x_i - \bar{x})^2}$ deviations from the mean get squared. This means that outliers (which are far from the mean) or values in tails get exaggerated by squaring in the calculation of the variance and standard deviation. The consequence of the sensitivity of the variance and standard deviation to skewness is that the *IQR* is a more useful measure of spread since it can be used for highly skewed distributions. You may well ask at this point: why do statisticians use them at all (and why are they so important)? The answer is that the standard deviation makes complete sense for an important *model* distribution, the ***Normal Distribution.*** In the next section, we will look at *model distributions* and the *Normal Distribution* specifically.

Resistance and Measures of Spread

The *range, standard deviation,* and *variance* are *not resistant* to skewness and outliers.

The *Inter-Quartile Range* is *resistant* to skewness and outliers.

IQR and the identification of outliers by "Lower and Upper Fences"

Up to now, we have identified outliers by noting whether these data are "outside" the main body of data. There is a stricter definition that uses the *IQR*. We calculate what are sometimes called ***fences;*** any data point that is *outside* these fences is deemed an outlier and shown as a dot in a box plot; data that are within the fences are included in the whiskers or the box of a box plot. Here are the definitions of the upper and lower fences.

Definition of Lower and Upper Fences:

Lower Fence $= Q_1 - 1.5 * IQR$ Upper Fence $= Q_3 + 1.5 * IQR$

Example. The ***Summary Table*** for the distribution of *TuitionFees* for public four-year colleges is shown below. From the numbers there, we can calculate that the $IQR = 5496 - 3030 = 2466$, and with the *IQR* and the numbers for the Q_1 and Q_3 we can calculate the lower and upper fences.

Far West Colleges Collection

	TuitionFees
	1214
	3030
	3656
	5496
	12506
	62

S1 = min( )
S2 = Q1 ( )
S3 = median ( )
S4 = Q3 ( )
S5 = max ( )
S6 = count ( )

sector = "Public 4-year"

Lower Fence $= Q_1 - 1.5 * IQR = 3030 - 1.5 * 2466 = -669$, and

Upper Fence $= Q_3 + 1.5 * IQR = 5496 + 1.5 * 2466 = 9195$.

There are no outliers on the left since all the data are within the lower fence, which is negative here; on the right end, the maximum value is greater than the upper fence and is a dot. The whiskers are drawn to the biggest (or smallest) values that are within the fences. Here is graphic showing the fences and the data as well as the box plot. Usually, what we will see displayed is just the box plot and not the fences.

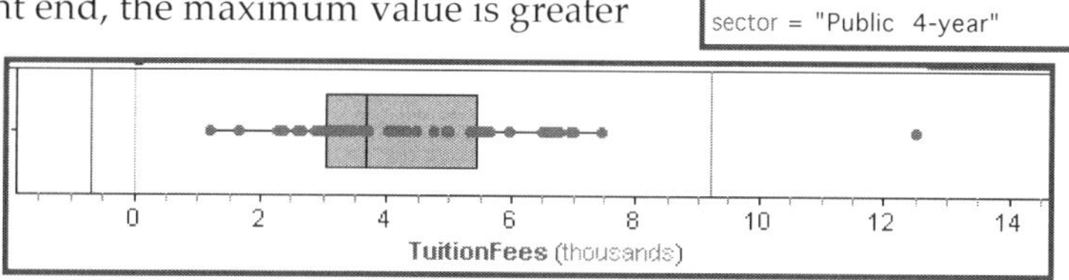

Summary: Measures of Spread

Measures of spread are numerical measures of the variability of data in distributions. There are four different measures that we use: the *range, the Inter-Quartile Range (IQR), variance* (s^2), and *standard deviation (s).*

- **Range** = *Maximum value – Minimum value*
- ***Inter-Quartile Range*** $IQR = Q_3 - Q_1$
- ***Variance*** $s^2 = \frac{1}{n-1}\sum_{i=1}^{n}(x_i - \bar{x})^2$
- ***Standard Deviation*** $s = \sqrt{\frac{1}{n-1}\sum_{i=1}^{n}(x_i - \bar{x})^2}$
- ***Interpretation of Measures of Variation*** The bigger the value of the standard deviation (s) or inter-quartile range (*IQR*), the more variability the data have. This interpretation is true for all of the measures. However, see the comments on *resistance* below.
- ***Resistance of Measures of Spread***
  - The *range, standard deviation* and *variance* are *not resistant* to skewness and outliers.
  - The *Inter-Quartile Range* is *resistant* to skewness and outliers.

  Therefore, the *range* may be a bad measure of variability, even though the concept is familiar. The *standard distribution* and *variance* are more appropriate for symmetrical distributions.
- ***Lower and Upper Fences***
  - Lower Fence = $Q_1 - 1.5 * IQR$
  - Upper Fence = $Q_3 + 1.5 * IQR$

  Whiskers are drawn to largest or smallest values within the fences; data outside the fences are shown by dots.

§1.7 Models for Distributions: The Normal Model

A model airplane or model train is a scaled-down version of the real airplane or train that hopefully preserves all of the features of reality. In mathematics and statistics, a model is often a function used to portray or replicate in succinct form some part of reality. We use models in statistics.

Distributions have *shape, center,* and *spread*, and we have found graphics (dot plots, histograms, box plots) to display and compare the shape, center, and spread of different distributions. We have also calculated measures so that we can compare the centers or locations of distributions (mean and median) and compare the spreads of distributions (standard deviation and inter-quartile range). But if we had a model of a distribution that we could use to compare with our actual real data distributions, we might be able to say: "Well, our distribution looks like a..." and then give a name that would be meaningful to others. By having a name (and perhaps some of the measures), these fortunate people would know pretty much what the distribution looked like. That is the idea of a ***model distribution***; a model distribution has a known shape, a known center, and a known spread, which are determined by a mathematical formula. The model that is the most important for this course is called the ***Normal Distribution***, although you may have thought of it as the ***bell curve.***

Example: Heights of Elderly English Women (EEW)

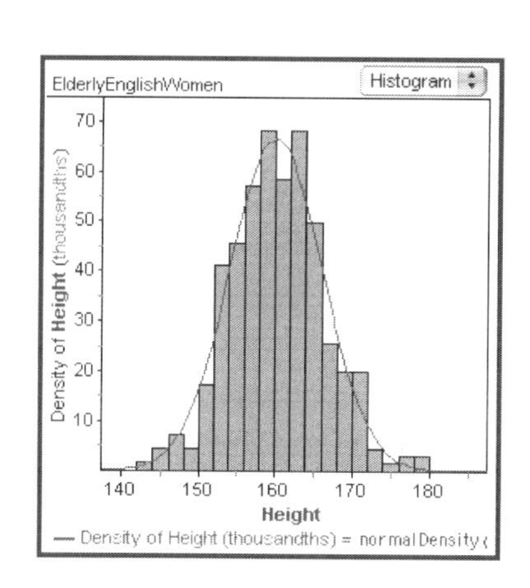

The histogram shows the distribution of the heights of $n = 351$ elderly English women who were part of a study of osteoporosis; they are short—their mean height is 160 centimeters. California and Penn State female students have a mean height of 165 centimeters, and Australian female high school students have a mean height of 164 centimeters. (To put all this in feet and inches, divide by 2.54 cm/in and get $160/2.54 \approx 63$ in. and $165/2.54 \approx 65$ in.—so 5′ 3″ and 5′5″). The smooth curve on top of the histogram is the ***Normal Distribution*** with mean 160 centimeters and standard deviation of 6 centimeters. From the graphic we can see that the Normal distribution model does not fit the histogram of the heights of the elderly English women *exactly*, but the model fits well.

Our Normal distribution smooth curve is one example of what is known as a ***density curve;*** it is a smooth curve and not a histogram because the normal distribution is a mathematical model—you can think of the density curve as the shape that you would get if you had a huge collection and a histogram of heights with an extremely large number of narrow bars, as in this example.

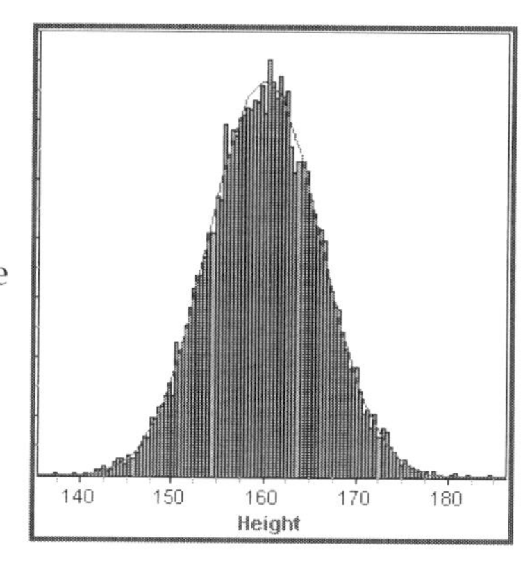

Histograms show an entire distribution; the bars include 100% of a collection, and typically we use them to determine the part or fraction or proportion of a distribution is above or below or between certain values of the variable, as you have done in the exercises. ***Density curves*** are designed for the same purposes, and so the area between the curve and the horizontal axis is 100%, or in terms of proportions, 1.

The Center and Spread of Normal Distributions

Our example collection of the heights of elderly English women (EEW for short) has mean 160 centimeters and standard deviation 6 centimeters. So, to this data we will fit a Normal model with mean $\mu = 160$ and standard deviation $\sigma=6$. We use Greek letters here (μ is Greek "m" and σ is Greek small "s") for the mean and the standard deviation because we are referring to the model distribution that we are fitting.

Notation for Normal Model Distributions	
μ is the mean	σ is the standard deviation

Center The ***Normal Distributions*** are perfectly symmetrical, so the mean μ is in the center of the distribution, with 50% of all of the distribution (think of the histogram again) on the left of the mean and 50% on the right of the mean. Hence, for our model for the heights of EEW, the mean is in the center of the graph.

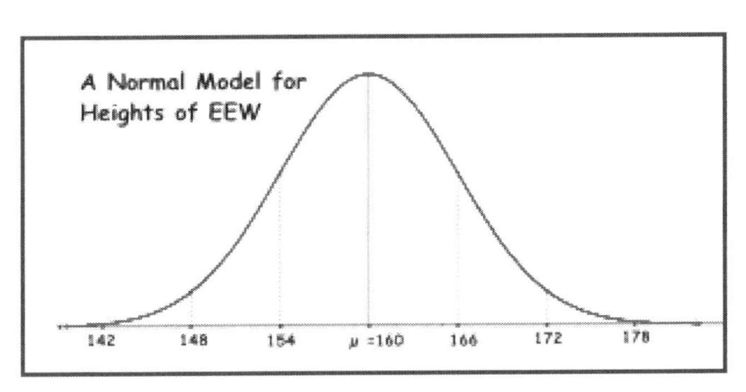

Spread What about spread? It is with this model that the standard deviation as we calculated it comes into its own. For these ***Normal Distributions*** the standard deviation measures the distance from the mean μ to the ***inflection point*** of the Normal density curve. The inflection point of the Normal curve is the point at which the slope changes from getting steeper and steeper to getting less and less steep. (To understand this, you can talk with little Udo, who has a Normal slide.) We can apply this to our model for the elderly English women. Since the standard deviation $\sigma = 6$ cm, we can draw on the horizontal scale the height that represents the mean plus one standard deviation, or $\mu + \sigma = 160 + 6 = 166$ cm. And, on the side of the mean that represents the EEW shorter than the mean, we can put $\mu + \sigma = 160 - 6 = 154$ cm. Then using this same distance (which is defined by the inflection point), we can specify the next standard deviation away from the mean on either side, and these will be $\mu + 2\sigma = 160 + 2 \cdot 6 = 160 + 12 = 172$, and $\mu - 2\sigma = 160 - 2 \cdot 6 = 160 - 12 = 148$.

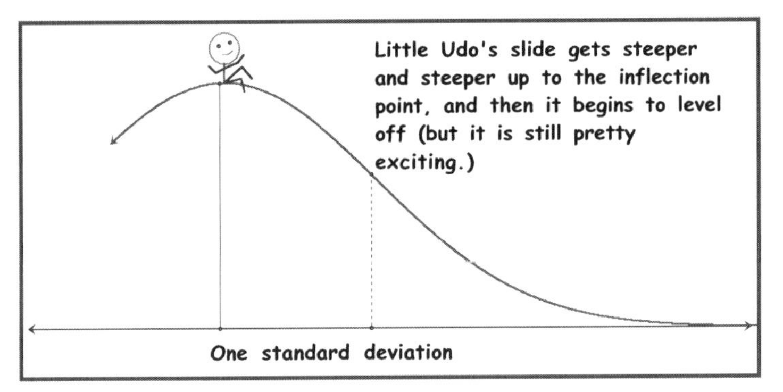

The Shape of the Normal Distributions and the Questions We Ask

For the *Normal Distributions* there is more to shape than symmetry. All Normal distributions have exactly the same shape, even though their pictures may make them look either tall and narrow or flattened out; the normal models have basically the same shape because their shape is determined by a mathematical formula.

To get an idea of what "the same shape" means, recall how we have been shading histograms to show the proportion of cases less than (or greater than) a particular value. We answered questions like this: "What proportion of EEW are shorter than 154 centimeters tall?" Or "What is the probability that an EEW is 172 centimeters or taller?" For the first question we looked at the heights of the bars of the histograms that included the women shorter than 154 centimeters, and we expressed that using our probability notation. If we did this for the EEW histogram, we would get $P(X < 154) \approx 0.15$, or about 15% of the EEW are shorter than 154 centimeters. For the question about the probability of a woman being 172 centimeters or taller we would do the same calculations, and finally get $P(X \geq 172) \approx 0.02$—that the probability that an EEW is 172 centimeters or taller is only about 2%.

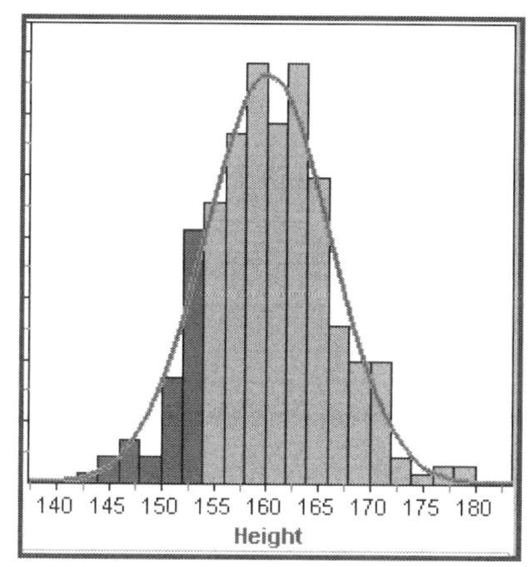

Think about what we are doing. We are using the space (actually the area of the histograms bars) to picture the answers to proportion or probability questions. *That is also what we will do with the Normal model.* Look at the graphic above; what can we say from the normal distribution model? We can say (because of the symmetry of the model) that 50% of the EEW have heights over the mean 160 centimeters and 50% less than 160 centimeters. We can actually say more because the model has a specific mathematical form. For any normal distribution we can specify that the proportion of data within one, two, and three standard deviations from the mean is approximately 68%, 95%, and 99.7%. This specification is called the ***Empirical Rule.***

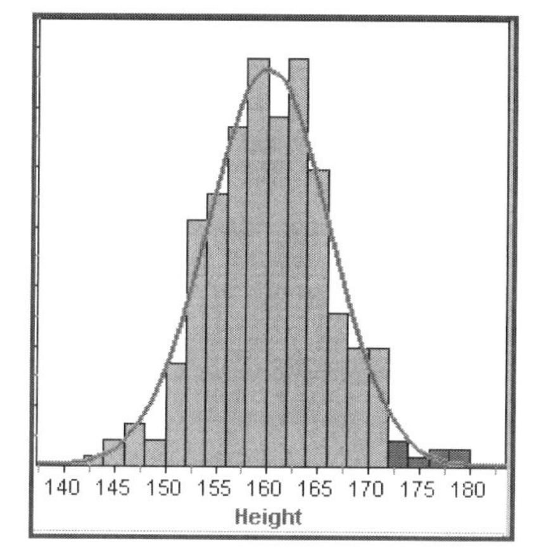

Empirical Rule for any Normal Distribution:

Approximately 68% of the distribution is between $\mu - \sigma$ and $\mu + \sigma$; i.e. within one sd of the mean.

Approximately 95% of the distribution is between $\mu - 2\sigma$ and $\mu + 2\sigma$; i.e. within two sd of the mean.

Approximately 99.7% of the distribution is between $\mu - 3\sigma$ and $\mu + 3\sigma$; i.e. within three sd of the mean.

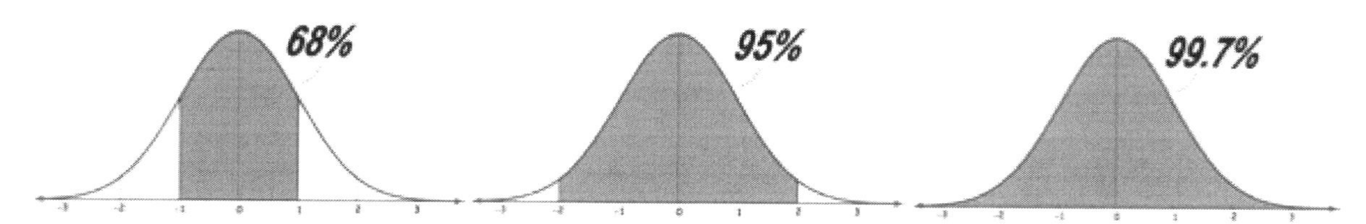

(We use the word "approximately" in the box above because the percentages are rounded; we will be able to get more exact numbers.) For our example of the elderly English women, this means that we can expect approximately 95% of the women to be between 148 centimeters tall ($\mu - 2\sigma$) and 172 centimeters tall ($\mu + 2\sigma$), according to the *Empirical Rule*. We can go farther than this.

Since 95% of the distribution is between 148 and 172 centimeters, it follows that 5% is outside this interval. That 5% we can equally divide between the women shorter than 148 centimeters and the women taller than 172 centimeters since the normal distribution is symmetrical. Therefore, we can say that the normal distribution model shows us that about 2.5% of the women should be 172 centimeters or taller. In symbols, our prediction from the model is that $P(X \geq 172) \approx 0.025$. This estimate from the normal distribution model is close to what we calculated from the data, and that shows that the model is a good fit.

Example 1: Using z Scores and the Normal Probability Chart

The proportion of EEW shorter than 151 cm? You should be able to see (using the same reasoning as just above) that about 2.5% of the EEW are shorter than 148 centimeters tall. However, if we want to know what proportion of the women are shorter than 151 centimeters, our *Empirical Rule* (68%-95%-99.7%) is not going to help us because 151 is somewhere between exactly one and two standard deviations shorter than the mean (see the graphic). The proportion of women shorter than 151 centimeters must be bigger than 2.5%, but what is it? We make a sketch (always!). In the sketch, the value of 151 centimeters is shown the thing we want. We want the proportion (or probability) of EEW who are shorter than 151 centimeters and that is shown the shaded region. We have a value (151 cm) and we want to know the area shaded in. Using our probability notation, we want to know $P(X < 151)$ and we are relying on the Normal model to describe the heights of EEW. To proceed, we need to calculate something called a *z score.*

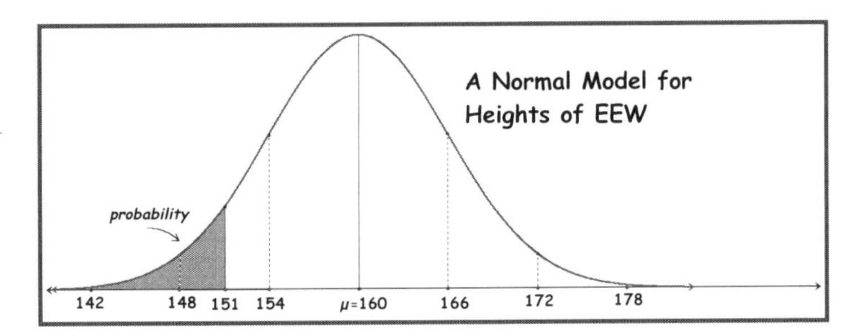

z scores. What we must do is to translate our question about the proportion of heights less than 151 centimeters into a question that can be answered with the ***Standard Normal Distribution;*** this is the Normal distribution that has mean $\mu = 0$ and standard deviation $\sigma = 1$. The 1, 2, and other numbers you see on the graphic are standard deviations (notice that the numbers -1 and 1 are at the inflection points). We do the translation using the formula for the ***z score***; we calculate

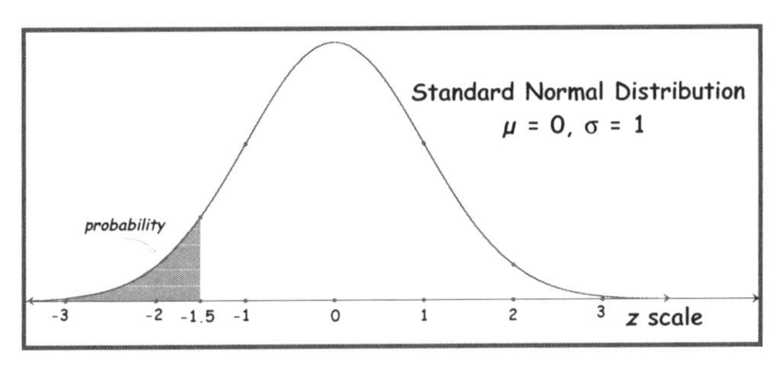

$z = \dfrac{x - \text{mean}}{\text{standard deviation}} = \dfrac{151-160}{6} = \dfrac{-9}{6} = -1.51$ since the Normal model that we are using for the heights of the women has μ =160 cm and σ = 6 cm. What this calculation tells us is that our height of x = 151 is $z = -1.50$ standard deviations from the mean. We can say that EEW who have a height of 151 centimeters are 1.5 standard deviations lower than the mean height of our model for the heights of the elderly English women.

Standard Normal Distribution

is the Normal distribution that has mean $\mu = 0$ and standard deviation $\sigma = 1$.

Definition and Formula for z score

The *z score:* $z = \dfrac{x - \text{mean}}{\text{standard deviation}}$ translates an x value on the variable being measured to a z value on the *Standard Normal Distribution.*

A *z score* shows the value of x in standard deviation units from the mean of a *Normal Distribution*.

Normal Probability Chart. The chart has the title **_Standard Normal Probabilities_** and can be found as the first chart in the **_Tables_** section after the end of the **_Notes._** The title refers to the numbers in the *body* of the table that give the proportion or the probability to the left of a given *z score*; the units digit and the tenths (the first number to the right of the decimal point) of *z scores* are given in the left-hand column. The second decimal of *z scores* is given in the row at the top of the chart. So, for our $z = -1.50$, the chart shows that the proportion of the shaded region is .0668. We conclude that our Normal model says that about 6.68% of the EEW have heights less than $x = 151$ cm.

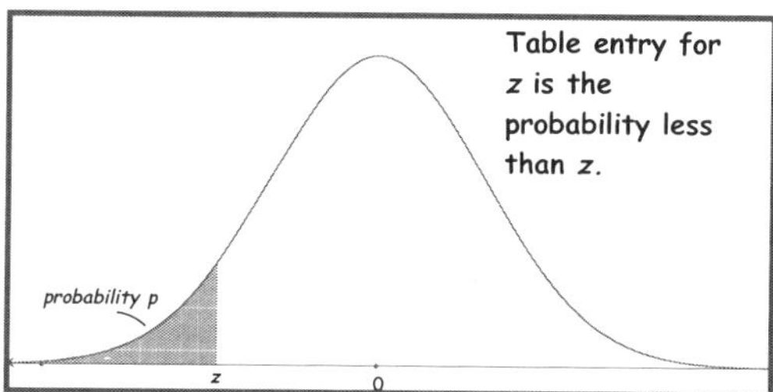

Shading and its meaning. We will often show a probability as a shaded area as in the graph above. This may be a new idea for you; you may well be more comfortable putting a numerical value on a scale (e.g., on the x axis). Our advice: get comfortable with shading; that is how we show probabilities in distributions.

Example 2: Given a Value—Find a Proportion

The type of problem we worked in the example (at the same time introducing the z score and the *Normal Probability Chart*) asks for a proportion when we have (or are given) a value; hence we call it a **_Given a Value—Find a Proportion_** problem. Our question is very similar: what proportion of elderly English women (EEW) has heights less than 168 cm? In our probability notation, we want to find $P(X < 168)$.

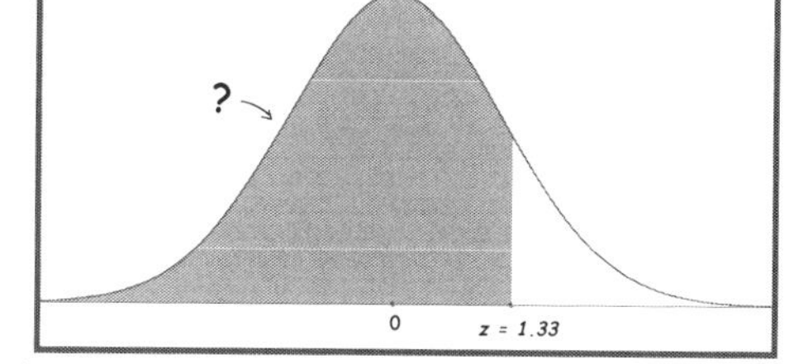

1. Sketch a Normal curve showing what you are trying to find; the "question mark" alerts us that we want a probability. (We do not have to place z on our graph exactly; since 168 is bigger than 160, our z will be positive.)
2. Calculate $z = \dfrac{x - \text{mean}}{\text{standard deviation}} = \dfrac{168 - 160}{6} = \dfrac{8}{6} \approx 1.33$ using $\mu = 160$ cm. and $\sigma = 6$ cm. (Now we can put $z = 1.33$ on our sketch.)
3. Consult the **_Normal Probability Chart_** for $z = 1.33$. Now we have to use the second decimal place at the top of the chart. For the row 1.3, and the column .03, (shown here) we see $P(X < 168) = P(z < 1.33) = .9082$.

z	.00	.01	.02	.03
0.0	.5000	.5040	.5080	.5120
0.1	.5398	.5438	.5478	.5517
0.2	.5793	.5832	.5871	.5910
0.3	.6179	.6217	.6255	.6293
0.4	.6554	.6591	.6628	.6664
0.5	.6915	.6950	.6985	.7019
0.6	.7257	.7291	.7324	.7357
0.7	.7580	.7611	.7642	.7673
0.8	.7881	.7910	.7939	.7967
0.9	.8159	.8186	.8212	.8238
1.0	.8413	.8438	.8461	.8485
1.1	.8643	.8665	.8686	.8708
1.2	.8849	.8869	.8888	.8907
1.3	.9032	.9049	.9066	.9082
1.4	.9192	.9207	.9222	.9236
1.5	.9332	.9345	.9357	.9370

4. Interpret the result in the context of the question. "According to the Normal model, we can expect that nearly 91% of elderly English women will be shorter than 168 centimeters."

Actually, when we use the Normal model, the answer to the question "what proportion of these women have heights *168 centimeters or less* (including the x = 168 cm)?" will be the same as the answer to the question "what proportion of women have heights *less than* 168 centimeters (excluding the x = 168 cm)?" This is because when we use the Normal model, we are using a continuous density curve. When we get our answer from a histogram or a dot plot, or from a collection of data, there *will* be a difference in the answers to the questions "what is $P(X \leq 168)$?" and "what is $P(X < 168)$?"

Example 3: Given a Value—Find a Proportion

Our question is: "According to the Normal model, what is the probability that an elderly English woman in our collection is taller than 170 centimeters?" We are given a value and want to find a proportion—which is here interpreted as a probability. Read the question carefully; here we are asked about heights *greater than* ("taller") 170 centimeters rather than "less than." In symbols we want $P(X > 170)$. That makes this question slightly different. Here is the solution in steps.

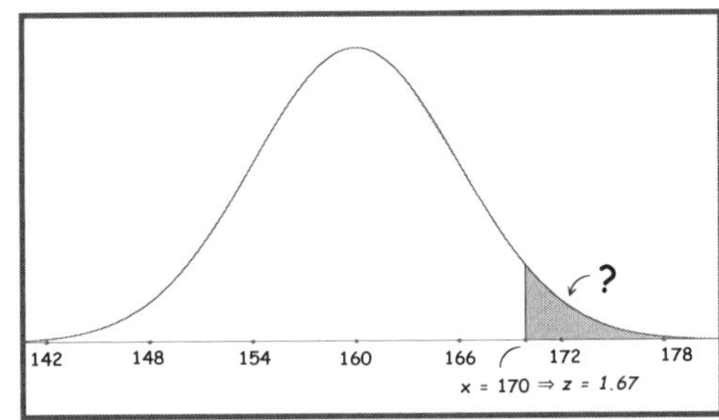

1. Make a sketch. The sketch incorporates what we want: "what is the proportion greater than 170 centimeters?" (We have shown the Normal sketch with the heights for 1, 2, etc. standard deviations; you may make the sketch with or without these shown. The sketch is meant to help.)
2. Calculate $z = \frac{x - \text{mean}}{\text{standard deviation}} = \frac{170 - 160}{6} = \frac{10}{6} \approx 1.67$ which we can now put on the sketch.
3. Consult the ***Normal Probability Chart***. The number that corresponds to z = 1.67 is .9525. This cannot be the answer to our question! That would be saying that 95% of elderly English women are *taller* than 170 centimeters. Remember that the chart gives the proportion *less than* a z; in other words, the chart is telling us that $P(z < 1.67) = .9525$. We want the other side—the "greater than" side. This looks like a "not" probability problem; indeed it is. The solution is: $P(z > 1.67) = 1 - P(z < 1.67) = 1 - .9525 = .0475$ or, in words, "the probability that a woman is taller than 170 centimeters" *is* 1 – "the probability that a women is *not* taller than 170 centimeters." Comparing the number we got with the size of the shaded area, we see that 4.75% makes sense. [Remember that in using the Normal model, $P(z < 1.67) = P(z \leq 1.67)$.]
4. Interpret in context. "According to the Normal model, we can say that about 4.75% of the elderly English women are taller than 170 centimeters."

Checking Hand Calculated Answers with Software

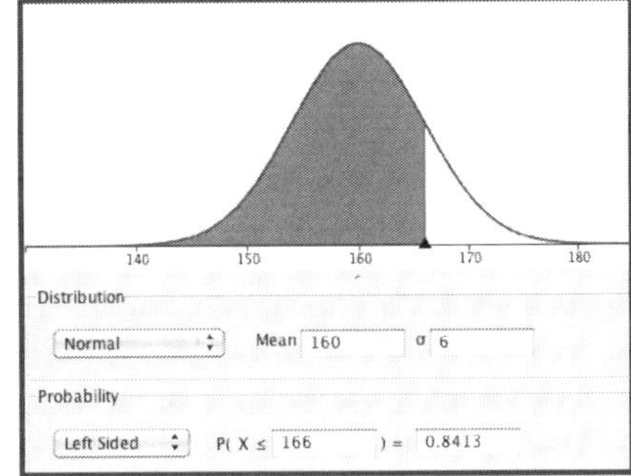

There are many software applications that can do all these calculations. Here is a solution using ***DistributionCalculator.ggb***, one of the easiest applications to use. The answer is disshow the $P(X < 168) = .9082$ is displayed by the shading under the curve.

Example 4: Given a Proportion—Find a Value

We can use a Normal model to answer what is really the "reverse" question. What values of our variable correspond to a given proportion of the data?

When we have this situation— we are given a proportion or probability, and what we want is a range or interval of values—we call this a ***"Given a Proportion — Find a Value"*** problem. Here is a simple example: "Using the Normal model with μ =160 cm. and σ = 6 cm, what are the values of the middle 95% of the distribution of the heights of elderly English women?"

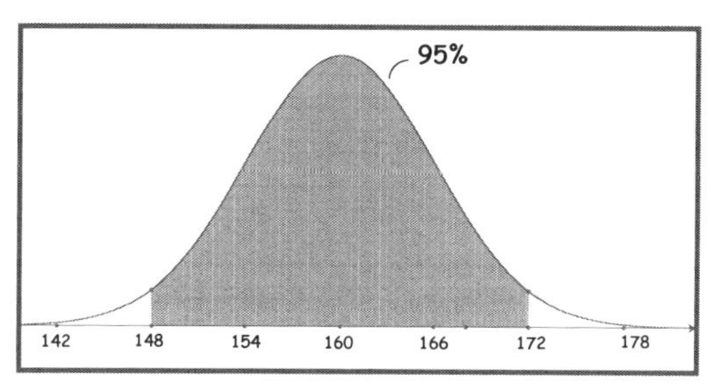

The answer to this question is not difficult; it can be derived from the *Empirical Rule*. Since about 95% of a Normal distribution is within two standard deviations on either side of the mean, we can say that about 95% of the women will have heights between 148 centimeters and 172 centimeters. Notice that the answer to this question is a range of values and not a probability; the probability (or proportion) was *given*. In symbols: $P(148 \le X \le 172) \approx 0.95$.

Example 5: Given a Proportion—Find a Value

Let us ask a more difficult question that cannot be answered with the *Empirical Rule*. We may want to know the *height of the tallest 5% of elderly English women.*

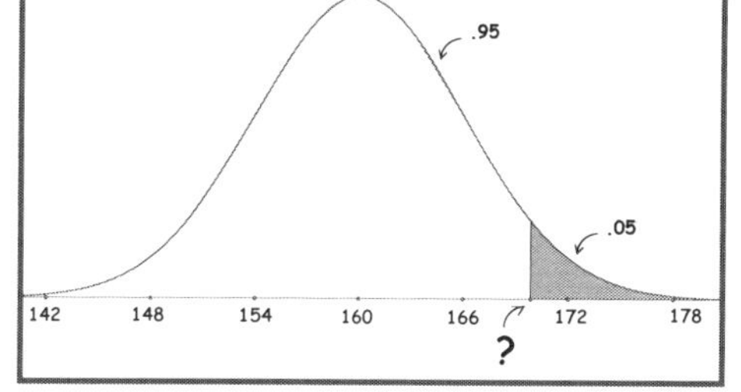

1. Make a sketch. Our sketch looks very much like the one for Example 3, but it is not the same; we see ".05" instead of a question mark pointing at the shaded region, showing that this region is what is *given* and *not* what is sought after. The question mark is pointing at a value for X, a height; this is what we want.
2. We work backwards; rather than calculate a *z score* using a value, we find the *z score* implied by our given .05. To find this we look at the body of the **Normal Probability Chart,** but remember that the chart gives cumulative proportions (proportions less than a *z*); the chart gives proportions to the left of a *z*. We need to look at what is to the left of our shaded area, and that is .95. So look at the *body* of the chart for .95. Alas, we do not see .9500, but we see two entries that are close to .95. One is .9505, and that entry corresponds to a *z score* of $z = 1.65$, and we see .9495, and that corresponds to $z = 1.64$. If our desired number were closer to either one of these, that it the one we would choose for our *z*. Since each of these entries are exactly the same distance from .9500, we get the midpoint, and our $z = 1.645$,

z	.00	.01	.02	.03	.04	.05	.06
0.0	.5000	.5040	.5080	.5120	.5160	.5199	.5239
0.1	.5398	.5438	.5478	.5517	.5557	.5596	.5636
0.2	.5793	.5832	.5871	.5910	.5948	.5987	.6026
0.3	.6179	.6217	.6255	.6293	.6331	.6368	.6406
0.4	.6554	.6591	.6628	.6664	.6700	.6736	.6772
0.5	.6915	.6950	.6985	.7019	.7054	.7088	.7123
0.6	.7257	.7291	.7324	.7357	.7389	.7422	.7454
0.7	.7580	.7611	.7642	.7673	.7704	.7734	.7764
0.8	.7881	.7910	.7939	.7967	.7995	.8023	.8051
0.9	.8159	.8186	.8212	.8238	.8264	.8289	.8315
1.0	.8413	.8438	.8461	.8485	.8508	.8531	.8554
1.1	.8643	.8665	.8686	.8708	.8729	.8749	.8770
1.2	.8849	.8869	.8888	.8907	.8925	.8944	.8962
1.3	.9032	.9049	.9066	.9082	.9099	.9115	.9131
1.4	.9192	.9207	.9222	.9236	.9251	.9265	.9279
1.5	.9332	.9345	.9357	.9370	.9382	.9394	.9406
1.6	.9452	.9463	.9474	.9484	.9495	.9505	.9515
1.7	.9554	.9564	.9573	.9582	.9591	.9599	.9608
[illegible]	[illegible]	[illegible]	[illegible]	[illegible]	[illegible]	[illegible]	[illegible]

3. Use the *z* score formula $z = \dfrac{x - \text{mean}}{\text{standard deviation}}$ and solve for what we want, *x*. We get:

$1.645 = \dfrac{x - 160}{6}$; multiplying, $6 \times 1.645 = x - 160$ and finally $9.87 + 160 = x$, we find $x = 169.87$ cm.

4. Interpret in context. "Our estimate according to the Normal model is that the tallest 5% of elderly English women have heights very near to 170 cm (169.87 cm)."

The Uses of Models and Specifically Normal Distributions

"Essentially, all models are wrong, but some are useful," wrote George E. P. Box.[1]

The *Normal distribution* for heights is a good example of this principle; as a mathematical model, the Normal curve never actually touches the horizontal axis. That means that according to the model, there is some extremely small probability that someone could be just one centimeter tall. Nonetheless, the Normal model is useful for our collection. Our example uses heights; however, other measures, such as foot lengths or arm spans, tend to have distributions that can be approximated by a Normal distribution. Normal distributions seem to fit biological measurements well (at least within the same species).

Here is another application; if you made repeated measurements of the same quantity (say, the weight of one single hefty Physics textbook) you would find that most of the measures of weight would cluster close to a peak (as the Normal distribution does), some of the measurements would be a bit off from this peak in one direction or another, and just a few of the measurements would be far off—either bigger or smaller.

The Normal model is also very useful in describing the distribution of averages of samples drawn randomly from the same population. It appears that Normal distributions describe situations where we have a great many small and independent "forces" operating, which is what we think is happening when we measure the same thing over and over again.

Of course, there are collections of data and there are situations where the distributions are not anywhere near to being Normal or even symmetrical; you have seen such skewed data. Many of the measures for the real estate data are extremely right-skewed; these data will not be well modeled by a Normal distribution. For skewed data, and for situations that do not lead to the Normal model, statisticians have other model distributions, and you will see some of these.

---

[1] Box, George E. P. and Norman R. Draper (1987), *Empirical Model-Building Response Surfaces,* New York: John Wiley and Sons, p. 424

Summary: Models of Distributions

- ***Models of Distributions*** are used to summarize the shape, center, and spread of actual distributions in a concise way. These model distributions are generally drawn from mathematical theory.
- ***Density Curves*** show the probability that a variable attains a range of values as the area between the curve and the horizontal axis if the variable has a (usually) model distribution. Since the sum of the probabilities for a sample space must sum to one, the area under a density curve must be one.
- ***Normal Distribution*** is a useful single-peaked model used extensively in statistics that:
  - Is perfectly symmetrical about its mean μ
  - Has its standard deviation σ located at the inflection points of the curve, such that
- ***Standard Normal Distribution*** is the normal distribution with $\mu = 0$ and $\sigma = 1$.
- ***Empirical Rule*** is a quick way of describing the shape of a Normal distribution; it says:
  - Approximately 68% of the distribution is between $\mu - \sigma$ and $\mu + \sigma$; i.e. within one sd of the mean.
  - Approximately 95% of the distribution is between $\mu - 2\sigma$ and $\mu + 2\sigma$; i.e. within two sd of the mean.
  - Approximately 99.7% is between $\mu - 3\sigma$ and $\mu + 3\sigma$; i.e. within three sd of the mean.
- ***z score*** $z = \dfrac{x - \text{mean}}{\text{standard deviation}}$ translates an x value to a z value on the *Standard Normal Distribution.*
  - A *z score* shows the value of x in standard deviation units from the mean of a *Normal Distribution.*
- ***Normal Probability Chart*** A chart showing the area (and therefore the probability) under the *Standard Normal Distribution* to the left of (or less than) a given value of z. Since the chart gives probabilities less than a value of z, probabilities greater than a value of z must be found by subtracting the probability found in the chart *from* 1, the total probability under the Normal density curve.
- The software application ***DistributionCalculator.ggb*** can be used to do the probability calculations and give a graphic showing the solution, where…
- ***Shading indicates the size of a probability.*** When working with density curves, a proportion (or probability) is shown by shading in an area of a curve. 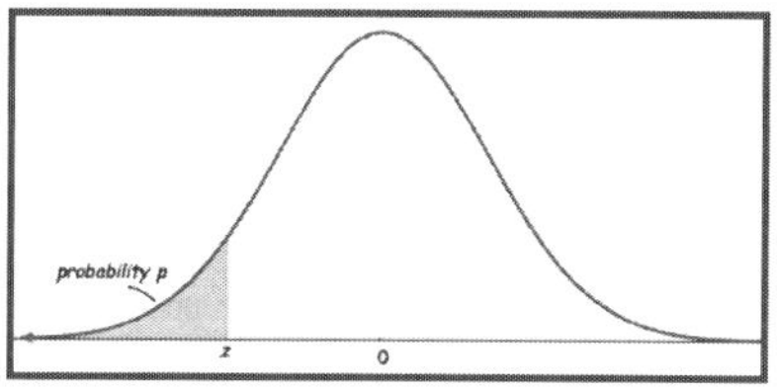
- ***Given a value, find a probability*** If a particular value is known for a variable modeled by a Normal distribution with mean μ and standard deviation σ, find the z score and consult the Normal Probability Chart, keeping in mind that the chart gives probabilities to the left of the z score.
- ***Given a probability, find a value*** If a particular probability is given for a variable modeled by a Normal distribution with mean μ and standard deviation σ, work backwards by finding the z score for the relevant probability and then use the formula for the z score to find the value.
- ***Normal Distributions*** are used to describe some empirical distributions, errors in measurement, and random phenomena.

§2.1 Comparisons, Relationships, and Interpretation

Review Example: Using the tools to calculate and interpret

This section begins by reviewing what we have done in the previous sections. The first examples are similar to the exercises for this section. At the same time, we will look at some statistical questions in some new ways.

Data on used BMWs, Audis, etc. In this data set, each case is a used car that had been placed on the website www.cars.com in Boston, Chicago, Dallas, or the San Francisco Bay Area. We chose to look at only five different models that compete in the "imported " sport sedan sector; the five are the Audi A4, BMW 3 Series, Mercedes C-class, Infiniti G35, and Lexus IS. The data were collected from the website in summer 2009. All of the cars were chosen of the makes listed just above that were advertised on the site for the four places listed (Boston, etc.). We can compare the cars or we can compare the places. The data look like this; the variables and their definitions are listed below.

Summer 09 BMWLexusInfiniti Student

	Make1	Place1	Year	Price	Miles	Distance	Age	Convert...	NoPrice	Body	Dist
605	BMW 3 S...	SF Bay A...	2008	31788	13194	33	1.33333	Not Conv...	Price Given	Sedan	33 mi.
606	BMW 3 S...	SF Bay A...	2008	30988	13656	33	1.33333	Not Conv...	Price Given	Sedan	33 mi.
607	BMW 3 S...	SF Bay A...	2008	30788	14573	33	1.33333	Not Conv...	Price Given	Sedan	33 mi.
608	BMW 3 S...	SF Bay A...	2008	30692	8040	11	1.33333	Not Conv...	Price Given	Wagon	11 mi.
609	BMW 3 S...	SF Bay A...	2006	30388	26787	33	3.33333	Not Conv...	Price Given	Sedan	33 mi.
610	BMW 3 S...	SF Bay A...	2006	29995	30862	27	3.33333	Convertible	Price Given	Convertible	27 mi.
611	BMW 3 S...	SF Bay A...	2006	29995	24907	27	3.33333	Not Conv...	Price Given	Sedan	27 mi.

Make1:	*Make of the car: Audi A4, BMW 3-Series, Mercedes C-Class, Infiniti G35, Lexus I.S*
Place1:	*Place the car was being sold: Boston, Chicago, Dallas, San Francisco Bay Area*
Year:	*Model year of the car being sold*
Price:	*Price that the seller was asking for the car*
Miles:	*Miles that the car being advertised had been driven*
Distance:	*Distance that the seller is from the zip code entered in the www.cars.com search engine*
Age:	*Age of the car; depends upon the model year and time of year the data were collected*
Convertible:	*Whether the car being sold is a convertible or not*
NoPrice:	*Whether or not the price of the car is given*
Body:	*Body style of the car: sedan, coupe, convertible, wagon, hatchback*

First Question: Examining two categorical variables *In which city (Boston, Chicago, Dallas, or San Francisco) are convertibles more likely to be found in the cars for sale?* How do we answer this question? This looks like the kind of question that we answered in §1.2 when we looked at probability; we want to *compare* the different places by the proportion of convertibles for sale. Using software, we would make a *summary table* such as the one shown. If we let C stand for the *event* that the car being sold is a convertible, we would then calculate the conditional probabilities for convertibles $P(C \mid Boston)$, $P(C \mid Chicago)$ etc. and compare these. Here are two examples for review:

Summer 09 Used Cars

		Convertible		Row Summary
		Convertible	Not Convertible	
Place1	Boston	17	338	355
	Chicago	60	631	691
	Dallas	36	558	594
	SF Bay Area	39	536	575
Column Summary		152	2063	2215

S1 = count ( )

$P(C \mid Boston) = \dfrac{17}{355} \approx 0.048$ and $P(C \mid SF\ Bay\ Area) = \dfrac{39}{575} \approx 0.068$. It will be a good exercise to calculate and compare the other two conditional probabilities; you may be surprised at the result.

The lesson here is that if we want to examine the relationship between two categorical variables then the tool that we have at our disposal is to compare conditional probabilities. We can think of this as comparing one of the categorical variables (likelihood of a convertible) by the categories of the other categorical variable (the place).

Second Question: A quantitative variable by categories of a categorical variable

If we look just at the model year 2006 cars being sold, are there differences in the distributions of miles driven between convertibles and cars that are not convertibles? (In other words, are convertibles driven more or less?) This looks like the kind of question we answered when we were discussing the center and spread of distributions. Here are the kinds of analyses we would do. We would make both a graph showing the distributions and get a summary table showing the calculations. Our results show that the distribution for *Miles* for the convertibles is to the left the distribution for the non-convertibles, indicating that the convertibles were on average driven fewer miles. The mean and medians (numerical measures of location) agree with this assessment, since the numbers are higher for the non-convertibles than for the convertibles. The measures of spread (standard deviation and iqr) indicate that there is also somewhat *less variability* in the miles driven for the convertibles being sold than for the other body styles.

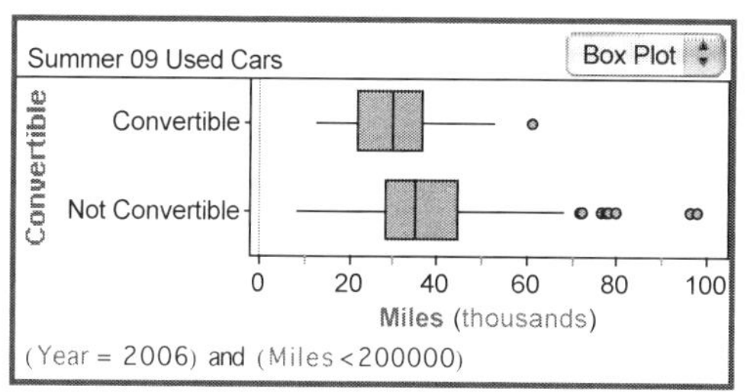

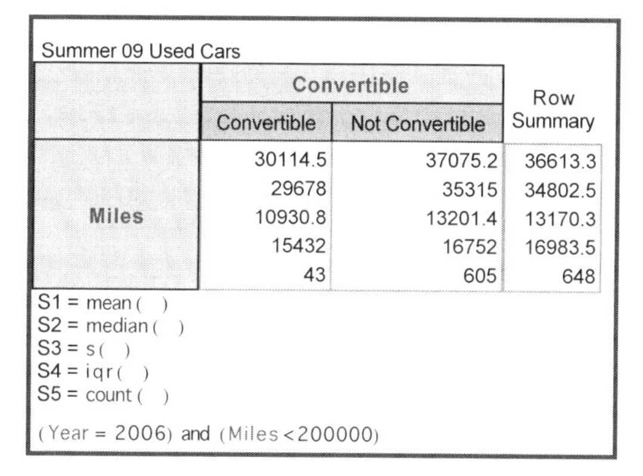

Summer 09 Used Cars

	Convertible: Convertible	Convertible: Not Convertible	Row Summary
Miles	30114.5	37075.2	36613.3
	29678	35315	34802.5
	10930.8	13201.4	13170.3
	15432	16752	16983.5
	43	605	648

S1 = mean( )
S2 = median( )
S3 = s( )
S4 = iqr( )
S5 = count( )

(Year = 2006) and (Miles <200000)

How do we decide what to calculate?

Think: Categorical and Quantitative Variables How did we decide that we should calculate means, medians, standard deviations, and IQRs in the *second question* and probabilities in the *first question?* The answer is that we looked at whether the variables were *categorical* or *quantitative.* In the *first question,* the two variables (*Place1* and *Convertible*) are both *categorical*; to handle the situation where we are comparing categories of one categorical variable (*Convertible*) according to the categories of a second categorical variable (*Place1*), we employ conditional probabilities. In the second question, one of the variables is *quantitative* (*Miles*), but the second variable (*Convertible*) is categorical. Since we have a quantitative variable, it makes sense to calculate means and medians and compare them according to the categories of the categorical variable.

We can also use the tools of probability when we have a quantitative variable; for example we could calculate the proportion of 2006 cars for sale that have been driven more than (say) 27,000 miles for the convertibles and non-convertibles. The results can be written $P(Miles > 27000 \mid C) \approx 0.56$ and $P(Miles > 27000 \mid NC) = 0.79$.

Summer 09 Used Cars

	Convertible: Convertible	Convertible: Not Convertible	Row Summary
Miles	0.55814	0.785124	0.770062

S1 = proportion(Miles >27000)

(Year = 2006) and (Miles <200000)

Think: Explanatory and Response Variables The phrase "according to the categories of the categorical variable" appeared just above. We call a variable ***explanatory*** when we have reason to think that a second variable (which we call a ***response variable***) will vary according to the values of this ***explanatory variable***. So, we suspect (though we do not know until we look at the data) that the likelihood that a car for sale is a convertible (the *response variable*) will vary according to the place (the *explanatory variable*) in which it is sold. That is, we think that the *miles driven* may vary among the various makes of car. Of course, there may be no substantial differences in the *miles driven* among the various makes of car, but we will not know until we look at the data. Here are some questions to consider:

- How do you decide which variable is the *explanatory* and which variable is the *response* variable? The answer depends on how we think one variable affects another or which variable is in some sense "prior." For example: we think that it may be possible that the convertibles are driven less over a number of years—so that the fact that a car is a convertible may affect its miles driven. However, the miles that a car is driven cannot affect whether it is a convertible or not.
- Can *quantitative* variables also be *explanatory?* Yes; a large part of this chapter is about quantitative variables that are explanatory. The age of a car we think affects the price of a car; in this example, age is a *quantitative* variable being used as an explanatory variable.
- Are there situations where the choice of one variable as *explanatory* and the other as *response* is either impossible or does not matter? Yes. If you had data on the exam scores on a collection of students who were all taking chemistry and history then we cannot say that chemistry affects history or history affects chemistry, even though there may be a relationship between the chemistry and history scores.
- Is the distinction between *explanatory* and *response* variables the same as the distinction between *independent* and *dependent* variables? Yes; in mathematical terms, the distinction is the same. The reason that different terms are used in statistics texts is that the terms *independent* and *dependent* have a different, and well-defined, meaning.

Explanatory and Response Variables

An ***explanatory*** variable allows us to predict in some way the values of a ***response*** variable.

Putting these bits of advice together, we offer these guidelines.

Guidelines for Calculation

- If the variables in the statistical question are both *categorical*, calculate *conditional probabilities* with the categories of the *explanatory variable* as the conditions.
- If one variable is *quantitative,* make a graphic that compares the distributions of the *quantitative variable* according to the categories of the *categorical variable.*
- If one variable is *quantitative,* calculate and compare measures of center and spread for the categories of the *categorical variable.*

Suppose the two variables are *both quantitative?* We have not considered that situation up to now.

Comparisons between Groups or...Relationships between Variables

Consider the example at the beginning of this section. The most natural way of thinking about the data in the first example—the one about the proportion of convertibles in the four places—is to think of four groups: the four different places the data were collected. We compare the proportions of convertibles being sold in these four places. However, there is another way to think to look at the same table.

Summer 09 Used Cars

		Convertible		Row Summary
		Convertible	Not Convertible	
Place1	Boston	17	338	355
	Chicago	60	631	691
	Dallas	36	558	594
	SF Bay Area	39	536	575
Column Summary		152	2063	2215

S1 = count ( )

We can *also* think of *Place1* and *Convertible* as two variables within one collection; after all, both of these are columns in a spreadsheet, and we think of the columns as variables and the rows as cases. If we think this way then the *summary table* shows the ***relationship*** between the variables. Whether we think of the categories as groups, or whether we think of them as values of a variable, the calculations we make are very often the same.

Although the calculations are the same, there may be differences in the way in which the data were collected that lead to differences in interpretation. Consider the cars example. We collected the data on the cars from four different places; the fact that we did leads us to think of comparing places. On the other hand, in those four places, *for each car* we recorded whether the car was a convertible and *for each car* we recorded the miles that the car was driven. The fact that both of these variables (*Convertible* and *Miles*) were recorded (or measured) *for each case* leads us to think of looking at the *relationship* between a car being a convertible and the miles that the car was driven. To analyze the proportion of convertibles by place, so far we have just one choice of the kind of calculation we will do: we will calculate conditional probabilities. However, we can *look* at these calculations from two different perspectives.

Calculation and then...Interpretation

You should already be aware that it is not sufficient just to quote the numbers; the numbers need to be related to the context of the data. However, it is also wise to keep in mind how the data were collected. Do the proportions of convertibles in Boston and San Francisco tell us that about 4.8% of the cars in Boston are convertibles and about 6.8% of cars in San Francisco are convertibles? It may be, but we cannot *infer* or generalize that from our data; recall that these data represent only cars being sold over the Internet and only five models of "imported sport sedans." It may even be that the proportion of convertibles *for sale* may be higher than the proportion of convertibles on the road. Keep in mind how the data were collected.

Interpreting calculations: how to complete the Writing Exercises successfully

The exercises for this section involve analyzing data and then writing about what you have found. You will find this a challenge because you probably have not done this in the past; you will make mistakes, and you will have to work at it and revise what you have done. Some advice:

- ***Understand the context: the cases, the variables.*** Make certain you understand clearly what your cases are, what the variables are, whether the variables are quantitative or categorical, and which variable is being considered as the explanatory and which variable is the response.
- ***Correct calculations.*** Make certain that the calculations that you are doing (or having software do) are the calculations that are appropriate for the questions being asked.

- ***Consider your audience.*** Your audience is *not* your instructor. It is better to think of your audience as the readers of a student newspaper. A common mistake is to be too technical. You do not need to explain the standard deviation formula; you *may* need to explain to your readers, in the context of the data, what the standard deviation tells you.
- ***Beware writing "it" and "they."*** Writing about technical matters demands that we be very specific. A common writing mistake with statistics is to use "it" and "they" vaguely. Here is an example: "It is right-skewed." What is right-skewed? Australian students? Used cars? Mammals? (What does a right-skewed Australian student or used car or mammal look like?) You are of course referring to the *distribution* of a variable. Distributions of variables can be right-skewed: Australian students or mammals cannot be. Be specific: "The distribution of the variable [name the variable being analyzed] is right-skewed."
- ***Over-Inference***. Another common mistake is to read into the data more than the data can bear. "The students who play lots of computer games come from dysfunctional families, and so..." when in fact there are no data about family backgrounds and you have no way of knowing what you have written.
- ***Start early and check your work.*** This assignment is not one you can do well by starting the night before the assignment is due. If you try that, there is a high probability that you will fail (that is, $P(\text{Fail} \mid \text{Start Late}) > 0.8$ or $P(\text{Fail} \mid \text{Start Late}) > 0.9$).

You will find these exercises challenging; however, if you master the idea, you will ultimately find it very interesting to be able to take some data, do some calculations to analyze the data, and then make sense of what the data show you about a statistical question. Good luck!

Summary: Types of Variables, Calculation, and Interpretation

- ***Calculations depend upon the types of variables involved.***
  - Analyzing two *categorical* variables leads to using conditional probabilities.
  - Analyzing a *quantitative* variable by the categories of a *categorical* variable generally leads to comparing the graphics and the summary measures of the quantitative variable according to the categories of the categorical variable.
- ***Calculations depend upon which variable is explanatory and which is response.***
  - An **explanatory variable** allows us to predict in some way the values of a **response variable**.
  - The choice of which variable is *explanatory* and which variable is *response* depends upon how we think the variables are related.
  - It may be neither of the two variables can be assigned the role of *explanatory* or *response*.
- ***Interpretation of calculations***
  - Must take account of the statistical question being asked
  - Must take account of how the data were collected
  - Must take account of the audience to which the interpretation is directed

§2.2 A Graphic, a Model, and a Measure

Cars lose their value with age; a bigger house is more valuable

These are the kinds of relationships between variables we will explore in the remainder of the chapter. Notice that we have *two quantitative* variables, and we are looking at the relationship between them. How can we handle this new situation? Answer: in a way similar to what we did before—we will look at graphics, a model, and a measure. When we looked at a single distribution, we looked at dot plots (or box plots, or histograms)—in other words, *graphics*—then we calculated measures such as the median, IQR, or the mean and standard deviation, and, finally, for certain kinds of distributions, we considered the *Normal distribution* as a model of the data.

Graphic: a Scatterplot Here is an example of the $n = 23$ Lexus IS cars for sale in Boston in summer 2009. A **Scatterplot** is essentially an $X - Y$ coordinate system where x is the horizontal axis and y is the vertical axis. Each dot represents the **ordered pair** of the values on each of the two variables for each case, where here the cases are cars for sale on the Internet through www.cars.com. Here the ordered pairs represent the *ordered pair* (*Age, Price*) so that the most expensive Lexus for sale is plotted at (1.33, 46985), that is, was 1.33 years old (a 2008 car) and the price was $46,985. The oldest Lexus for sale was plotted at (7.33, 14995). However, it is the pattern of the data for all of the points plotted that interests us, and from the pattern we can see what we know: cars lose their value as they age. The higher the age, we see that the lower the price is, in general. Of course, there is some variability depending on other factors about the cars.

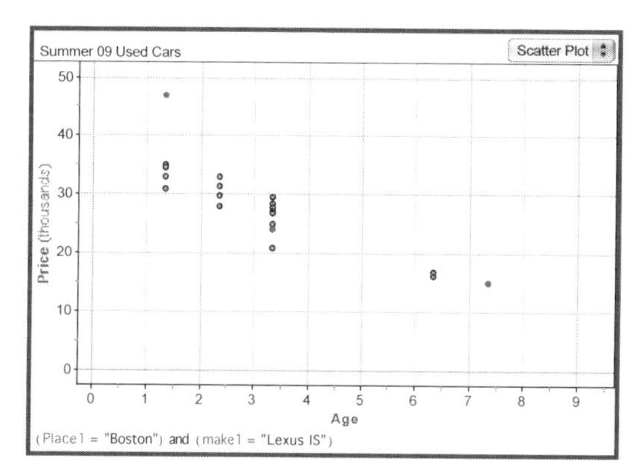

Scatterplots: Direction, Form, and Strength Here is another example of data plotted on a *scatterplot*. Here the cases (the dots) are houses that have been sold in San Mateo County, and the variables are a measure of the living area of the house for sale *SqFt* and the *ListPrice* of the house. The first thing we notice about the data is something we think we know: as the size of the house increases, so does the value (or listed price) of the house. There are many more houses than cars, but that fact is incidental.

Direction With *scatterplots* we look first at the **direction** of the plot. A plot like the one relating *SqFt* and *ListPrice* shows a **positive** relationship; since as the x value increases, the y value also increases. A plot like the one relating *Age* and *Price* for used cars shows a **negative** relationship; as the x value increases, the y value *decreases*. It is possible for a plot to show a relationship that is neither positive nor negative; here is an example. Our question is: do older or younger houses sell more quickly so that their *Time in Market* is less?

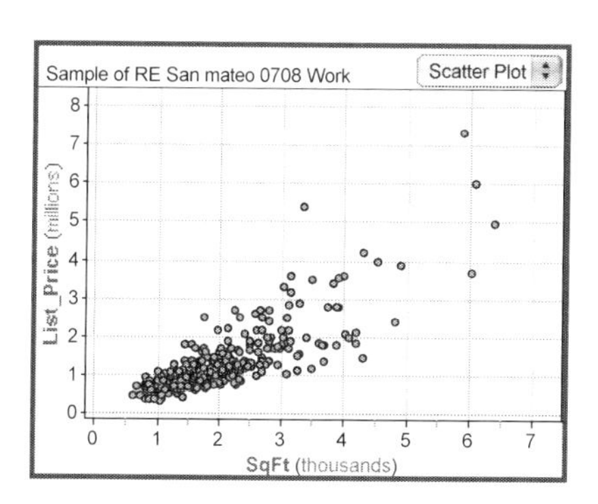

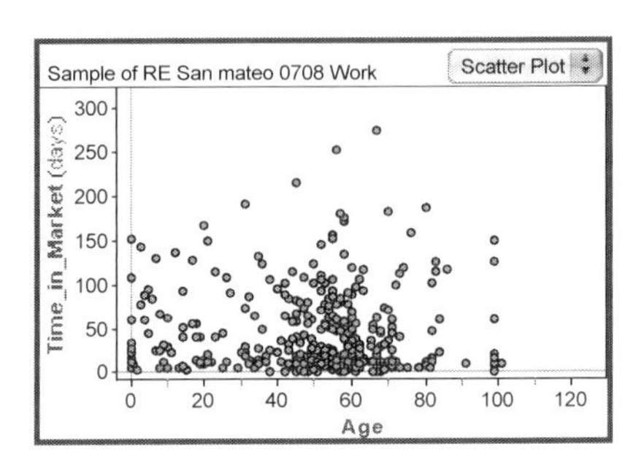

If there were a tendency for younger houses to spend less time on market, we would see a *positive* relationship (low age ~ short time on market/high age ~ long time on market), and if younger houses tended to take longer to sell, we would see a *negative* relationship. What we actually see for our sample of houses is that the *Age* of a house appears to have very little to do with the *Time in Market*. In such instances, we would say that there appears to be ***no relationship*** between the variables. The terms *positive* and *negative* follow the terminology you encountered in algebra when studying the slopes of straight lines.

Direction

We call a *relationship* **positive** if as the values of x gets larger, the values of y also get larger.

We call a *relationship* **negative** if as the values of x gets larger, the values of y get smaller.

Form The second thing we look at is the ***form*** of the relationship, and in studying the *form*, we are primarily interested in whether the plot shows a ***linear*** or straight-line pattern or not. Why? A straight-line relationship between two variables is extremely simple—much simpler than a quadratic or an exponential relationship; so if it is possible to use a straight-line model for the data, we will. The *scatterplot* of the negative relationship between *Age* and *Price* for the Lexuses in Boston (the plot on the previous page) appears linear, as does the positive relationship between *SqFt* and *ListPrice*, at least for houses with less than about 500 square feet of area. Not every relationship between variables is linear in form, however. Here is the scatterplot of the relationship between *Age* and *Price* for Porsche 911 sports cars for sale in the San Francisco Bay Area and in Los Angeles. The plot shows a curve rather than a straight line. (We sometimes call relationships that exhibit a curve ***curvilinear;*** in this example, you may recognize a pattern that may be *exponential decay*, at least for the cars younger than about twelve years old.) If the relationship in a scatterplot shows *no* relationship between the variables (such as for *Age* and *Time_in_Market* above) and resembles a cloud of points, we would *not* use either the term linear or the term curvilinear.

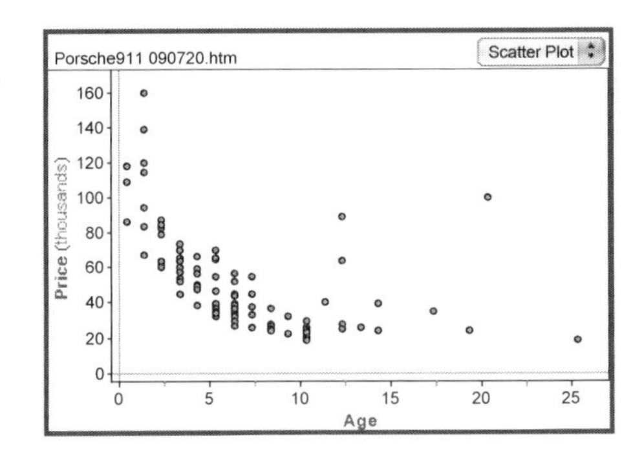

Strength The third aspect of scatterplots to look at is the most difficult to assess visually from a graphic; this third aspect looks at the ***strength*** of the relationship between the variables. By *strength* we mean (roughly, here) the amount of scatter around the basic pattern, where the basic pattern may or may not be linear. Here are two examples of a strong relationship, one very linear, the other not. The linear one is the relationship between the *ListPrice* and *SalePrice* of our sample of San Mateo County houses; the price at which a house is sold is strongly associated with its list price.

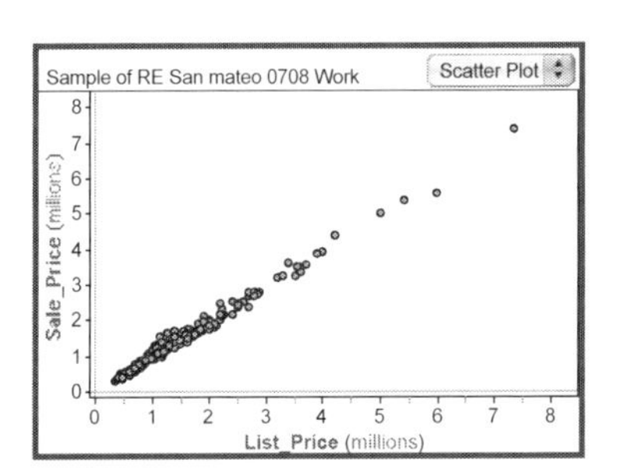

The non-linear example shows the *strong* relationship between the length (measured in millimeters) and the mass (measured in grams) of a fish popularly called snook (*Centropomus undecimalis*) found in warm waters in Florida, the Gulf Coast of the Southern states, and in Southern California. Longer snook have greater mass (hence a positive relationship), but the relationship is not linear; the slope relating the mass to the length of the longer snook is steeper than the slope for the shorter snook.

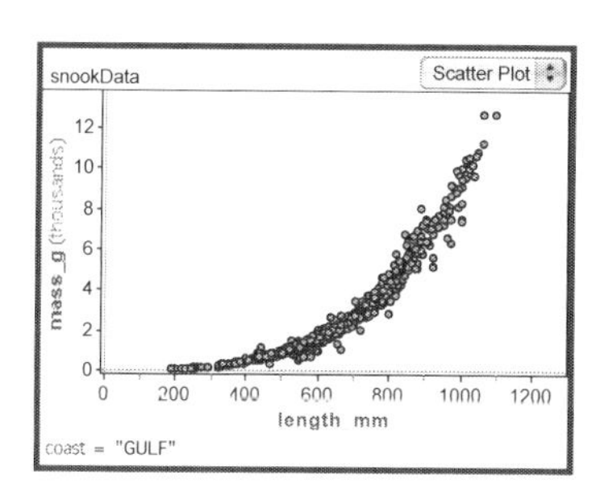

The assessment of strength for scatterplots that are neither strong nor weak seems a bit subjective. Are they "moderately strong" or "moderately weak"? What we need is a measure of strength of relationship. After the next paragraph, we introduce such a measure.

Explanatory and Response Variables Again In the scatterplots shown above, did you notice that the variable that we think of as *explanatory* is regarded as the x variable and the variable we think of as *response* is regarded as the y variable? We think that the *Age* of a car influences the *Price* that the car can fetch in the market. Asking a high price for the car cannot change its age; it may delay the sale of the car while it grows still older, but it cannot make the car younger. Of course, there are situations when it is not clear which variable should be regarded as the *explanatory* and which the *response*. In those situations, the placement of the two variables on a scatterplot may not matter. However, for any specific pair of variables, we usually have a "natural way of thinking" about the relationship between the two variables, and the way we think of the relationship usually indicates which variable should be the *explanatory* and which variable the *response*.

The Correlation Coefficient: A Measure Based on a Model

As it happens, we do have a measure of the strength of association between two variables. This measure also shows the direction of the relationship, and it is based upon a linear model for the data. It is called the ***correlation coefficient***. The symbol for this measure is r. The *correlation coefficient* measures the ***strength of linear association*** between two quantitative variables. It is calculated so that it varies between $r = -1$, which indicates a perfectly linear negative association, and $r = 1$, which indicates a perfectly linear positive association. Hence, $-1 \le r \le 1$ and values between these limits indicate a negative or positive linear association with some amount of scatter—or, it is important to note, a relationship that happens not to be linear. Here are some examples (we will actually look at the data we have been analyzing shortly).

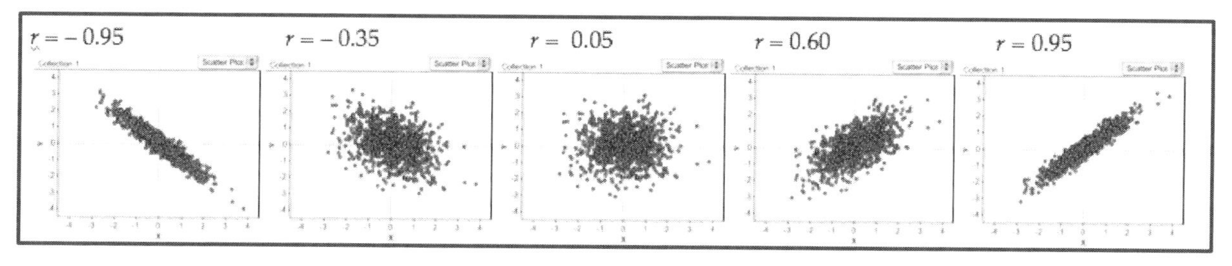

Correlation Coefficient, r

- Measures the *linear* association between two *quantitative* variables
- Has values in the interval: $-1 \le r \le 1$
- Is calculated by the formula: $r = \dfrac{1}{n-1}\sum_{i=1}^{n}\left(\dfrac{x_i - \bar{x}}{s_x}\right)\left(\dfrac{y_i - \bar{y}}{s_y}\right) = \dfrac{1}{n-1}\sum_{i=1}^{n} z_x \cdot z_y$

Using the Formula: Calculating the Correlation Coefficient

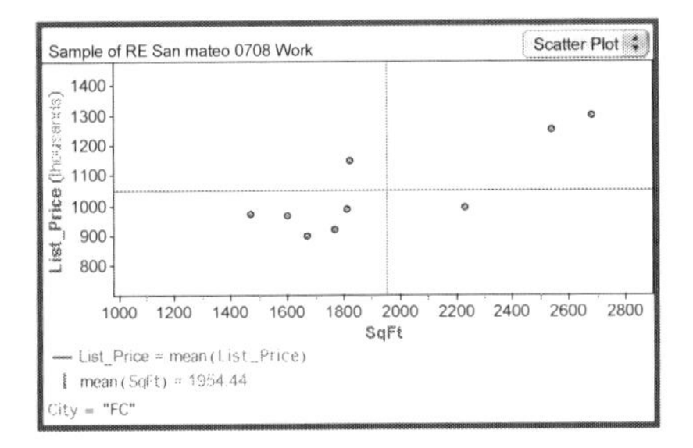

We will show an example that has to do with just nine houses for sale in Foster City in our real estate sample; we are looking at the relationship between the size of the house (measured by the variable *SqFt*) and the *ListPrice* of the house. Here is the plot for just these $n = 9$ houses. The vertical and horizontal lines that we have drawn on the plot show the means for the two variables *SqFt* and *ListPrice*; we will see that having these lines there gives us some insight to what the formula is actually doing. The formula uses the means and standard deviations for the two variables (shown in the *Summary Table*), so that here for *SqFt*, $\bar{x} = 1954.4$ and $s_x = 427.1$ and for *ListPrice*, $\bar{y} = 1048864$ and $s_y = 146217$. For these $n = 9$ cases, the *Correlation Coefficient* $r = 0.821$.

Sample of RE San mateo 0708 Work

	SqFt	List_Price
	1954.44	1048864
	427.116	146217
	9	9

S1 = mean( )
S2 = s( )
S3 = count( )

City = "FC"

It may help to follow the calculations below by having the complete data in a table. The data for the nine houses is shown in the two columns on the left, and the results of the calculations of the formula are shown in the remaining columns, with the sums at foot of the table. Here is what the calculation looks like in "formula form."

SqFt x	ListPrice y	z(x)	z(y)	Product
1470	975000	-1.13420116	-0.50516698	0.57296097
1820	1150000	-0.31475932	0.69168428	-0.21771407
1600	968000	-0.82983705	-0.55304103	0.45893393
1810	985888	-0.33817194	-0.43070231	0.14565144
2540	1248888	1.37094962	1.36799415	1.87545105
1770	920000	-0.43182244	-0.88132023	0.38057385
2230	995000	0.64515827	-0.36838398	-0.23766597
1670	898000	-0.66594868	-1.03178153	0.68711355
2680	1299000	1.69872635	1.71071763	2.90604112
	Sum	0.00009365	0.00000000	6.57134588

$$
\begin{aligned}
r &= \frac{1}{n-1}\sum_{i=1}^{n}\left(\frac{x_i - \bar{x}}{s_x}\right)\left(\frac{y_i - \bar{y}}{s_y}\right) \\
&\approx \frac{1}{9-1}\left[\left(\frac{1470-1954.4}{427.1}\right)\left(\frac{975000-1048864}{146217}\right)+\cdots+\left(\frac{2680-1954.4}{427.1}\right)\left(\frac{1290000-1048864}{146217}\right)\right] \\
&\approx \frac{1}{8}\left[(-1.1342)(-0.5052)+(-0.3148)(0.6917)+(-0.8298)(-0.5530)+\cdots+(-0.6659.)(-1.0318)+(1.6987)(1.7107)\right] \\
&\approx \frac{1}{8}\left[(0.5729)+(-.02177)+(0.4589)+(0.1457)+(1.8755)+(0.3806)+(-0.2377)+(0.6871)+(2.9060)\right] \\
&\approx \frac{6.5713}{8} \approx 0.8214
\end{aligned}
$$

ZZzzzz: a picture of what the formula is doing The graphic below (if studied carefully) shows what is happening inside the formula. We have drawn the means of the two variables in the graph: the dashed vertical line is the *SqFt* mean and the dashed horizontal line is the *ListPrice* mean. Then we can show the distances of each data point from the mean of x (here *SqFt*) and from the mean of y (here *ListPrice*) in *standard deviation* units. These distances are actually z scores: that is $z_x = \frac{x_i - \bar{x}}{s_x}$ and

$z_x = \frac{y_i - \bar{y}}{s_y}$. In the formula, it is these zs that get multiplied before the terms are added.

- ***Point A*** is the first point (the first house) and ***Point B*** the second point (house) in the list above; check that the z_x and z_y scores on the graph agree with the table.
- Points that are in ***quadrant I*** will contribute a positive product, that is $z_x z_y = \left(\frac{x_i - \bar{x}}{s_x}\right)\left(\frac{y_i - \bar{y}}{s_y}\right) > 0$)

  in the sum in the formula, and points that are in ***quadrant III*** will also have a positive product, since the two z scores will both be negative. This means that if our data are mainly in quadrants I and III, our sum will be positive, and our r will be positive. In our example, most of our points are in these two quadrants. Check out that the product of the z scores for point A is 0.573.
- Points that are in ***quadrant II*** (or ***quadrant IV***) will contribute a *negative* product $z_x z_y < 0$ for the sum in the formula (as does our point B), and so if most of the data are in these two quadrants, the correlation coefficient r will be negative.

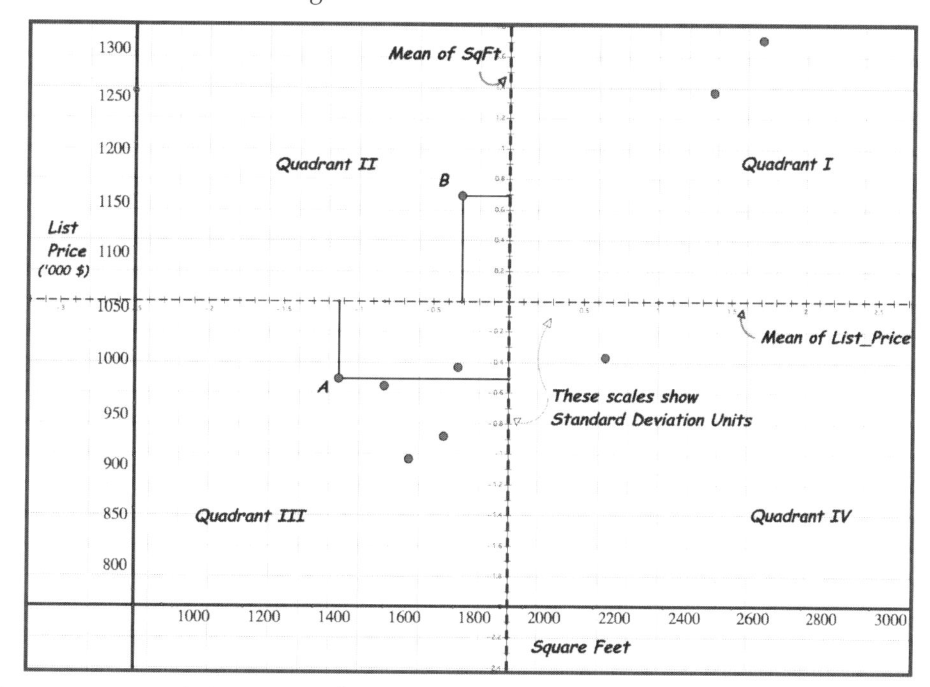

- If the data are scattered about in *all* the quadrants, the positive and negative products—when they are added—will tend to cancel each other out, and we will end up with a number near to zero. The number may be positive and may be negative, but it will be near to zero.

Correlations in Data: Used car data We return to our data sets to see how the measure of linear association actually works. Here are three examples from the used car study of Lexuses and Infinitis.

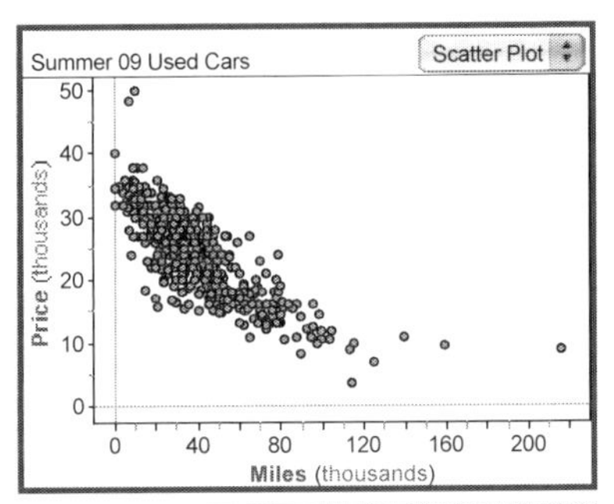

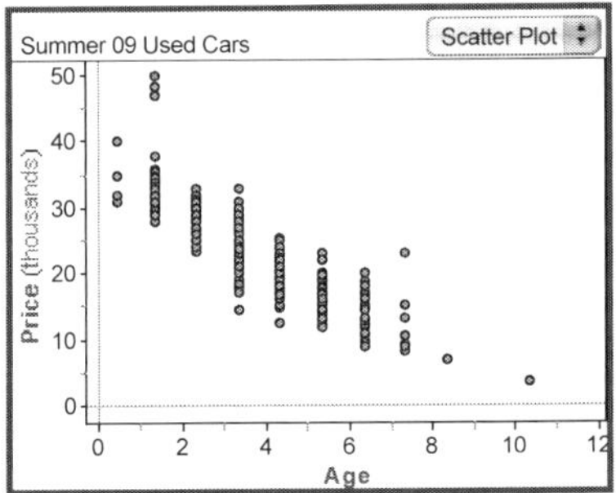

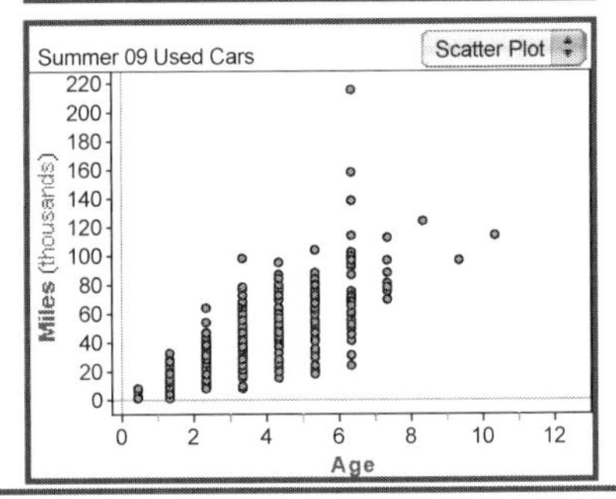

The first graph shows the relationship between *Age* and *Price* for all of the cars (not just the Lexuses from Boston). The association looks fairly linear and negative as we would expect (as the age of a car increases the value decreases) and with a bit of scatter; obviously, even with cars that are the same year model, there are differences in price asked. So about what value should the correlation coefficient be?

The second graph shows the relationship between *Miles* (miles that the car has been driven) and *Price*, whose association we would expect to be negative. In the center portion the relationship appears to be quite linear.

The third graph shows the association between *Age* and *Miles* driven, which we expect to be a positive association.

What we see in the graphs should be reflected in the numbers for the *correlation coefficients* between these variables. The values of the correlation coefficients r are shown in the summary table below.

Summer 09 Used Cars

	Age	Miles	Price
Age	1	0.756543	-0.883391
Miles	0.756543	1	-0.799797
Price	-0.883391	-0.799797	1

S1 = correlation ( )

Notice that the correlation of a variable (such as *Age*) with itself is $r = 1$. Secondly, notice that the correlation coefficient does not depend upon which variable we choose as x and which variable we choose as y. Look at the number for *Age* and *Miles* and then look at the number for *Miles* and *Age* in the summary table. The value of r does not depend on the choice of *explanatory* and *response* variables. We can see this in the form of the formula for the correlation coefficient, $r = \frac{1}{n-1}\sum_{i=1}^{n}\left(\frac{x_i - \bar{x}}{s_x}\right)\left(\frac{y_i - \bar{y}}{s_y}\right) = \frac{1}{n-1}\sum_{i=1}^{n} z_x \cdot z_y$ since before we sum, there is a product, and reversing the order of the product does not change its value.

Real Estate Data Here is the summary table for the correlations for the real estate data. You should be able to check that r =-0.061 for the association between *Age* and *Time_in_Market* is sufficiently weak to fit what you see in the scatterplot for the two variables above. The r = 0.995 for *List_Price* and *Sale_Price* should accord with the graphic for those two variables.

Sample of RE San mateo 0708 Work

	Age	List_Price	Sale_Price	SqFt	Time_in_Market
Age	1	-0.205108	-0.185829	-0.348588	-0.0608161
List_Price	-0.205108	1	0.994569	0.822379	-0.104747
Sale_Price	-0.185829	0.994569	1	0.807327	-0.138747
SqFt	-0.348588	0.822379	0.807327	1	-0.0396667
Time_in_Market	-0.0608161	-0.104747	-0.138747	-0.0396667	1

S1 = correlation ( )

Cautionary Comments on Correlation There are some facts about correlation that are sometimes misunderstood. Here are some of those facts, followed by the potential misunderstandings.

The correlation coefficient r is for quantitative variables only. The word "correlation" has entered into modern speech to such an extent that people easily say: "There is a correlation between gender and political party preference." In statistics, the word *correlation* is reserved for the association between quantitative variables, and neither gender nor political party preference is quantitative. It is quite correct to say, "There is an *association* between gender and political party preference" but not a *correlation.*

The correlation coefficient r measures a linear association between two variables. Recall the snook data where the relationship between the length and the mass appears to be without much scatter at all, but the relationship is curved and not linear. For these data, $r = 0.911$, whose proximity to 1.00 you may find impressive. However, if these data were "straightened out" but had the same amount of scatter about a straight line, the correlation coefficient would be about 0.99. The lesson is that a value of r measures either scatter about linearity or lack of linearity, or both! Studying the scatterplot should show what is happening.

Correlation does not imply causation. Many of the examples in this section have been chosen because the relationship between the variables "makes sense" in that we have some ideas about the connection between the variables. As cars grow older, people know that the cars are likely to have been driven much, that the car is more likely to cause trouble, that the car looks out of date, and so people are less willing to pay what they would for a younger car. So we say that *Age* affects *Price.*

It may be that there are some other variables affecting both of the variables we are analyzing and that there is not a direct connection between the variables. If we had a collection of students who were all taking (say) statistics and chemistry, and we looked at the correlations between these test scores in statistics and chemistry, we would probably find a fairly strong positive correlation; that positive correlation does not mean that taking chemistry helps students taking statistics or vice versa. There are other variables.

Cautions about Correlation

- The *correlation coefficient r* is only appropriate for quantitative variables.
- The *correlation coefficient r* measures linear association between two variables.
- *Correlation* between variables does not imply that one variable causes the other.

Summary: Analyzing the relationship between two quantitative variables

- A ***scatterplot*** is an X – Y coordinate system on which the ordered (x, y) pairs of the two quantitative variables are plotted to show the *direction, form,* and *strength* of the relationship between the two variables.
  - A ***positive direction*** indicates that as x is larger, y is also larger.
  - A ***negative direction*** indicates that as x is larger, y is smaller.
  - ***Form*** refers to whether the data in the plot appear to approximate ***linear*** form (a straight line) or not.
  - ***Strength*** refers to the extent to which the data in the plot appear to adhere to a particular form (hence, "stronger") or show scatter with respect to a particular form ("weaker").
- The ***correlation coefficient,*** whose symbol is ***r,*** is a numerical summary of the relationship between two quantitative variables that:
  - Is based upon a *linear* model for the data, and
  - May take on the values in the interval $-1 \le r \le 1$, where
  - The value $r = -1$ indicates a (perfect) negative linear relationship, the value $r = 1$ indicates a (perfect) negative linear relationship, and values near zero either no relationship or possibly one non-linear.
  - Is calculated using the formula $r = \frac{1}{n-1}\sum_{i=1}^{n}\left(\frac{x_i - \bar{x}}{s_x}\right)\left(\frac{y_i - \bar{y}}{s_y}\right) = \frac{1}{n-1}\sum_{i=1}^{n} z_x \cdot z_y$.
- ***Correlation and causation*** A *correlation coefficient r* (even one that is strong—near to $r = -1$, or $r = 1$) does not necessarily indicate a ***causal relationship*** between the two quantitative variables.

§2.3 Best-Fitting Lines: Making the Most of the Model

More Statistical Questions about Used Cars and Used Houses

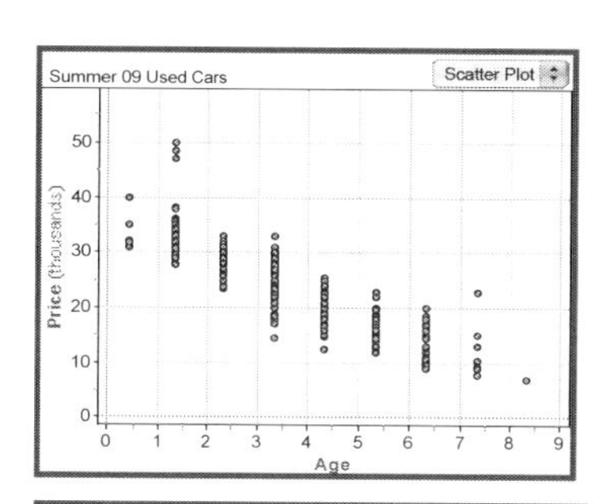

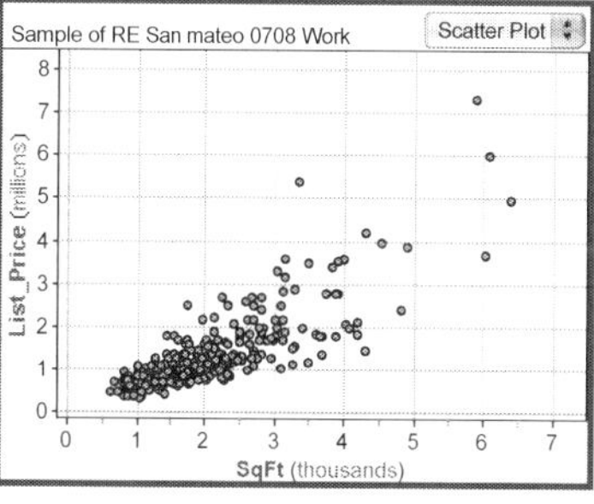

In the last section we looked at graphs showing how cars lose their value as they get older (not a big surprise) and the larger the house is, the more valuable it is (also not surprising).

In these scatterplots, the form of the data is fairly linear, although there is some scatter. For the relationship between the *Age* and the *Price* of used Lexuses and Infinitis, $r = -.88$, and for the relationship between *SqFt* and *List_Price* of houses in San Mateo County, we can calculate $r = .82$. However, we can go beyond the correlation coefficient r, a single measure of linear relationship. We get an estimate of *how much* a used Lexus or Infiniti loses each year it ages. We use our data to get an idea of *how much* a square foot of house in San Mateo County costs. We can get these things by fitting a straight-line model—a linear model—to the data.

Remembering about straight lines Recall from algebra courses that a straight line can be expressed by the equation $y = mx + b$, where m is the slope and b is the y-intercept. Remember also that slope can be thought of as $m = \frac{\text{rise}}{\text{run}} = \frac{\text{change in } y}{\text{change in } x}$, so that $y = \frac{3}{4}x + 2$ means that for every 4 units running along the x-axis, the line rises by 3 units. Or, for every one unit running on the x-axis, the line rises ¾ of a unit. For statistics, the notation is slightly different.

Notation for equations of lines in statistics

In statistics we express a straight line by the equation $y = a + bx$, where b is the slope and a is the y-intercept.

Example of a linear model for prediction and interpretation: used Infiniti cars The plot here shows a line that we have drawn to model the relationship of the *Price* of a used Infiniti G-35 to its *Age*. Our line has the equation $\hat{y} = 36201 - 3516x$, where y is *Price* and x is the *Age* of the used car. We use the symbol $\hat{y}$ (pronounced "y-hat") because we use our equation to predict the value of y (in this instance, a price) from a value for x (here, age). Our prediction of the price $\hat{y}$ for the age $x = 3$ we get from the equation $\hat{y} = 36201 - 3516x$. When we put $x = 3$ into this equation, we get: $\hat{y} = 36201 - 3516(3) \approx 36201 - 10548 = 25653$

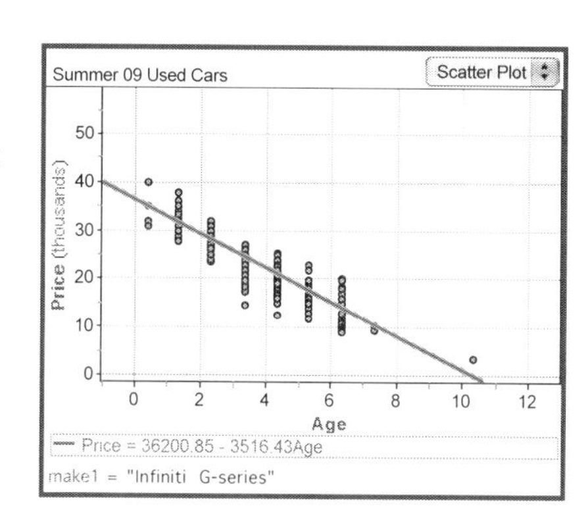

We could say that we expect that in three years our new Infiniti G-35 has lost \$10,550 and is now worth about \$25,650. This predicted value $\hat{y}$ for $x = 3$ is *on* the line shown on the plot; indeed, *all* the predicted values are on the line we have drawn. We can do the same calculation for any other value of *Age* and get a predicted value for the *Price*, as long as the value for *Age* is reasonable. Notice, however, that our linear model predicts that after about eleven years, the Infiniti G-35 will be worth nothing; what this shows is that our linear model is probably not good for values of *Age* beyond nine years or so. We have no data for the years beyond about eight years old, except for one car at age eleven years.

For a three year-old Infiniti G-35, we lose \$10,550. Using our equation $\hat{y} = 36201 - 3516x$ we can also say that for *every year that Age increases for an Infiniti G-35 the Price will change by the value of the slope – 3516*. To be specific, we can say that for every year that an Infiniti G-35 ages, it loses about \$3,516. This sentence is an *interpretation* of the slope of our model.

We have used our linear model for both ***prediction*** and ***interpretation.*** We have used the equation to *predict* a value for price, and we have *interpreted* the ***slope*** to say how the price declines as a car ages. We may also sometimes be able to interpret the ***y-intercept***, but not always. The y intercept of a line is the value of y where $x = 0$, so we could say that our prediction for the *Price* of a *new* Infiniti G-35 is about \$36,200. However, notice that our data is about *used cars;* we do not have any data on *new* Infinitis—just used ones, so we should probably be a bit cautious about making this prediction.

Another Example of Prediction and Interpretation: Size and Price of House. Here is the plot showing the relationship between the size (*SqFt*) and the listed price (*List_Price*) for houses sold in San Mateo County in 2007–2008. We have fitted a linear model with the equation, $\hat{y} = 25331.67 + 728x$, where x is *SqFt* and y is *List_Price*.

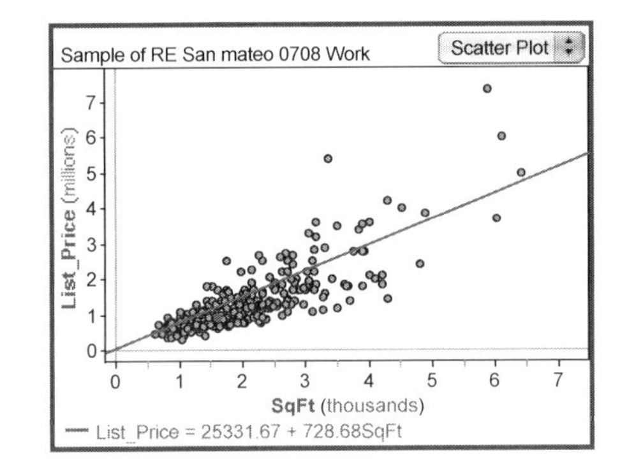

There was a house sold in Redwood City in the sample that had *SqFt* equal to $x = 1610$. Using our equation $\hat{y} = 25331.67 + 728(1610) = 25331.67 + 1173174 = 1198506.47$, our predicted list price for this house is \$1,198,506. How does this compare with the actual value of the list price (which we also have in the data)? The actual value of *List_Price* for this house is \$949,000; what they got (in this case) was lower than what we predicted. When we compare the two, we calculate what we call the ***residual,*** which is the difference between the actual value and the predicted value. For this case we get the

$$residual = y_i - \hat{y} = 949000.00 - 1198506.47 = -249506.47 \text{ dollars}$$

The value is negative because the actual value is less than the predicted value and thus below our linear model, whose equation is $\hat{y} = 25331.67 + 728.68x$. Another house had $x = 2131$ square feet of living area; if you calculate the predicted value $\hat{y}$ for *List_Price* and compare it with the actual value of \$2,250,000, you will get a *residual* of \$671,851.25. This *residual* in this example is positive, showing that the actual value is above the predicted value on the line.

Definition of residual

$$\textbf{Residual} = \textbf{Actual value} - \textbf{Predicted value} = y_i - \hat{y}$$

Our *interpretation* of the slope tells as *that for every square foot in size that a house is bigger, we can expect that the list price will be about $729 more.* Notice that in our interpretation, it is important to mention the units for the x variable and for the y variable. We can generalize this interpretation.

Interpretation of Slope

The slope tells us the amount of change in the y variable for every *unit* change in the x variable. The interpretation is to be done in the context of the variables, using the units in which the variables are measured.

Explanatory and Response Variables Again In both our examples we have chosen one variable to be the *explanatory* variable—for the used cars example, the explanatory variable, denoted by x, was *Age*. That implies that the *response* variable was *Price*, denoted by y. It would be possible to get a linear model to predict the *Age* of a used car (making *Age* the y variable) from the *Price* (making *Price* the x variable), but that choice would make less sense to us. As a car grows older, we think the value tends to decrease; we do not think that making a car worth less (for example, by abusing the car in some way) will make the car older in age.

When we calculated the *correlation coefficient,* which also depends on a linear model, it did not matter which variable we chose to be x (the explanatory variable) and which variable we chose to be y (the response variable); the result was the same whatever choice of explanatory and response we made because r just looks for a linear relationship between the variables. In fitting a model for prediction and interpretation, it *does* make a difference because in our prediction and in our interpretation, we look at *how* one variable (the explanatory) affects the other (the response.) It should also be clear from looking at a scatterplot the variable chosen to be the response variable; the response variable is the one on the y-axis and the *explanatory* variable is the one on the x-axis.

Are there instances in analyzing data where it is difficult to choose which variable should be the *explanatory* and which variable the *response*? Or are there perhaps instances both choices make some sense? Yes, such situations exist. You may have run into that difficulty when you looked at the length and height of roller coasters; which variable has priority: length or height? Lesson: the application of statistics follows the way we think about things.

How do we get the straight line? Is there a "best-fitting" straight line?

There are many, many possible lines but one best-fitting line Here is a scatterplot from the used car data again; this plot shows the relationship between the *Price* (the *response* or y variable) of the car being sold and the *Miles* that car had been driven (the *explanatory* or x variable). Once again, our general idea is that there is a negative association—the greater the number of miles the car had been driven, the lower the price. In the plot we have drawn in a number of lines to represent the relationship; some of the lines obviously do not fit the data well at all. Each of the lines drawn here has a different slope and different y-intercept. We could draw an infinite number of different prediction lines, some of them good but many bad as models. There *is* a best-fitting line, but how do we find it?

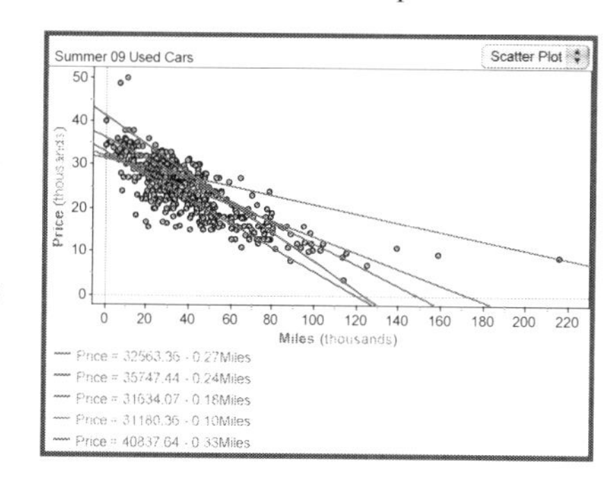

Some wrong but popular ideas followed by the correct idea. Here are two wrong ideas.

- "Of all the possible lines, the best-fitting line is the one that hits (or includes) the largest number of data points." No! In fact, the best fitting line may hit *none* of the data points and be better than a line that hits some of the data points.
- "You can get a good line by connecting the points at the ends of the graph and then finding the equation of that line." No! A look at our example above should convince you that this is not a good strategy; any line that includes the car with nearly 220,000 miles will result in a poorly fitting line.
- You can "eye-ball" a line that will fit the data. Perhaps so, but there are some simple formulas that will guarantee a line and an equation for that line that is better than any estimate that comes from just looking at the plot.

So, how do we decide which line, of all the infinite number of possible lines, is the best one? The theoretical answer is that we inspect the sizes of the *residuals* for *every* single data point for *every* possible line. Of an infinite number of possible lines, the one where the sum of squared residuals the smallest is the *best* line. Since there are an infinite number of lines to check, this looks impossible.

As it happens, we do not have to actually do this (to Caspar's relief!) because there is a formula for the slope and the y-intercept of the line that makes the *sum of squared residuals* the smallest. However, even though we are spared the toil of inspecting every possible line, the principle is a very important one and merits its own box showing the formulas for the slope b and y-intercept a.

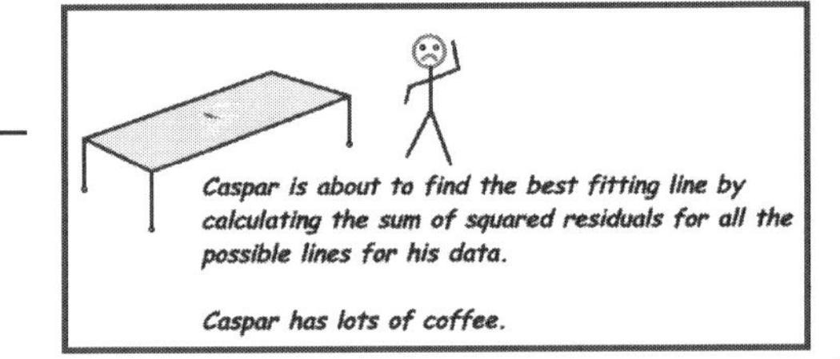

The Best-Fitting Line

- The *best-fitting* straight line $\hat{y} = a + bx$ to predict $\hat{y}$ from x is the line that makes the sum of squared residuals the smallest, or
- In symbols: the *best-fitting line* makes $\sum_{i=1}^{n}(y_i - \hat{y})^2$ as small as possible, and so
- The best-fitting line is also called the **Least Squares Regression Line** or **Least Squares Line**

The Least Squares Regression Line is $\hat{y} = a + bx$ where

- the slope is: $b = r\dfrac{s_y}{s_x}$
- the y-intercept is: $a = \bar{y} - b\bar{x}$

and where $\bar{x}$ and $\bar{y}$ are the means and s_x and s_y the standard deviations of the *explanatory* and *response* variables x and y, and r is the correlation coefficient for x and y.

Example: Using the formulas We will actually get the best-fitting (or *least squares*) lines for two cars (Lexus and Infiniti) so that we can compare the *two* cars and specifically their *rates of depreciation.* The scatterplot with the two *least squares regression lines* is shown on the next page; one of the lines is to predict the *Price* from *Age* for the Lexuses, and the other line is for the Infiniti G-35 cars. Here are the means, standard deviations for *Age* and *Price,* and the correlation coefficient for the relationship between them. To get the *slope* of the least squares line for the Infiniti, we calculate, using the number shown in the table for the Infiniti:

Summer 09 Used Cars

		Age	Price
Make1	Infiniti G-series	3.58847	22621.2
		1.57391	6515.4
		-0.902579	-0.902579
		308	308
	Lexus IS	2.90331	27571.6
		1.52305	6426.06
		-0.851356	-0.851356
		181	181
Column Summary		3.33487	24453.6
		1.58861	6903.76
		-0.883391	-0.883391
		489	489

S1 = mean( )
S2 = s( )
S3 = correlation(Age, Price)
S4 = count( )

¬((Price = "missing") or (Age = "missing"))

$$b = r\frac{s_y}{s_x} = -0.902579\frac{6515.4}{1.57391} \approx -3736.3402$$

Then, using this value for the slope, we calculate the y-intercept, using the formula:

$$a = \bar{y} - b\bar{x} = 22621.2 - (-3736.3402)(3.58847) \approx 36028.94 .$$

Our equation for predicting the *Price* from *Age* for the Infiniti G-35 cars sold is $\hat{y} = a + bx = 36028.94 - 3736.34x$. For our interpretation of the slope, we can say that *for every year that an Infiniti G-35 ages, we expect the price to decrease by $3,736.* (The calculation of the regression equation for the Lexus IS cars is left as an exercise.)

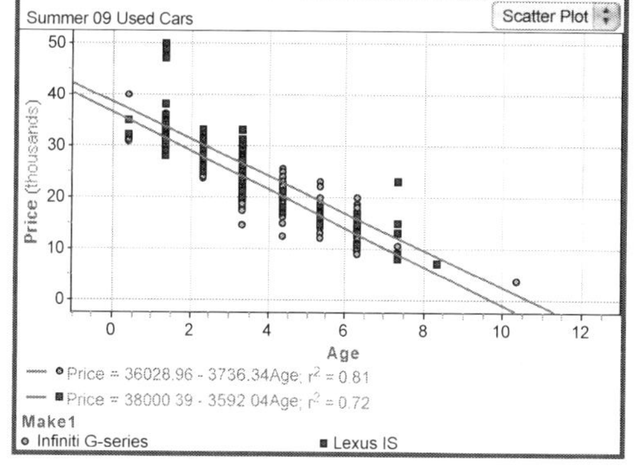

Pondering the calculations. Our value for the y-intercept was not exactly what software got with its calculation; we were two cents off. This is an instance of ***rounding error***; in our calculations we used numbers with far fewer decimal places' accuracy than what software typically uses.

Notice that we use the *correlation coefficient r to get the slope and the y.* That means that in order to calculate our best-fitting line, we must first calculate the correlation coefficient as well as the means and standard deviations for the variables x and y-intercept of the *Least Squares Line.* Notice that we have to calculate the slope b before we can calculate the y-intercept $a = \bar{y} - b\bar{x}$ because the formula for the y-intercept uses b.

Pondering the meaning: Comparing the lines. Look at the lines in the plot and look at the equations. The lines are nearly parallel—not exactly parallel but almost, and you can see that the slopes are very similar. (If the lines were actually parallel, the slopes would be equal.) We can identify which car is which by noting that the y-intercept for the Lexus is about $38,000 whereas the y-intercept for the Infiniti is about $36000; therefore, the top line represents the best-fitting line for the Lexus and the lower one, the best-fitting line for the Infiniti.

What meaning can we draw from the fact that the slopes of the two lines are nearly equal? In the context of our used cars, our evidence is that the value that these two competing cars lose per year is just about the same; in car jargon, we say that their rates of depreciation are about equal.

That the slopes are nearly equal also means that one car is consistently more expensive. However, the difference is not large—about two thousand dollars, on average. Our data come from just one collection of cars being sold at a particular time; our collection is merely one ***sample*** of many such samples that we might have downloaded from www.cars.com. Every sample that we choose will be different; there is variation from sample to sample. Hence, we cannot be completely certain that the difference of two thousand dollars between the Lexus cars and the Infiniti cars is just a feature of the particular sample we have chosen or whether it is something that we would see if we took sample after sample. It is possible that if we had access to *all* the Lexus IS cars and Infiniti cars sold over the Internet, we would find essentially no difference or that we would find the kind of difference that we have found in our sample. Can we generalize from our sample to say something about Lexus and Infiniti cars? These questions we will begin to take up in Unit 3 and especially in Units 4 and 5.

Summary: Fitting a Straight Line Model to Data

- The overall idea is to fit a ***linear model***, $\hat{y} = a + bx$, which we use for ***prediction*** and ***interpretation.***
  - We get a *predicted value* $\hat{y}$ using our linear model by inserting a value for x in $\hat{y} = a + bx$.
  - We *interpret* the slope b of the equation $\hat{y} = a + bx$ as the change we expect in y for a unit change in x.
  - We have chosen one of the two quantitative variables as the ***explanatory*** variable (so, in symbols, x) and the other quantitative variable as the ***response*** variable, y.
- The ***residual*** for any given data point is the difference between the actual and the predicted or, in symbols:

$$\textit{residual} = \text{Actual value} - \text{Predicted value} = y_i - \hat{y}$$

- There are many possible linear models for any given set of data but only one model that we declare to be the ***best-fitting line.***
  - The *best-fitting* straight line $\hat{y} = a + bx$ to predict $\hat{y}$ from x is the line that makes the sum of squared residuals the smallest, or
  - In symbols: the *best-fitting line* makes $\sum_{i=1}^{n}(y_i - \hat{y})^2$ as small as possible, and so
  - The best-fitting line is also called the ***Least Squares Regression Line*** or ***Least Squares Line***
  - The ***Least Squares Line*** has the equation $\hat{y} = a + bx$ where
  - the slope is: $b = r\dfrac{s_y}{s_x}$,
  - the y-intercept is: $a = \overline{y} - b\overline{x}$, where
  - $\overline{x}$ and $\overline{y}$ are the means and s_x and s_y the standard deviations of the *explanatory* and *response* variables x and y, and r is the correlation coefficient for the relationship between x and y.

§2.4 "Is the Model Good and Useful?"

A measure of how well a model fits: R^2

Recall the quote from George E. P. Box in §1.7: "All models are wrong, but some are useful." We have a way of assessing whether it is a good model or not. A general measure of how well the model fits is called the ***coefficient of determination,*** and its symbol is $\boldsymbol{R^2}$. Here are some facts about $\boldsymbol{R^2}$.

R^2: Facts and Interpretation

- For simple regression models, $R^2 = (r)^2$; that is, R^2 is the square of the correlation coefficient r.
- $0 \le R^2 \le 1$
- *Interpretation:* R^2 shows the proportion (or percentage) of variation in the *response* variable *explained* by the model being used. (See the examples just below for the wording.)

Here is the plot from the end of the last section on the relationship between the *Age* and the *Price* of used Lexus and Infiniti cars. (Note that the top-most line is the least squares line for the Lexus cars, and the second line is for the Infiniti cars, even though the equations are listed in alphabetical order in the output.) At the foot of the plot, you can read that for the Infiniti cars, $R^2 = r^2 = 0.81$ and for the Lexus cars, we see $R^2 = r^2 = 0.72$. For the regression model for the Infiniti cars, we would say:

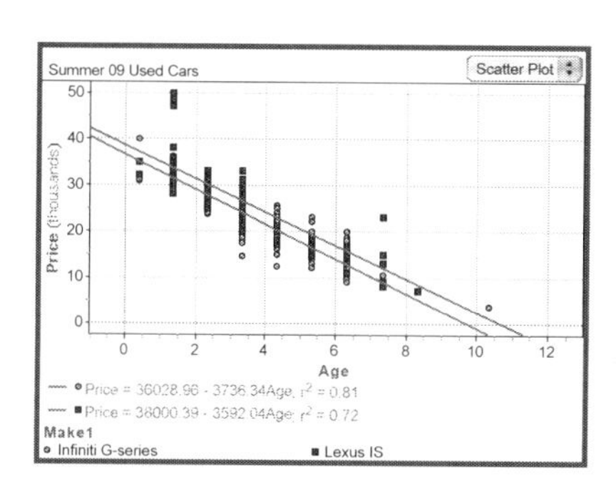

> 81 % of the variation in the *Price* of used Infiniti G-Series can be explained (or accounted for) by the linear model on the *Age* of the car.

For the Lexus cars, we would say:

> 72 % of the variation in the *Price* of used Lexus ISs can be explained (or accounted for) by the linear model on the *Age* of the car.

The linear model on the variable *Age* does a slightly better job of explaining the variation in the *Price* of used Infiniti cars than it does for the Lexus IS; however, to explain 70% or 80% of the variation in the response variable is doing pretty well. Notice also that there are some alternate verbs that can be used rather than "explained."

R^2 is also known as the ***Coefficient of Determination,*** but it is also convenient to refer to it as *"big R-squared"* to prevent confusion with the correlation coefficient r. It is easy to confuse R^2 (and its interpretation) with the correlation coefficient r or with residuals, $y_i - \hat{y}$, or with the least squares regression equation $\hat{y} = a + bx$ since all of these involve the letter "r" in their names or in their symbol.

Here is a further warning: be prepared to find the remainder of this section challenging since the ideas are new and different. However, the ideas are also important and are used widely in statistics.

Interactive Notes: How and why R2 works. For this subsection, it will be helpful to have software opened and follow with the exposition. All of the graphics will be shown, but seeing the graphics will give the argument a certain force.

Our example will use the data for the women's heptathlon of the 2008 Summer Olympics in Beijing. (For a description of the data and the event, see Exercise 4 of §2.3.)

- Open the file ***Heptathlon2008A*** and get a scatterplot for the relationship between *HundredMeterHurdles* and *TwoHundredMeters.*

The scatterplot shows a positive linear relationship but with some scatter. We expect that a linear model that relates the time for the *HundredMeterHurdles* is appropriate since there does not appear to be a curved relationship. We also note scatter around the basic linear pattern. Here is the *Summary Table* showing the means, standard deviations, and the correlation coefficient r for this relationship. Notice that we have all the information here that we need to calculate the *Least Squares Regression Equation.*

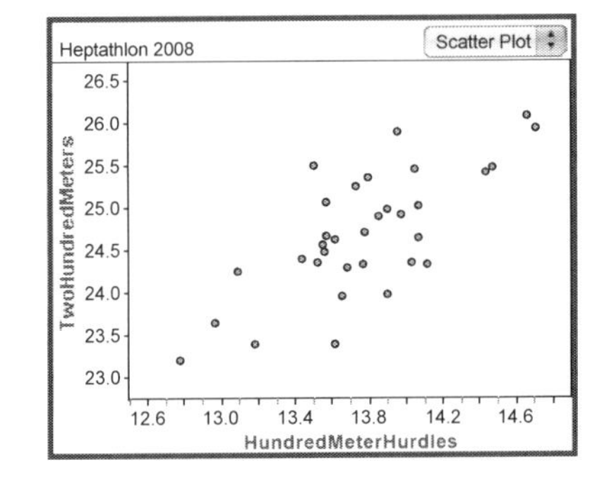

We *also* have all the information we need to calculate R^2, which is just $R^2 = r^2 = (0.723836)^2 = 0.5239$. We can interpret this R^2 and say that about 52% of the variation in the *TwoHundredMeters* times can be explained by the linear regression on the times for the *HundredMeterHurdles*. It does make some sense that the faster athletes should be faster in both events, the slower ones slower in both events. It also makes sense that there should be some other variation; one of the events is a purely running event, while the other event involves some jumping as well.

Heptathlon 2008

	HundredMeterHurdles	TwoHundredMeters
	13.7738	24.6815
	0.428952	0.728124
	34	34
	0.723836	0.723836

S1 = mean( )
S2 = s( )
S3 = count( )
S4 = correlation(HundredMeterHurdles, TwoHundredMeters)

Now we shall proceed to see why we say that R^2 shows the "the proportion (or percentage) of variation in the *response* variable *explained* by the model being used"—in this instance, that 52% of the variation in the *TwoHundredMeters* times is explained by our model on the *HundredMeterHurdles*.

- Using software get a graphic showing the "squared residusls."

The plot at the right shows the mean of the response variable, *TwoHundredMeters* times, and many squares. Each square is a quantity $(y_i - \overline{y})^2$, where the y_i is the time for one of the athletes in the *TwoHundredMeters*, and $\overline{y}$ is the mean time for all the athletes. So the best "heptathlete" for both these events (the least time for both events) was Hyleas Fountain (of the USA), whose time was 23.21 seconds for the *TwoHundredMeters* (and 12.78 seconds for the *HundredMeterHurdles*, which means that she is the "lower-left-most" dot on the scatterplot). The mean for all n = 34 heptathletes was 24.68 s., so for Ms. Fountain the calculation of $(y_i - \overline{y})^2$ will be $(23.21 - 24.68)^2 = (-1.47)^2 = 2.161$. Ms. Fountain is 1.47 seconds below the mean time for the *TwoHundredMeters* (which is a good place to be!), and the square of this number is 2.16. For each heptathlete, there is a square like this.

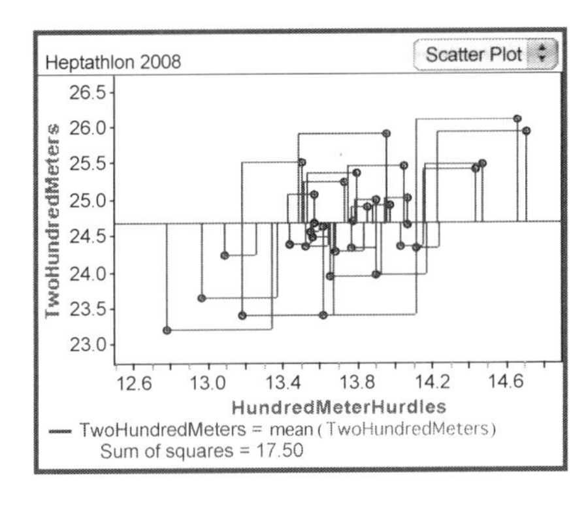

For the athletes whose times are near the mean time in the *TwoHundredMeters* of 24. 68 seconds, the square is very small, and for the frontrunners and for the stragglers, the squares are big.

Total Sum of Squares What we do with these squares is add them to get a *Total Sum of Squares,* whose formula is $\sum_{i=1}^{n}(y_i-\bar{y})^2$. We should recognize $\sum_{i=1}^{n}(y_i-\bar{y})^2$ as part of the formula for the variance and the standard deviation, $s=\sqrt{\frac{\sum_{i=1}^{n}(y_i-\bar{y})^2}{n-1}}$. This sum tells us how much *variation* we have in the data as a whole. Usually we divide this by $n-1$ so that we can compare collections, but we will see we do not have to do that here.

For our example, the *Total Sum of Squares* $=\sum_{i=1}^{n}(y_i-\bar{y})^2=17.50$.

- With the graph selected, go to ***Graph>Least-Squares Line.*** You should see the plot shown here; if you see this plot on the computer, you should be able to see that there are two sets of squares shown in two different colors.

Sum of Squared Residuals The plot is looking really complicated because there is now an additional set of squares. If you are doing this with software, you should just about be able to see that this new set is composed of the squares that come from the *residuals* between the data points and the *Least Squares Line.* Look at the dot for Hyleas Fountain (the best runner). Remember that her square for her distance from the *mean*, that is $(y_i-\bar{y})^2$ was large; now look closely and you will see a tiny square for her dot that represents her squared residual, that is $(y_i-\hat{y})^2$. Here is the calculation with comments.

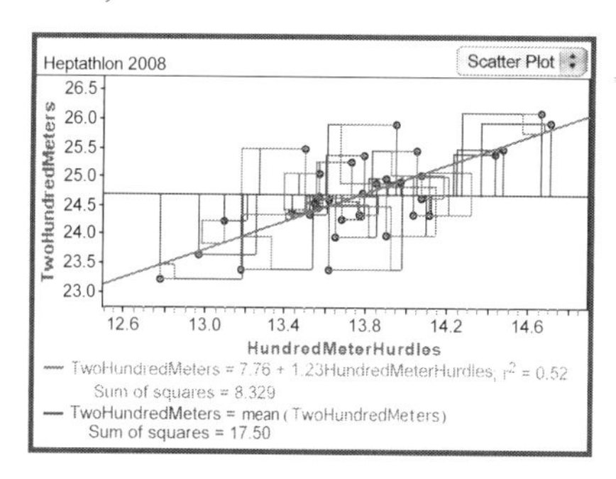

- Hyleas Fountain's time for the *HundredMeterHurdles* was 12.78 seconds.
- The *Least Squares Regression Equation* is $\hat{y}=7.76+1.23x$, as can be checked from the graph.
- Therefore, Ms. Fountain's *predicted TwoHundredMeters* time should be $\hat{y}=7.76+1.23(12.78)\approx 7.76+15.72=23.48$ seconds.
- Hence, Hyleas Fountain's residual is $y_{HyleasFountain}-\hat{y}=23.21-23.48=-0.27$, which means that she was even slightly faster than her predicted time.
- The squared residual for Ms. Fountain is $(y_i-\hat{y})^2=(-0.27)^2\approx 0.073$

What we have done and where this is going. Let us review what we have done. First, we calculated the sum of the squares from the mean of y, which in symbolic form is:

$$\textit{Total Sum of Squares}=\sum_{i=1}^{n}(y_i-\bar{y})^2=17.50$$

We can think of this number as the *total amount of variation* in the data, or we can think of this number as a kind of *baseline* amount of variation. Then we calculated a similar—but different—sum of squares between the actual values and the predicted values, in symbolic form:

$$\textit{Sum of Squared Residuals } \sum_{i=1}^{n}(y_i - \hat{y})^2 = 8.329$$

We can think of this as the variation still left over when we applied the least squares regression line. Notice that the *Sum of Squared Residuals* is smaller, and we would like to reduce the amount of variability, so smaller is good. That means that by applying the least squares line, we have disposed of, or gotten rid of, some of the original variation measured by the *Total Sum of Squares.* By applying the least squares line, we have accounted for (or, "explained") of some of the *baseline* amount of variation. If you look carefully at the plot again you should be able to see that the squares for the squared *residuals* $(y_i - \hat{y})^2$, the squares around the *Least Squares Line,* appear to be smaller than the squares around the mean, $(y_i - \bar{y})^2$. If the squares are smaller then, when we add them all up, we should get a smaller sum; and since software often reports the sum of squares, you can see that the sum is smaller.

The sum of squares for the least squares line, the *Sum of Squared Residuals* $\sum_{i=1}^{n}(y_i - \hat{y})^2 = 8.329$, whereas the *Total Sum of Squares* $= \sum_{i=1}^{n}(y_i - \bar{y})^2 = 17.50$. We have reduced the amount of variation by applying the linear model in the form of the *Least Squares Line*. The next step is to calculate—in percentage terms—how much we have reduced the original variation.

Calculating the proportion of explained and unexplained variation. Of course, we have not completely reduced the baseline variation to zero; the *Sum of Squared Residuals* is not zero. (If there were no variation, there would be no scatter around the least squares line.) There is variation that is still *unexplained* (or as yet *unaccounted for*) by the model. How much *unexplained variation* remains compared with what we started with? We calculate the ***Proportion of Unexplained Variation*** by making a fraction of the *Sum of Squared Residuals* to the *Total Sum of Squares.* Here is how it works out in our example:

$$\textbf{\textit{Proportion of Unexplained Variation}} = \frac{\textit{Sum of Squared Residuals}}{\textit{Total Sum of Squares}} = \frac{\sum_{i=1}^{n}(y_i - \hat{y})^2}{\sum_{i=1}^{n}(y_i - \bar{y})^2} = \frac{8.329}{17.500} \approx 0.476$$

Now comes what may appear at first to be a linguistic trick, but it is not. Recall the rule that says that the probability of "not *A*" is equal to 1 minus the probability of the event. In our notation, we write $P(\text{Not } A) = 1 - P(A)$. What we have calculated so far is the proportion of (still) unexplained variation. If we want the ***Proportion of Explained Variation,*** we will subtract what we have just calculated from the number 1 (the 1 stands for "all the variation") and get

Proportion of Explained Variation = 1 − Proportion of Unexplained Variation

$$R^2 = \textit{Proportion of Explained Variation} = 1 - \frac{\sum_{i=1}^{n}(y_i - \hat{y})^2}{\sum_{i=1}^{n}(y_i - \bar{y})^2} = 1 - \frac{8.329}{17.500} \approx 1 - 0.476 = 0.524.$$

Compare this with what we calculated before; you will find it just below the first heptathlon plot. There we calculated $R^2 = r^2 = (0.723836)^2 = 0.5239$. We can get R^2 by squaring the correlation coefficient or by calculating sums of squares. If the second appears to you more cumbersome, you should know that it has the advantage of showing *why* we *interpret* R^2 in the way that we do.

R^2 **The calculation from sums of squares**
$R^2 = \textit{Proportion of Explained Variation} = 1 - \frac{\sum_{i=1}^{n}(y_i - \hat{y})^2}{\sum_{i=1}^{n}(y_i - \bar{y})^2}$

Flawed Models: Ways the model may not be adequate

We have been using a "straight-line" or linear model for the data. However, there are was in which the model building may be flawed. First, the data may not be linear. If that is so, the linear model that we fit will be simply wrong. We may be able to make good predictions for a part of the data that appear linear, but ultimately, something other than a linear model is called for. The first part of this section illustrates that. Secondly, it may be that there are a few observations that make our estimates of the slope of the linear model problematical. That is the subject the second sub-section here. Thirdly, it is possible to misuse our model to make unwarranted predictions; that is the third sub-section.

Flawed Models: Curves and Bends and Gearhead Cars Actually, "gearheads" tend to enjoy driving fast through curves and bends. However, when we use the correlation coefficient and the least squares line, we are using a *linear model*—we want a straight road. If our data is not linear, if it has curves and bends, then the conclusions that we draw from the linear model will not be useful. Here is an example from our *Road and Track* data. One of the most common measures with cars is the time in seconds that it takes for a car to accelerate from zero miles per hours to sixty miles per hour.

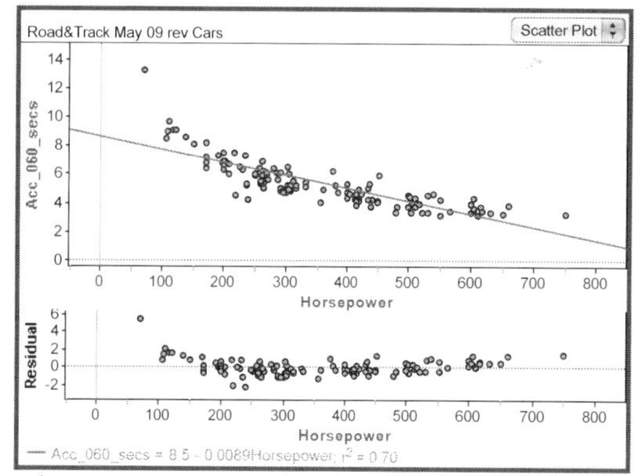

The plot showing how the acceleration time for cars depends upon horsepower—is displayed here. The relationship is negative—the more power a car has, the smaller the time it takes to accelerate to 60 mph—but there is also a certain amount of curvilinearity evident. Look at the cars with less than 150 horsepower; all of the residuals for these cars are positive. You can see that in the original scatterplot, but software can make a plot of residual values that exhibits the same feature. Notice also that in all of the residuals for the cars that have more than 650 horsepower, the residuals are also positive. For the cars having 200 to 500 horsepower, there is a mix of positive and negative residuals, and the linear model is more appropriate.

In general, you can detect ***curvilinearity*** by looking at the shape of the graph, but curves and bends can also be seen in the residuals; if the residuals exhibit a definite pattern, the linear model breaks down in some way. If the residuals show a random scatter then the linear model is a good model.

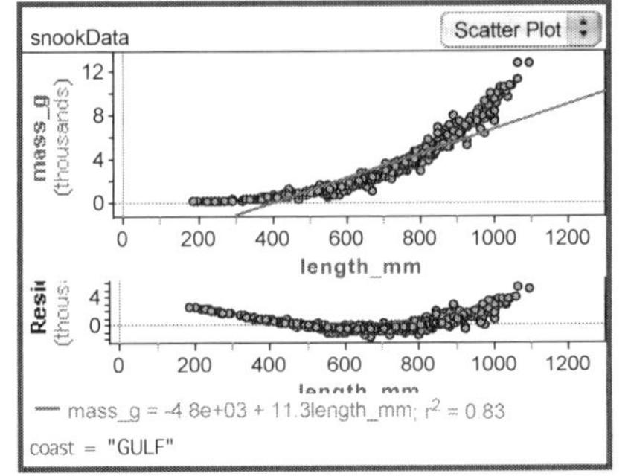

Here is another example of curves and bends in data (the Snook data); notice again how you can detect the curve in the scatterplot and also how the residuals form a pattern.

If the data are not linear, the data can still be analyzed. However, the techniques employed (such as ***transforming*** the data using logarithms) and using alternate models are for the next course in statistics. For this course, it is sufficient to *detect* curvilinearity in data.

> ***Curvilinearity***
>
> *Curvilinearity* (lack of a linear relationship) is seen in a definite pattern to the residuals; residuals around the least squares regression line should exhibit random scatter for a linear model.

Flawed Models: Influential Observations One or two observations may have the potential to determine the slope and the y-intercept of the *Least Squares Equation.* Such observations are called ***influential observations.*** Here is one example. In the 2008 Olympic heptathlon, one competitor (Yana Maksimova of Belarus) fouled out in the long jump and got zero points for that event. As a result, her overall score of 4,806 was very low. Here are two plots showing the relationship between the total score (which determines the rankings in the event) and performance on two events, the *HundredMeterHurdles* and the *HighJump,* with Ms. Maksimova highlighted.

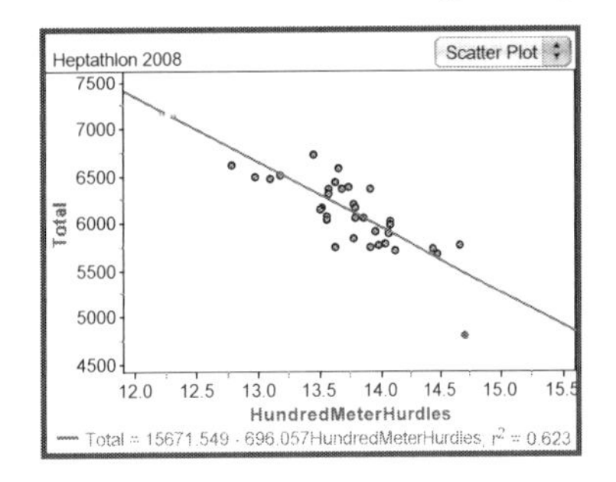

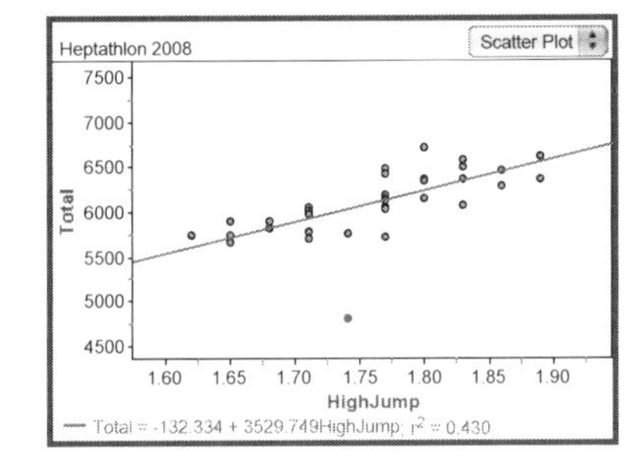

What if? Now suppose that Ms. Maksimova had *not* fouled out in the long jump and so had a higher *TotalScore.* Then, if she had a *TotalScore* of about 6,000, the scatterplots might look like this.

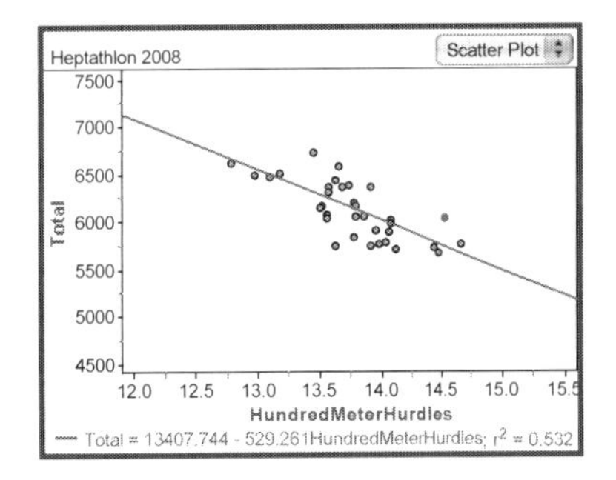

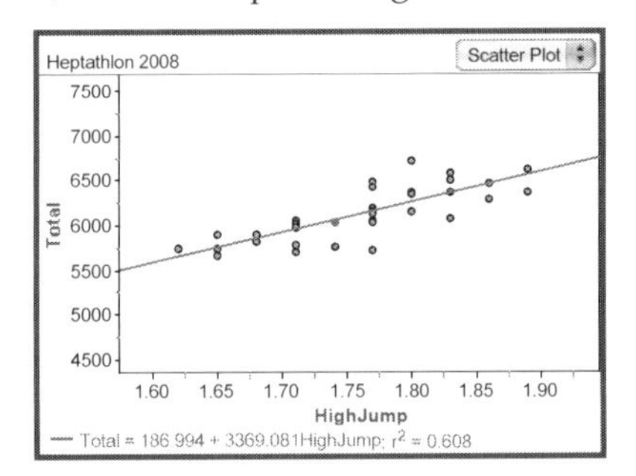

So what has happened? Or not happened? If Ms. Maksimova's *TotalScore* had been higher then our least squares model looking at the effect that *HundedMeterHurdles* had on *TotalScore* would look much different. The slope would be different, and the R^2 is also different. Raising the *TotalScore* changes the slope of the *HighJump* least squares equation but much less so. The change for the *HundedMeterHurdles* is about 24% of its original value (about $\frac{167}{696} \approx 0.24$), but the change in the slope for the *HighJump* is only about 4.5% of the original value (about $\frac{160}{3529} \approx 0.045$). Notice that Ms. Maksimova is an *outlier* on the *response* variable (*TotalScore*) in both plots. However, her performance in the *HighJump* was neither an outlier or toward the "end" of the data, as it was for the *HundedMeterHurdles.* If an observation is also an outlier or near the end of the data on the *explanatory* variable then that observation has the potential to be influential. Whether a data point actually *is* an *influential observation* depends upon its relationship with the remainder of the data. With some software (such as Fathom), it is a simple matter to run an analysis with and without a single data point to explore whether the data point is *influential.*

Influential Observations

A data point is an *influential observation* when small changes in the values of the data would substantially change the slope and *y*-intercept of the linear model fitted.

Flawed Models; Extrapolation One of the purposes of the *Least Squares Regression Equation* is to make predictions. One of the mistakes that can be made is to make predictions for values of the *explanatory variable* outside of the range of the data we have. This mistake is called ***extrapolation.*** Here is an example that looks at the relationship between *Horsepower* and *Acc_060_secs* for just the cars with four cylinders.

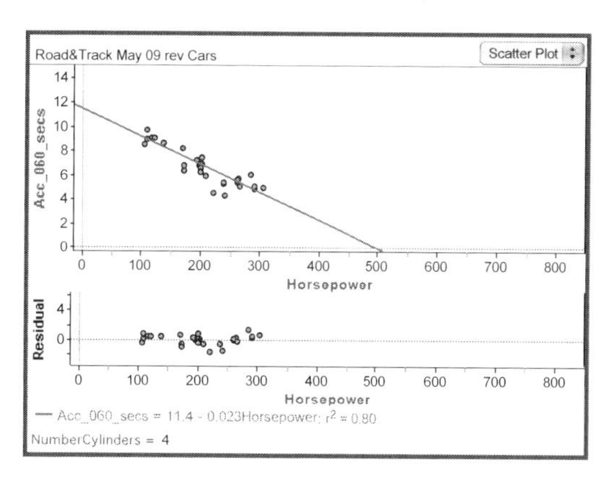

For just the four cylinder cars, the plot looks linear, and the least squares line has the equation $\hat{y} = 11.4 - 0.023x$, where x is *Horsepower* and y is the time to accelerate to 60 mph, or *Acc_060_secs.* It does not make sense at all to insert $x = 37$ into the equation and get $\hat{y} = 11.4 - 0.023(37) = 11.4 - 0.23 \approx 10.55$ because the data from which we derived the least squares line do not include cars with horsepower in the region of thirty-five to forty horsepower. (There were cars made with horsepower in this range, but their zero to sixty times tended to be in the neighborhood of twenty-five to thirty seconds!) For the same reason, we should not use this equation to predict the zero to sixty time for a four-cylinder car with 450 horsepower. There are cars with 450 horsepower, but these cars are not four-cylinder cars, and so a different linear model will be appropriate.

In fact, with our data, it makes sense to look at linear models by the number of cylinders car have. Here is the plot. (Note: The variable *Cylinders* includes three-cylinder cars and five-cylinder cars in the categories "Four Cylinders" and "Six Cylinders" respectively.) As you can see from the plot with the lines shown, the linear relationship differs depending on the number of cylinders.

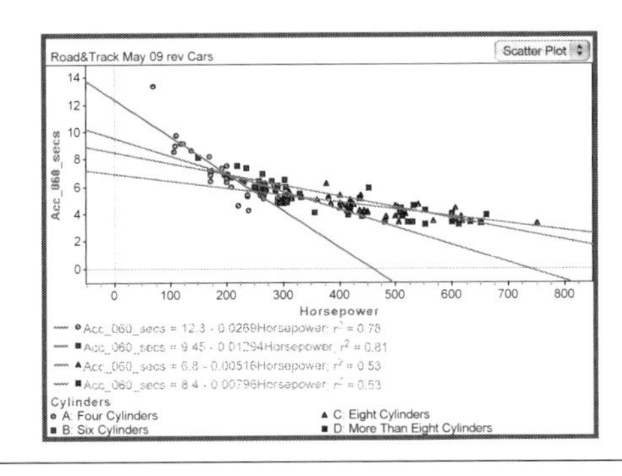

Extrapolation

Extrapolation is the practice of using a linear model to predict outside the range of data from which the linear model was derived. *Extrapolation* is to be avoided.

Summary: Evaluating a Linear Model

Much data analysis involves fitting a model to data. But then the question immediately comes up as to whether our model—simple and beautiful as it may be, such a straight line—is any good. Our models attempt to make sense of this variability in our data—to "explain" it, to "account" for it. This section introduces a measure and perspectives for assessing how good linear models are.

- The ***Coefficient of Determination,*** whose symbol is R^2, is a measure to assess how good a linear model fits the data.
  - For simple regression models, $R^2 = (r)^2$; that is, R^2 is the square of the correlation coefficient r.
  - $0 \le R^2 \le 1$, so that the numbers can be interpreted as proportions;
- ***Interpretation of R^2*** : R^2 shows the proportion (or percentage) of variation in the *response* variable *explained* by the model being used.
- The ***calculation of R^2*** may also be done by considering:
  - The total variation in the response variable is taken as variation that is initially unexplained and measured by the *Total Sum of Squares* $= \sum_{i=1}^{n}(y_i - \bar{y})^2$ and,
  - The amount of variation still remaining after applying the model, measured by the *Sum of Squared Residuals* $\sum_{i=1}^{n}(y_i - \hat{y})^2$; then
  - $R^2 = 1 - \dfrac{\text{Sum of Squared Residuals}}{\text{Total Sum of Squares}} = 1 - \dfrac{\sum_{i=1}^{n}(y_i - \hat{y})^2}{\sum_{i=1}^{n}(y_i - \bar{y})^2}$
  - The *interpretation of R^2* follows from

    R^2= *Proportion of Explained Variation = 1 – Proportion of Unexplained Variation.*
- ***Flawed Models*** It may be that our model is not good for a number of reasons.
  - There may be ***Influential Observations,*** observations whose position in the data would substantially change the slope and *y*-intercept of the linear model fitted if the data were changed or omitted. Hence, the model we are using may have a misleading slope or *y*-intercept.
  - The data may be ***curvilinear*** rather than linear. *Curvilinearity* is reflected in a definite pattern to the residuals around the least squares regression line rather than a random scatter about the line we expect to see if the linear model is appropriate. In this instance, the idea for the model itself is flawed.
  - We may be attempting to ***extrapolate*** prediction beyond the domain of data from which the model was constructed.

§3.1 Can We Trust Data (to Reflect Reality)?

Example: the Presidential Approval Rating

If you go to http://www.gallup.com/poll/113980/Gallup-Daily-Obama-Job-Approval.aspx, you will find something very much like:

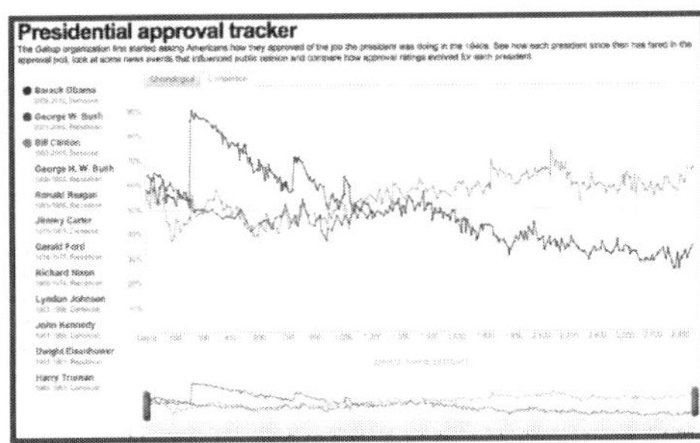

> Gallup tracks daily the percentage of Americans who approve or disapprove of the job Barack Obama is doing as president. Daily results are based on telephone interviews with approximately 1,500 national adults; margin of error is ±3 percentage points.

There are approximately 217,000,000 adults in the USA who are eligible to vote. How can we trust findings that are based upon fifteen hundred out of 217,000,000 of the people involved? In the following sections, we will see:

- We *can* trust results based upon a miniscule sample of the voting population, if the data are collected and analyzed carefully. The next sections cover what "collected and analyzed carefully" means.
- We *cannot* trust results if data are collected badly, even if the number of cases being studied is large.
- The meaning of the term "margin of error" mentioned above

Some terminology We start with some essential terminology, making a distinction between a population and sample. By a population we mean whatever it is that we are ultimately interested in describing and understanding. A sample is a representative part of that population. In the example of the Presidential Approval Rating, the population of interest is the adults in the USA who are eligible to vote.

Population and Sample	
Population:	the collection that a statistical analysis of data seeks to describe or understand
Sample:	a collection that is a representative part or subset of a population
Census	if a "sample" includes all of the population, it is called a census

By a population we mean a *description* of something and not a number. The number would be the size of the population. The Gallup organization wants to know something about the adult voters of the USA, and not of Canada or Mexico. They could be interested in knowing something (incomes, for example) about those who work in the construction industry in California. Some other organization could be interested in knowing something about fir tree production in Oregon. Voters in the USA, construction workers in California, fir trees in Oregon are all different populations, with different descriptions.

Typically the thing we want to know about a population is either impossible (or very hard) to find out about for the entire population, but *is* possible for a sample of the population. It is possible for Gallup to contact fifteen hundred people and ask them about the president's job performance; it would not be feasible to ask *all* the voters.

Generalizing from a sample to a population: an example. We make the distinction between population and sample because we want to be clear about what we are searching for. If we do not clearly know what we want, we probably have a good chance of not finding it! However, sometimes we have data, and it is not easy to define a population to which the results can be reasonably generalized. An example is the student data from statistics classes that we have been analyzing. It comes from students taking statistics at one community college in the San Francisco Bay Area. Here are some possible descriptions of populations to which the results of the data might possibly be generalized but with an increasing amount of risk of being wrong.

- Students taking statistics at one community college
- Students at one community college
- Students at community colleges in the San Francisco Bay Area
- Students at all colleges and universities in the San Francisco Bay Area
- Students at all community colleges in Northern California
- Students at all colleges and universities in California
- Students at colleges and universities in North America

Moreover, the risk of being wrong as we generalize to a "wider" population may depend on what we are measuring. If we are looking at the proportion of students who speak more than one language, we may guess that this proportion may differ for students outside the San Francisco Bay Area (but perhaps not), and so it would be risky to generalize to all colleges and universities in Northern California. What about the heights of students? Can we generalize data on the heights of students from one college to students in all of North America?

Best practice. The best practice will be to start with the idea of a population and devise a good method to get a sample. That is what we consider next. However, as "consumers" of statistical analyses, we have to do the harder thing of thinking how a sample that has already been analyzed can be generalized or not.

How to get bad samples (aka non-random samples)

It may seem strange to start negatively, but bad samples are common, and it is well to deal with the bad before the good. There are a huge number of ways for data to come from bad samples, and as "consumers" of statistics, it is our job to look critically at how data are got. We start with some common examples of bad data.

Voluntary Response (or Self-Selected) Samples and Non-response. A web site or a publication asks for votes or "likes" or asks for opinions is a good example of a voluntary response or self-selected sample. A self-selected sample is one in which the data come *solely* from people voluntarily responding. Voluntary response is therefore not the same as *non-response* from people who are chosen to participate. A sample of people may be chosen to answer a questionnaire, and always there are some who do not respond; this is called *non-response.* A large proportion of *non-response* is problematical, but that large proportion does not make the data collection itself a voluntary response sample. We will reserve the words "voluntary response" for data collection schemes that depend wholly or solely on the voluntary response in a population. For voluntary response data collection (or for large *non-response*), the questions to ask are: (1) "Who is likely (and unlikely) to respond to the invitation to answer questions?" and (2) "Does the answer to question (1) represent any population in which we have an interest?"

Convenience Samples. Here is an (extreme?) example of a convenience sample: A college student is asked to collect data on some topic where the population is meant to be first-year college students at that student's college. So, the college student pesters all his or her friends until the friends give the required information about the topic. There may be an element of non-response in this procedure (some of the friends may refuse), but if the student is persistent enough, the non-response may not be a problem, although the student may lose some friends. We must still ask the same two questions: (1) "Who (or what) is likely to be included in the sample?" and (2) "Does the answer to question (1) fairly represent any population in which we have an interest?" Unless the student's friends really are a representative sample of all first-year college students at that college (unlikely!), this sample is not a good sample; it is a bad sample because it is likely to be unrepresentative.

Convenience samples may not always be as useless as our example suggests. The data collected from statistics students at one college in Northern California is a convenience sample. The answer to question (1) is clear: the data come from students in a statistics course at one community college. The little exercise in the sub-section above ***(Generalizing from a sample to a population: an example)*** comparing this sample to various populations in which we may have an interest is an attempt to answer question (2). One treads warily here.

Judgment (or quota) samples. One response to the extreme example of a convenience sample may be to say: "I would not be so naïve to think that my friends represent all first-year students. I would make certain that my sample had equal numbers of males and females, for example, and I would try to make the ethnic composition of my sample to reflect what I know of the ethnic composition of the first-year students at the college." This is a good move, and the sample produced will be a better sample than the one chosen naïvely in the example above. However, it is still not a good sample. Although the one choosing the sample may specify that half the sample be females (or 60% or whatever percentage at that college is), the choice of the individuals is still up to the one choosing the sample. Whenever there is human choice of the elements (the individuals or the cases), the sample has a chance of being unrepresentative. Odd as it may seem, the best way to choose a sample is to eliminate human choice in the final selection of the elements. Samples chosen by chance and not human choice are called ***random samples***. The three types of "bad samples" listed above are not the only types of bad sampling. Here is a summary; in all of these types of sampling, human choice is ultimately involved.

Types of Non-Random Sampling

- ***Voluntary Response Sampling***: Samples in which the elements in the sample (the individuals or cases) depend solely upon the choice of the individuals themselves
- ***Convenience Sampling*** Samples in which the choice of the elements in the sample (the individuals or cases) is determined primarily by which elements happen to be easily accessible to the researcher
- ***Judgment or Quota Sampling*** Samples in which the representativeness of the sample is furthered by having the elements fulfill quotas judged by the researcher to characterize the population

How to get good samples (aka random samples)

There are two good reasons for employing random sampling. One: in the long run, it is the only type of sampling that will guarantee truly representative samples. The second is that random sampling is an essential requirement for the techniques statisticians use to generalize. The "margin of error" cannot be calculated if the samples involved are not random, for example. But what are random samples?

Random is not random; so what is random? In that sentence, obviously the word "random" is used in several senses. The very first "random" refers to what people think of as random: "haphazard," "without pattern," "lacking regularity," "arbitrary," or perhaps even "unbidden" or "unwanted." If you are asked to choose from a group of people or things apparently randomly, your choices will not be random in the mathematical sense, so that kind of "random" is not "random" in a mathematical sense. Choices made by humans will always follow some pattern. We may not be able to know or predict what that pattern will be—it will be different for people from different cultures—but there will always be some pattern.

So what is random? It is helpful to reduce everything to numbers. Here is an initial working definition: if we have a finite set of numbers, a ***random*** choice will give each number an equal probability of being chosen. Here are some examples: on a ten-sided die there are ten choices, and the action of rolling the die gives each choice an equal chance of appearing. The equal probability is 1/10 or 0.10. In a lottery bin containing sixty-four numbers where a number is chosen "blindly" (without looking at the numbers) after thoroughly mixing the numbers in the bin, each of the sixty-four numbers has an equal chance of being chosen, with probability 1/64 or 0.015625.

Working Definition of Randomness

Given a finite set of numbers, a ***random*** choice gives each number the same probability of being chosen.

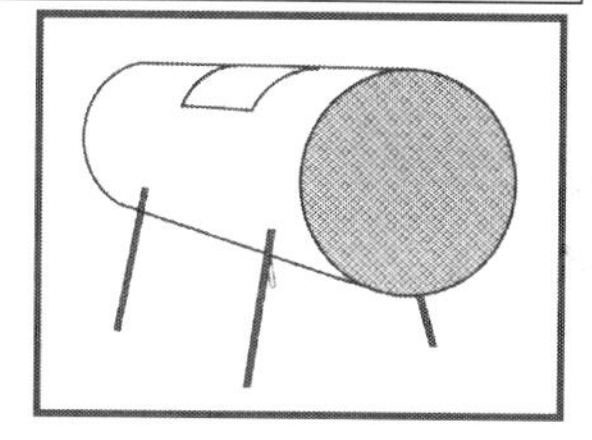

How this works in practice is that each case in a population is assigned a number. A device choosing numbers randomly (called a ***random-number generator***) that essentially works like the ten-sided die (or picking from a lottery bin) gives the result that each number chosen has the same probability as any number. Using a random process will guarantee a random sample from a population.

Simple random samples. The simplest kind of random sample involves two steps. First, there must be a list of the entire population, with a unique number assigned to each case in the population. Here is an example that comes from one of the exercises. We start with the collection of all of the houses that were sold in San Mateo County in 2005–2006. This is the population (a part of which is shown here), and in that population each of the houses is given a number: the assigned numbers are listed in the column labeled *Caseno*. Other variables are measured and recorded as well. A simple random sample from this population is chosen by getting a random-number generator (it could actually be a ten-sided die) choose however many numbers we want.

San Mateo Real Estate Y0506

	Caseno	Address	City	Area	Beds	Sq_Ft	Age	List_Price
698	698	944 WALNUT ST	SCL	353	3	1570	80	795000
699	699	1026 ELM ST	SCL	353	3	1710	78	899000
700	700	1951 ELIZABETH ST	SCL	353	4	2060	47	999950
701	701	2806 SAN CARLOS AV	SCL	355	3	1210	99	829000
702	702	28 ARROYO VIEW CI	BEL	360	3	1420	8	889500
703	703	2603 CARLMONT DR	BEL	360	3	2060	9	1099800
704	704	2604 CARMELITA AV	BEL	361	2	740	59	669000
705	705	2700 CORONET	BEL	361	3	1160	53	749000
706	706	2841 SAN JUAN BL	BEL	361	3	1350	45	839000
707	707	2511 BUENA VISTA AV	BEL	361	4	1890	49	895000

Simple random sampling is the basis for all other sampling and involves two essential features, which are listed in the box below.

Simple Random Sample (or SRS)

- To draw a ***SRS*** of sample size *n* from a population, you must have:
  1. A list of all the cases in the population, and
  2. A chance process that will choose randomly from the list.
- In drawing a ***SRS,*** every possible sample of size *n* has an equal chance of being chosen.

Random Cluster Samples: an example Often it is not possible for researchers to make a list of the entire population. Or, even if it is possible to make a list, choosing a simple random sample may create too much work collecting the data. For example, suppose the population in which we have an interest is all college students in California. A simple random sample of students may include some students from Yreka (near the Oregon border) and other from Calexico (on the border with Mexico) and some everywhere in between. The travel costs to collect data on these students might be enormous. It would be better and cheaper if the sample data were clustered. The way to accomplish this is to randomly sample *clusters* of cases, and then use *all* the cases in the clusters. The example in the exercises with the real estate data is to draw a random sample of area numbers and then take *all* of the houses within the randomly areas. A common question that is asked about random cluster samples is: "What are the clusters?" The answer is that clusters can be *any* grouping of the cases where each case belongs to just one of the groups or clusters. In the real estate example in the exercises, the clusters are areas. In the student example above, it would make sense to group students by the college or university they attend, with provisions made for students who attend more than one school at the same time.

Stratified Random Samples: an example Often researchers want to make certain that certain groups are represented in the sample in the same proportions as they think they exist in the population. One way to guarantee that this will be so is to randomly select from within groups. A researcher who is sampling students within one university may wish to randomly sample within the categories of the major administrative divisions of the university. Here are the "schools" that constitute Stanford University (see http://www.stanford.edu/academics/). The schools would become the groups, (or, as they are called, the "strata" in "stratified"), and from within each of these schools (business, earth sciences, etc.) a simple random sample of students would be chosen. It is the random selection makes it not merely a judgment sample. As with "clusters" in random cluster sampling, these "strata" are any groupings that make sense.

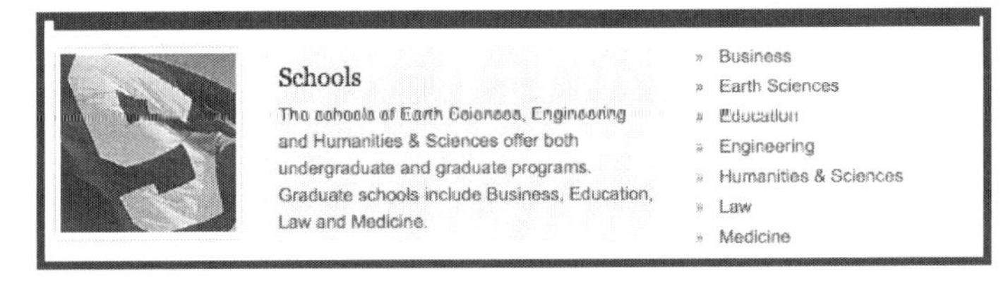

Random Cluster Sample

- To draw a ***random cluster sample*** from a population, you must have:
  1. A list of the clusters to be sampled
  2. A chance process that will choose randomly from the list of clusters
- Then *all* of the cases within the chosen clusters constitute the random sample.

Stratified Random Sample

- To draw a ***stratified random sample*** from a population, you must have:
  1. A list of all the possible cases in the population, grouped in the strata
  2. A chance process that will choose randomly from the cases *within* each of the strata
- Then the random sample is all of the cases randomly chosen from each of the strata (or groups).

Reality. Listed above are only three kinds of random samples. In reality, random samples are chosen in much more complicated ways. The sampling design may be a mixture of cluster and stratified sampling or a type of sampling that depends on the random arrangement of cases in the populations. Here is one example. (See: http://www.gallup.com/poll/101872/How-does-Gallup-Polling-work.aspx.)

> The majority of Gallup surveys in the U.S. are based on interviews conducted by landline and cellular telephones. Generally, Gallup refers to the target audience as "national adults," representing all adults, aged 18 and older, living in United States.
>
> The findings from Gallup's U.S. surveys are based on the organization's standard national telephone samples, consisting of directory-assisted random-digit-dial (RDD) telephone samples using a proportionate, stratified sampling design. A computer randomly generates the phone numbers Gallup calls from all working phone exchanges (the first three numbers of your local phone number) and not-listed phone numbers; thus, Gallup is as likely to call unlisted phone numbers as listed phone numbers.

In this example, what we have called the population Gallup calls the "target audience." The term "national adults" denotes the collection (of people, in this instance) that Gallup wants to describe or study. They do not have a list of the "national adults," but they do have a list of working telephone exchanges. They choose telephone numbers randomly by having a computer generate random telephone numbers within those exchanges. The important word here is "random": Gallup chose samples using a random process.

Experiments give data also, and randomization is still important

The examples in the first part of this section have all been from *observational studies*, but data also come from *experiments*. Here is an example of an experiment on the effect of negative political advertising.

Example: what effect does negative political advertising have? Do negative political ads turn voters against the political process, so that they are less likely to vote, or do negative political ads energize voters, so that they are actually more likely to vote? One political scientist from the Ohio State University conducted an experiment in a Florida mayoral election in 2003 to get some evidence[2]. Here is a brief description of the experiment (which can be found at http://www.jstor.org/stable/4148088, page 203):

> In brief, a random sample of voters was chosen for either the treatment (negative ads) or the control group (no ads) in a mayoral election. Subjects in the treatment received negative campaign ads (from an independent expenditure group) in the mail in the days immediately preceding the election. The resulting decision to vote was then measured by consulting official election records.

[2] Niven, David, "A Field Experiment on the Effects of Negative Campaign Mail on Voter Turnout in a Municipal Election," *Political Research Quarterly*, Vol. 59, No. 2 (Jun., 2006), pp. 203–210

We can make a diagram of this experiment that also shows some of the terminology associated with experiments.

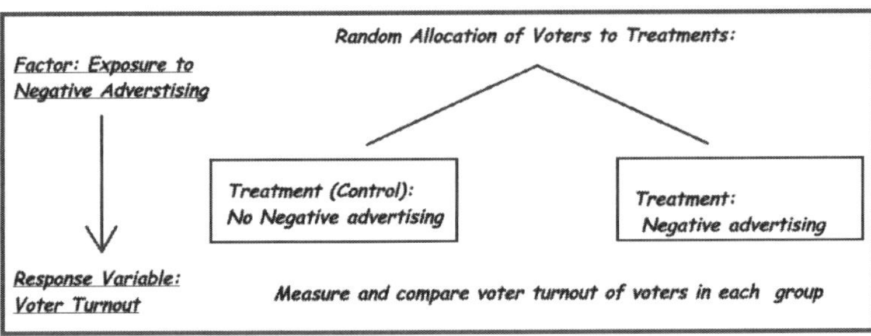

Terminology of experiments: With this experiment, we can illustrate the vocabulary that goes with such experiments.

- ***Factor:*** In experiments, the explanatory variable or variables are called *factors*. Here the factor is whether or not the voter received negative campaign ads. The experimenter determined by random choice which voters in the sample of voters received negative political mailings and which voters did not. (Think of flipping a coin for each voter—heads: negative ads; tails, no ads.)
- ***Treatment:*** Here there are just two treatments, and they are the values of the explanatory variable or factor. The two treatment groups are: (1) received negative political advertising, and (2) did not receive negative political advertising. The word "treatment" reminds one of medical experiments, and often the default "treatment," representing either a standard procedure or no action at all, is labeled the "control group" or just "control." We will regard the control group as one of the treatments.
- ***Response variable:*** The response variable for this experiment is whether or not the voter actually voted: "the resulting decision to vote was then measured by consulting official election records." If negative ads depress voter turnout then the turnout should be lower in the treatment group that had negative ads; if negative ads encourage voter turnout then the voter turnout should be higher.
- ***Experimental Units:*** The cases for an experiment are typically either called "experimental units" or "subjects." The word "subjects" is typically used if the experiment involves humans. However, an experiment may have experimental units that are not humans, just as cases need not be humans.

Example: what effect does fertilizer have on begonias? Here is a diagram of an experiment. Some horticulturists were concerned that some gardeners were using too much fertilizer on their begonias. They bought thirty-six begonia plants and randomly allocated the plants to three treatment groups (no fertilizer, 100 ppm fertilizer, and 300 ppm fertilizer; ppm means "parts per million"), and then gave all the begonia plants the same amount of water, the same amount of sunlight, and the same amount of weeding for some weeks. At each week, they measured the growth of the begonia plants. So for this experiment, as shown in the diagram, the *experimental units* are begonia plants, the *factor* is the amount of fertilizer given to the plants, and the *response variable* is the amount of growth in plants. In case you are wondering, the researchers found that giving begonias big amounts of fertilizer stunted growth.

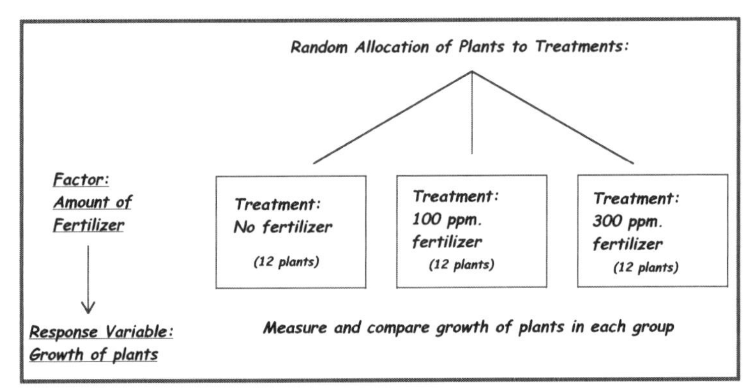

Essential terminology about experiments:

- **Factor:** The explanatory variable in an experiment, whose value for each case is determined; there may be more than one factor.
- **Treatment:** A value or combination of values of a factor or factors
- **Experimental units:** The cases that are used in the experiment; they are sometimes also called "subjects."
- **Control group:** A treatment against which the other treatments are compared; in medical experiments, the standard or no treatment

The role of randomization: The horticulturists who ran the begonia experiment were careful to treat all the begonia plants in the same way—the same amount of watering, etc. The important principle is that all of the cases should be as much alike as possible *except* for the value of the factor. The reason for this is logical: if the researchers *do* find a difference in the growth of the begonia plants after some weeks, they can legitimately conclude that the differences were caused by the amount of fertilizer given to the plants and not some other variable, such as the amount of water given, or the amount of sunlight, or the soil conditions. The researchers can confidently rule out some other cause.

On the other hand, if they find no difference in growth then they can legitimately conclude that the amount of fertilizer has no effect on growth, since all other things were kept equal. But can you be certain that the begonia plants were themselves "equal"? Perhaps some of them are initially healthier than others. What would happen if all the really healthy begonias got assigned to the "no fertilizer" treatment group? Then any greater growth of the no fertilizer treatment group that is found may have been because the plants in that group were healthier in the first place. That is where randomization comes in: using randomization to assign the begonias to the three treatment groups effectively makes it very unlikely that one particular treatment group gets all the healthy plants and another one unhealthy plants, since the probabilities of being assigned to each of the three groups should be equal. Where randomization is used to assign the experimental units to the treatment groups, the entire experiment is often called a ***randomized trial.*** This terminology is especially common in medical research.

The role of randomization for the negative political advertising experiment: The researcher first of all randomly selected 1400 voters from the list of voters. Then, he randomly assigned these sampled voters to the treatment groups, with 700 being randomly assigned to the group that did not receive any negative mailings. Randomly assigning the subjects (experimental units) made the two groups approximately equal in composition.

Blind, double blind, and the quest for equality. In the negative political ads experiment, neither the people who received the negative political ads nor those who did not realized that they were part of an experiment or that they were in one of the researcher's treatment groups. This is an example of a ***blind*** experiment, where the subjects are unaware ("blind") of the treatment group they are a part of. With people as experimental units (subjects) being aware of being in a "treatment" or "control" group may itself make a difference, which means that "not everything else is equal." In medical experiments that are blind, patients who are part of an experiment may be kept unaware of which treatment group they have been assigned.

Sometimes this is not possible; one of the exercises concerns an experiment on comparing different diets, and, obviously, the subjects knew which diet they were following, so the experiment was ***unblinded***. If in a medical experiment the medical personnel are also unaware of which treatment which patient has, that study is ***double blinded***.

Very elementary experimental design:

- In an experiment, the researcher is able to set the values of one explanatory variable (the ***factor***) and keep all other variables as equal (or constant) as possible, so that the ***treatment groups*** differ only by the values of the ***factor***.
- In order to keep the treatment groups as equal as possible, studies sometimes keep human subjects from knowing which treatment group they are a part of; this is known as a ***blind*** or ***blinded*** experiment.
- Where variables cannot be controlled to make the treatment groups equal then ***randomization*** is employed to assign experimental units to treatment groups, and the experiment is sometimes called a ***randomized trial.***

Where the experimental units are not people (such as in the begonia experiment), the issue of the subjects' awareness of their treatment group as a threat to the equality of the treatments does not arise.

How experiments differ from observational studies and why.

In an experiment, the researchers deliberately set values to one variable vary and keep everything else the same. In the begonia experiment, the amount of fertilizer was set by the horticulturists at 0 ppm, 100 ppm, or 300 ppm. In the negative political mailings experiment, the researcher was able to determine that some people received negative mailings, and others did not. Since those who received and those who did not were randomly determined, the randomization works to make the two groups as equal as possible. In both examples, the researcher was able to set values of the explanatory variable (the factor) for the experimental units. It is this ability to set the values of the factor that makes an ***experiment*** distinct and different from an ***observational study***. In a purely observational study, the researcher has no ability to set the values of any variable; the values of the variables are merely observed. The Gallup polls on the presidential approval rating are observational studies because the researchers do not have the ability to set any of the variables for the cases they select. One of the exercises draws samples from the population of houses sold in San Mateo County in 2005–2006. That is also an observational study since the data were simply observed, without any intervention by the researcher to set the values of any of the variables.

The advantage of an experiment over an observational study is that differences in a response variable can be directly linked to differences in the factor whose values have been set, since all other variables have been held constant or equal. If the begonia study had been an observational study where the growth of begonia plants was recorded (observed) as well as the amount of water, sunlight, fertilizer, weeding, etc. (all explanatory variables), it would be unclear whether any variation in growth (the response variable) was because of the fertilizer, the watering, the differences in sunlight, or some other variable.

Experiments and Observational Studies:

- In an ***observational study***, the researchers simply collect the data as it is, without the ability to intervene to set the values of one of the explanatory variables.
- In an ***experiment***, the researcher is able to intervene to set the values of one or more of the explanatory variables, often employing randomization to allocate or assign cases to treatment groups.
- ***Experiments*** offer a more direct basis for inferring effects (variation in response variables) from causes (variation in explanatory variables) than do ***observational studies***, since variation in response variables can logically be attributed to many combinations of variables.

Trusting data: what we mean by it and the first steps toward it

By trusting data, we mean: "Do the data say anything beyond the data themselves?" If Gallup find that 42% of the fifteen hundred people in their sample approve of the president's performance, must they say that all we know is that 42% from that sample approve, or can Gallup generalize beyond those fifteen hundred to all voters? No and yes.

The answer is no if we (or Gallup or anyone else) are sloppy about how the fifteen hundred people are chosen. If they use a non-random sample, whether it is a voluntary response sample or a convenience sample or a judgment sample, then the results cannot be trusted. Generalizations from sloppy sampling cannot be trusted, even if the sample size is large. If we (or Gallup or anyone else) have not thought through about what the intended population is—the target audience, as Gallup terms it—then generalizing is risky. The first steps to being able to generalize are to be clear about the population to which we are generalizing and to use more sophisticated sampling. More sophisticated sampling means randomization.

We also saw in this section that there is numerical data that comes from experiments as well as from observational studies, and with these kinds of data, randomization also has a role to play.

However, getting data using randomization are just the first steps in what statisticians do to have confidence in their data to be able to generalize beyond as sample. The technical term for generalization that statisticians use is inference. In the next sections we tackle inference head-on.

Inference

Inference is the term that statisticians use to refer to generalizing from a sample to a population they seek to understand or describe.

Summary: Trusting Data

Population and Sample

Population: the collection that a statistical analysis of data seeks to describe or understand

Sample: a collection that is a representative part or subset of a population

Census if a "sample" includes all of the population, it is called a census.

Types of Non-Random Sampling

Voluntary Response Sampling: Samples in which the elements in the sample (the individuals or cases) depend upon the choice of the individuals themselves, wholly or in part

Convenience Sampling Samples in which the choice of the elements in the sample (the individuals or cases) is determined primarily by which elements happen to be easily accessible to the researcher

Judgment or Quota Sampling Samples in which the representativeness of the sample is furthered by having the elements fulfill quotas judged by the researcher to characterize the population

Working Definition of Randomness:

Given a finite set of numbers, a ***random*** choice gives each number the same probability of being chosen.

Simple Random Sample (or SRS):

To draw a **SRS** of sample size n from a population, you must have:

1. A list of all the cases in the population, and
2. A chance process that will choose randomly from the list

In drawing a ***SRS,*** every possible sample of size n has an equal chance of being chosen.

Random Cluster Sample:

To draw a ***random cluster sample*** from a population, you must have:

1. A list of the clusters to be sampled
2. A chance process that will choose randomly from the list of clusters

Then all of the cases within the chosen clusters constitute the random sample.

Stratified Random Sample:

To draw a ***stratified random sample*** from a population, you must have:

1. A list of all the possible cases in the population, grouped in the strata
2. A chance process that will choose randomly from the cases *within* each of the strata

Then the random sample is all of the cases randomly chosen from each of the strata (or groups).

Essential terminology about experiments:

Factor: The explanatory variable in an experiment, whose value for each case is determined; there may be more than one factor.

Treatment: A value or combination of values of a factor or factors

Experimental units: The cases that are used in the experiment; they are sometimes also called "subjects."

Control group: A treatment against which the other treatments are compared; in medical experiments, the standard or no treatment.

Very elementary experimental design:

In an experiment, the researcher is able to set the values of one explanatory variable (the ***factor***) and keep all other variables as equal as possible, so that the ***treatment groups*** differ only by the values of the ***factor***.

When humans are used as the subjects in a study, the subjects are sometimes kept unaware of whether they are part of a treatment or control group. If this is done, the study is known as a ***blind*** or ***blinded*** experiment. If, in addition, the administrators of the experiment can be kept unaware of which subjects are in which treatment group, that study is called double blind.

Where variables cannot be controlled to make the treatment groups equal then ***randomization*** is employed to assign experimental units to treatment groups; the experiment is sometimes called a ***randomized trial.***

Experiments and Observational Studies:

In an ***observational study***, the researchers simply collect the data as it is, without the ability to intervene to set the values of one of the explanatory variables.

In an ***experiment***, the researcher is able to intervene to set the values of one or more of the explanatory variables, often employing randomization to allocate or assign cases to treatment groups.

Experiments offer a more direct basis for inferring effects (variation in response variables) from causes (variation in explanatory variables) than do ***observational studies***, since variation in response variables can logically be attributed to many combinations of variables.

Inference is the term that statisticians use to refer to generalizing from a sample to a population they seek to understand or describe.

§3.2 Trusting Data, Part 2: Sampling Distributions

How tall are Hobbits?

Imagine being transported in both time and place to the Shire, land populated by Hobbits. You are surrounded by a group of Hobbits despite the fact that they are "shy of 'the Big Folk.'" But how tall are they on average? Can you trust what you see? Even if you were brave enough to collect measurements and use all the stats you have learnt to calculate means and standard deviations and make dot plots and box plots, how do you know that you have not stumbled upon an especially short or an especially tall sample of Hobbits? There is the story (pictured here)[3] of Bandobras Took (Bullroarer), son of Isengrim the Second, who was said to be four-foot-five (about 135 centimeters) tall and was able to ride a horse.[4] Was Bandobras an "outlier" in height, as implied by the story? Or, in his time, was he typical? How can we ever know the answer to this question?

The answer: The idea of a sampling distribution. There is a surprising answer to having doubts about the trustworthiness of the samples that we have: the answer is to have an idea of *all* the possible samples that we could possibly have. With this idea, we will be able to judge if "our" sample is trustworthy. How can this be done? If we do not trust our one sample, how can we have an idea of all of them? Read on. Height data for Hobbits is very difficult to find, and so we have to abandon this rather intriguing question. Instead, we will concentrate on something more accessible: the size of houses for sale.

Warning: Difficult Ideas Ahead Many students find the idea of a *sampling distribution* and the way that we use it in statistics quite difficult. This is correct: the ideas are not immediately intuitive and cannot be reduced to a number of steps to solve problems. Advice: ask many questions to try to picture what a sampling distribution is. (The exercises are meant to lead you through the kinds of questions to ask.)

The ideas will be developed in the context of observational studies, so we will look at samples from a population; however, the same ideas, with appropriate changes in terminology, also apply to randomization in experiments. In the context of population and samples, take care to distinguish between a ***population distribution, a sample distribution,*** and a ***sampling distribution.*** (Here: "*ling*" is deliberately underlined.)

The Idea of a Sampling Distribution for Quantitative Variables

We are going to use an example where we know about the population; this is completely unrealistic because usually we do *not* know about the population. However, using an example where we *do* know everything about the population will help us better understand the concept.

---

[3] Bandobras Took during the fight with Golfimbul * Artist: Gregor Roffalski, alias Sigismond, * Copyright: Public Domain

[4] Tolkien, J. R. R., *The Fellowship of the Ring,* New York, Ballantine Books, 1965, p. 20

When we randomly sample from a population, we can never be completely certain how representative our sample is of the population. The plot here shows the distribution of the variable *Sq_Ft* (the number of square feet in the living area of the house) for the entire *population* of houses that were sold in San Mateo County in the years 2005–2006 (there were 1,890 houses sold in that year). The plot just below that shows the distribution of the same variable *Sq_Ft* for a random sample of $n = 40$. Since $n = 40$, there are far fewer dots in the dot plot. Is the sample like the population? One thing we can do is to compare the means, since we are fortunate to have the population. For the population, the mean for the variable *Sq_Ft* is $\mu = 1949.56$ ft^2 (we use the Greek letter μ because we are referring to the population mean). For the sample, the sample mean is $\bar{x} = 1965.08\ \text{ft}^2$; it is not the same as the population mean but not that far away. How do we know that the sample mean is not far from the population mean? It is because we can compare our sample mean $\bar{x} = 1965.08\ \text{ft}^2$ to the *collection of means from all possible random samples.* That little italicized phrase gives the idea of a *sampling distribution*. The idea is to collect together the sample means from *all* the possible random samples of size $n = 40$ drawn from the population of 1,890 houses in our population. Read the summary but do not stop there! The paragraphs below unpack the meaning.

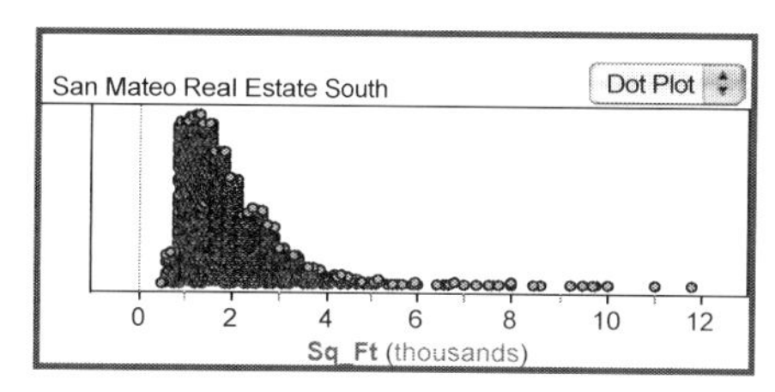

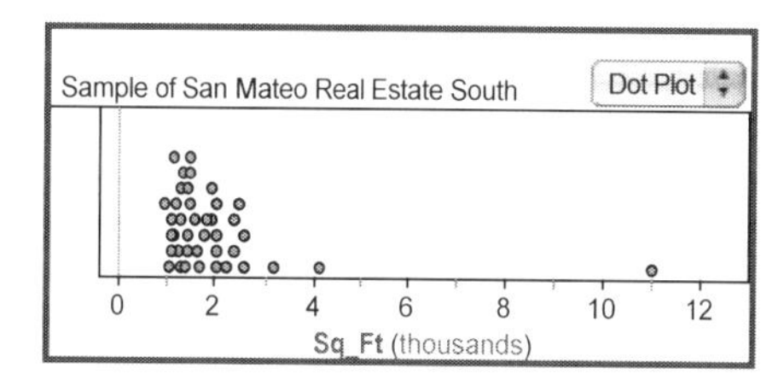

Idea of a Sampling Distribution of Sample Means

The ***sampling distribution of sample means*** is the distribution of the sample means for a specific variable of *all possible* random samples of a given size n drawn from a population.

Let us try to unpack the meaning of this. First, the number of possible different samples of size $n = 40$ is much greater than the size of the population. Our population has just 1,890 houses. The number of unique random samples of $n = 40$ (the count) that can be drawn from this population is approximately $9.254066615 \times 10^{82}$ if we each time we sample we sample *without replacement*. For our example, the population has 1,890 cases, a single sample has $n = 40$ cases, but the sampling distribution has a count of something close to that huge number, $9.254066615 \times 10^{82}$. Sampling distributions are huge.

Secondly, notice that the sampling distribution is a distribution of sample means, or in symbols, $\bar{x}$ s. The cases for the population were houses, the cases for our single sample of $n = 40$ were also houses, but the cases for the sampling distribution are sample means of the variable *Sq_Ft*. With a *sampling distribution*, we are interested in looking at all the possible samples (all $9.254066615 \times 10^{82}$ of them). For each sample (each of the $9.254066615 \times 10^{82}$), we calculate some measure, such as a sample mean, or a sample proportion, or a sample median, and make the sampling distribution out of one of these measures calculated for the different samples. We can therefore see how we would go about getting a sampling distribution for our real estate example. Our process has a number of steps.

➪ Step 1: Randomly sample a sample of $n = 40$ from the population of 1,890 houses.

➪ Step 2: Calculate the sample mean $\bar{x}$ for the sample just collected; store the result in a safe place.

➪ Step 3: Repeat steps 1 and 2 at least $9.254066615 \times 10^{82}$ times to include every possible sample.

➪ Step 4: Collect together all the different sample means into a collection.

Sound impossible? Yes, it is impossible; actual sampling distributions are derived from mathematical theory, but the process gives you an idea of what goes into a sampling distribution. Moreover, we will *simulate* a sampling distribution, actually a part of a sampling distribution, using these steps.

Building a Sampling Distribution Using software In §2.4, you worked "interactively." In this section it will be best to read the notes while (at the same time) following the bulleted instructions.

- The file ***SanMateoRealEstateSouth*** contains data on *all* the houses that were sold in the southern part of San Mateo County, California, from June 2005 to June 2006. It contains the *population* of houses sold.
- Use software to a ***dot plot*** of the variable *Sq_Ft* and summary statistics including the mean and the standard deviation of the population. It may resemble what is on the left of the graphic below.
- Next, use software to get a sample of n = 40 from the collection ***SanMateoRealEstateSouth*** . This is Step 1 of our steps.
- For the sample, use software to get a dot plot and summary statistics for the sample that you have drawn. What you should have should resemble the (greatly reduced) picture below. However, depending on the software used, the format may differ and the sample will be different from the sample chosen here, with the mean and standard deviation will be different. Randomness at work!

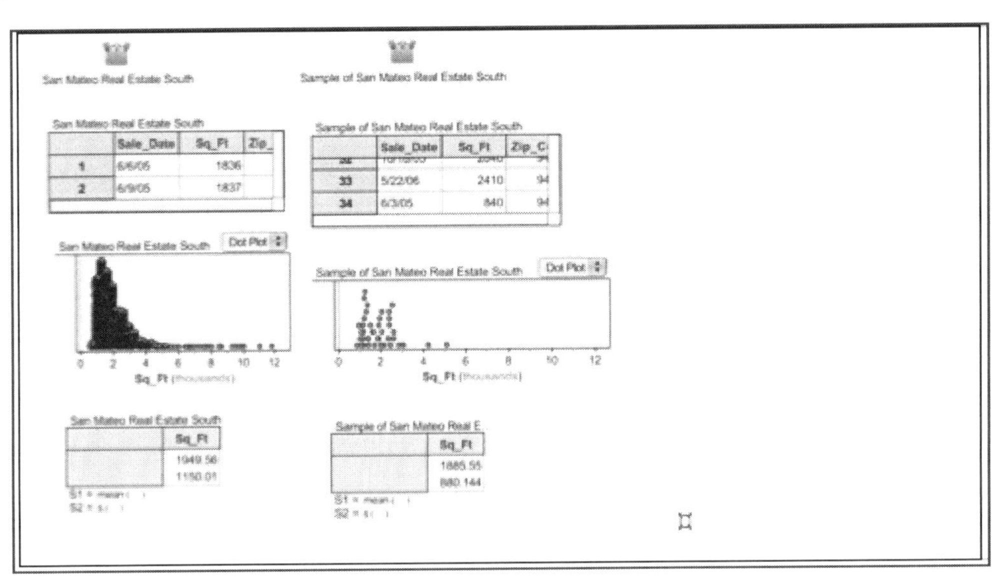

We have now completed (with the help of software) Step 2 of our four steps; we have calculated the sample mean $\bar{x}$ for our sample. Now it is time to take Step 3: getting more and more samples, each time calculating the mean. We will have software do the work.

- Use software to repeat the process of getting samples and calculating the mean for each of the samples drawn. Do this for thousands of times; the software should have a facility for doing this. Arrange the output so that there is a clear distinction between population, sample and the simulated sampling distribution.

Population ***Sample*** ***Sampling Distribution of Sample Means***

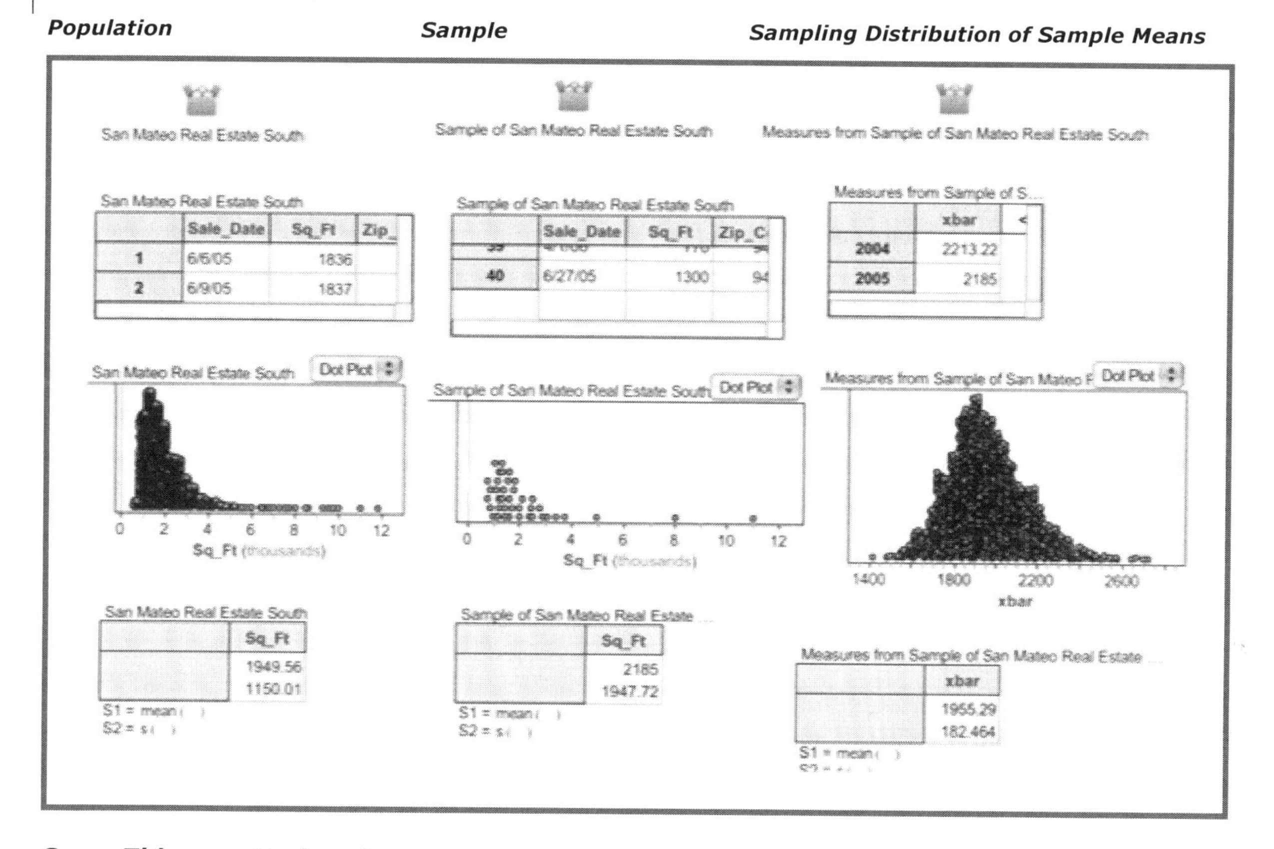

Some Things to Notice about What We Have Done

We have not derived the complete *sampling distribution of sample means*. To get the complete sampling distribution, we would have to continue until we had got *all* possible random samples, which we said would be at about $9.254066615 \times 10^{82}$ different samples; software would probably choke on this task, and we would also choke at the time it would take. We said that the facts about sampling distributions are derived mathematically. The simulation we have illustrates some facts about *sampling distributions of sample means*.

Compare the three distributions in your output (or the display above) and notice that:

- The population and the sample are both right skewed, but the sampling distribution looks symmetric. Indeed the sampling distribution looks like it may be a Normal distribution.
- The variable for the population and for the sample is the same, *Sq_Ft*, but for the sampling distribution the variable being measured is *xbar*. Look at the horizontal axes of the dot plots.
- The means for all three distributions are similar, but the standard deviation for the sampling distribution is much smaller than the standard deviation for the population and the sample.

There is another major, important difference between the population and sample on the one hand and the sampling distribution. Not only is the variable different, but the *cases* are not the same things.

For the population and the sample, the cases are houses, but for the sampling distribution, the cases are *not* houses. The cases for the sampling distribution are *means from samples of n = 40.* Each dot in the dot plot for the *population* distribution is a house, and each of the $n = 40$ dots in the *sample* distribution is one of the houses in that sample, but each dot in the dot plot of the *sampling* distribution is a sample mean of some sample.

Since the cases for the sampling distribution are different from the cases for the sample, the means have to be interpreted in the correct context; for the *sample* (not *sampling*) distribution shown above, we can say that that the average house size for that sample was $\bar{x} = 2185$ ft^2. However, the mean that you see for the *sampling distribution* (which is 1955.29 ft^2) is the ***mean of all the x-bars,*** or we could say the ***mean of all the sample means***. We assign a new symbol to the mean of all the x-bars; $\mu_{\bar{x}}$, which is read in short, "***mu of x-bar***" or "***mean of x-bar***." The standard deviation for a sampling distribution, since it is a standard deviation of all the sample means, is denoted $\sigma_{\bar{x}}$. Along with the three graphs, one for a population distribution, one for a sample distribution, and one for a sampling distribution, we have three sets of symbols for mean and standard deviation.

Notation for means and standard deviations

	Population Distribution	*Sample Distribution*	*Sampling Distribution*
Mean	μ	$\bar{x}$	$\mu_{\bar{x}}$
Standard Deviation	σ	s	$\sigma_{\bar{x}}$

Be prepared to memorize these symbols and their meanings; there is a certain logic to the usage that will help you. Non-Greek symbols are used for what we can easily calculate (things from samples), whereas Greek symbols are used either for theoretical quantities (such as $\mu_{\bar{x}}$ and $\sigma_{\bar{x}}$) or quantities that are not accessible to researchers, such as the population mean μ and population standard deviation σ.

How Random Samples Behave: Facts about Sampling Distributions

Look back at the output for our *part-of-a-sampling-distribution* that we had software simulate; it is only a part of a sampling distribution because we have only about two or three thousand samples, whereas the real sampling distribution would have $9.254066615 \times 10^{82}$ cases) Think, as always, about *shape, center,* and *spread.* The *shape* of our sampling distribution appears nearly Normal, despite the shape of the population distribution being very right-skewed. The center of the sampling distribution looks very close to the center of the population distribution; compare the means. However, the *standard deviation* looks much smaller than the standard deviation of the population distribution.

Facts about Sampling Distributions of Sample Means Calculated from Random Samples

Shape:	The shape of the sampling distribution of $\bar{x}$ will be approximately Normal if the sample size n is sufficiently large, even if the population distribution from which the samples were drawn is not Normal.
Center:	The mean of the sampling distribution of $\bar{x}$ is equal to the mean of the population distribution. That is, $\mu_{\bar{x}} = \mu$.
Spread:	The standard deviation of the sampling distribution of $\bar{x}$ is $\sigma_{\bar{x}} = \frac{\sigma}{\sqrt{n}}$, where σ is the population standard deviation.

Sampling Distributions are less variable than population distributions. It makes sense that the *spread* of sampling distribution is smaller than the spread of the population distribution. The standard deviation of the sampling distribution is the population standard deviation divided by the square root of the sample size $\sigma_{\bar{x}} = \frac{\sigma}{\sqrt{n}}$ and since $\sqrt{n}$ will be bigger than one, the standard deviation of the sampling distribution $\sigma_{\bar{x}} = \frac{\sigma}{\sqrt{n}}$ will be smaller than the standard deviation σ of the population. Hence the sampling distribution of sample means will be less variable than the distribution of the population since our sample size n will be bigger than one. Notice that the larger the sample size that we use, the less variation the sampling distribution has. This is important in practice.

The shape of sampling distributions may be Normal despite the population distribution being not Normal. The *shape* of the sampling distribution will be "approximately Normal" if the sample size n is sufficiently large, despite the *population distribution* being decidedly not at all Normal in shape. We can say more than this: we can say that the larger the sample we decide to draw, the closer the *sampling distribution* will be to a Normal distribution. Look at the shape of the distribution of our start-of-a-sampling-distribution (our simulation) with about two thousand different samples of $n = 40$. The dot plot looks Normal, but there is a hint of right skewness to it. If we increased the sample size to $n = 120$, the plot that we would get would conform more to a Normal distribution. The bigger the sample size, the more Normal is the *sampling distribution*. This characteristic of the shape of the sampling distribution is called the ***Central Limit Theorem.***

Central Limit Theorem

If a simple random sample of size n is drawn from any shape population with population mean μ and finite standard deviation σ, then if n is sufficiently large, the ***sampling distribution*** of the sample mean $\bar{x}$ is approximately a Normal distribution with mean $\mu_{\bar{x}} = \mu$ and standard deviation $\sigma_{\bar{x}} = \frac{\sigma}{\sqrt{n}}$. The Normal approximation is better the bigger the sample size n and the closer the population distribution is to Normal.

Sampling distributions are useful: Reasonably Like and Rare. We can use the facts about sampling distributions to calculate what x-bars we are reasonably likely to see when we randomly sample. We saw that with the standard Normal distribution, 95% of the distribution lay between 1.96 standard deviations less than the mean to 1.96 standard deviations more than the mean. (To check this again, remember that 95% in the middle of a Normal distribution leaves 5% for the tails, and hence 2.5% for each tail; look for .0250 in the *body* of the Standard Normal Chart and read off the z score.) We use this fact to define ***reasonably likely*** and ***rare*** sample means $\bar{x}$ from a random sample of size n.

Reasonably Likely and Rare

Sample means $\bar{x}$ that are *within* the interval $\mu - 1.96\frac{\sigma}{\sqrt{n}} < \bar{x} < \mu + 1.96\frac{\sigma}{\sqrt{n}}$ are called ***reasonably likely.***

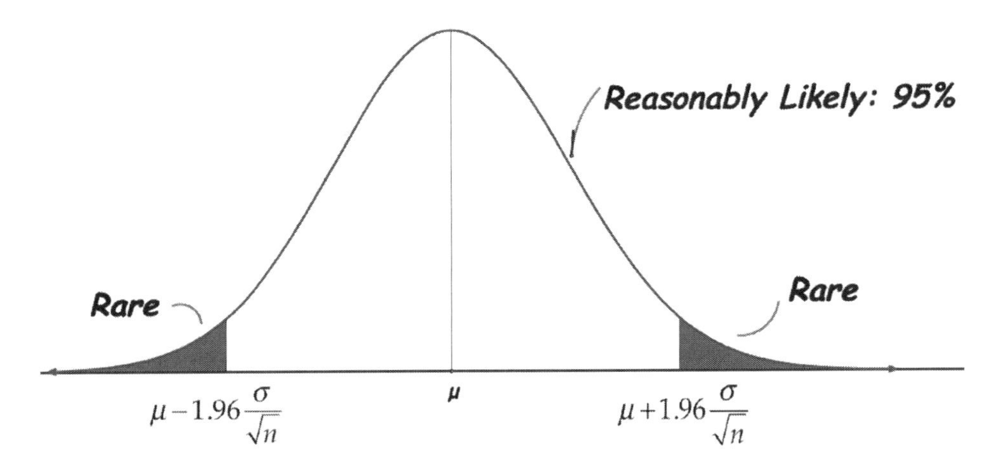

Sample means $\bar{x}$ that are *outside* the interval $\mu - 1.96\frac{\sigma}{\sqrt{n}} < \bar{x} < \mu + 1.96\frac{\sigma}{\sqrt{n}}$ are called **Rare.**

In probability notation, $P\left(\mu - 1.96\frac{\sigma}{\sqrt{n}} < \bar{x} < \mu + 1.96\frac{\sigma}{\sqrt{n}}\right) = 0.95$.

Examples with the San Mateo Real Estate Data

What sample means are reasonably likely? We used the real estate data for the South Region of San Mateo County as our population. For the variable *Sq_Ft* we saw that the population mean μ = 1949.56 ft^2 and that the population standard deviation $\sigma = 1150.81$ ft^2. Using the facts in the box on the previous pages, we can specify the center, spread, and shape of the *Sampling Distribution* for sample means $\bar{x}$ for random samples of $n = 40$:

Population Parameters San Mateo South Houses

San Mateo Real Estate South

	Sq_Ft
	1949.56
	1150.01

S1 = mean ( )
S2 = s ( )

Center: The mean will be $\mu_{\bar{x}} = \mu = 1949.56$.

Spread: The standard deviation will be $\sigma_{\bar{x}} = \frac{\sigma}{\sqrt{n}} = \frac{1150.01}{\sqrt{40}} \approx 181.83$.

Shape: The sampling distribution will be approximately Normal.

Then we can calculate the ***reasonably likely*** sample means for sampling from this population with a sample size of $n = 40$ by calculating:

$\mu - 1.96\frac{\sigma}{\sqrt{n}} = 1949.56 - 1.96\left(\frac{1150.01}{\sqrt{40}}\right) \approx 1949.56 - 1.96(181.83) = 1593.17$ ft^2, and

$\mu + 1.96\frac{\sigma}{\sqrt{n}} = 1949.56 + 1.96\left(\frac{1150.01}{\sqrt{40}}\right) \approx 1949.56 + 1.96(181.83) = 2305.95$. What this calculation means is that if we sample randomly from the population of San Mateo houses, we should expect to get sample means $\bar{x}$ between about 1593 ft^2 and 2306 ft^2 with probability 0.95.

Another way of putting this is that if we sampled repeatedly—sampled over and over again—we expect 95% of the sample means to be between about 1593 ft^2 and 2306 ft^2. We expect to get sample means that are *outside* the interval $1593 < \overline{x} < 2306$ with probability 5%. Using probability notation, we can write $P(1593 < \overline{x} < 2306) \approx 0.95$. Having made this calculation, we can answer questions such as this: *Is getting a random sample of $n = 40$ with $\overline{x} = 1800$ ft^2 reasonably likely or rare?* Since $\overline{x} = 1800$ is *inside* the interval $1593 < \overline{x} < 2306$, getting such a sample is ***reasonably likely.*** Getting a sample with $\overline{x} = 2400$ would be ***rare*** since this sample mean is outside the interval $1593 < \overline{x} < 2306$.

What is the probability that we will get a sample mean greater than 2185 ft2? This was the last sample mean $\overline{x}$ that we collected in our simulation of two thousand sample means. (If you repeat the simulation, you will get some other sample mean as your "last one"; this is just a convenient one to focus on—there is nothing special about it.) Since we have a normal distribution for our sampling distribution, this question looks like a *"given a value, find a probability"* Normal distribution type of problem. We want to know the probability $P(\overline{x} > 2185)$. The first step is to make a sketch of the situation, which is shown in the graphic here. Notice that our $\overline{x} = 2185$ is in the *reasonably likely* region and *not* in the *rare* region, and so we should expect to get a z score that is smaller than 1.96. The z score is $z = \dfrac{2185 - 1949.56}{181.83} \approx 1.29$. Then $P(\overline{x} > 2185) = P(z > 1.29) = 1 - P(z < 1.29) = 1 - 0.9015 = 0.0985$.

Hence we conclude that the probability that with random sampling, we would get a random sample with sample mean bigger than 2,185 square feet, or $\overline{x} > 2185$ is just under 10%.

What is the xbar for the lowest 5% of xbars that we could get with random sampling? Because the sampling distribution of the *xbars* is very approximately a Normal distribution, we can find such a value. This question looks like a *given a probability, find a value* type of Normal distribution problem. We want a value—in this situation a value for an *xbar*—that divides the lowest 5% of *xbars* from the remaining 95%.

Looking at the chart for the Normal distribution, we find that for the lowest 5%, we get a z score of $z = -1.645$, and we can use the formula for the z score to find the *xbar* that divides the lowest 5% from the rest. We calculate from: $z = \dfrac{\overline{x} - \mu_{\overline{x}}}{\sigma_{\overline{x}}}$. So, $-1.645 = \dfrac{\overline{x} - 1949.56}{181.83}$ and from this

$(-1.645)(181.83) = \overline{x} - 1949.56$, and so solving for $\overline{x}$, we get $\overline{x} = 1949.56 - 181.83(1.645) \approx 1650.89$ ft^2. The lowest 5% of all *xbars* we expect to get will have $\overline{x} \le 1650.89$.

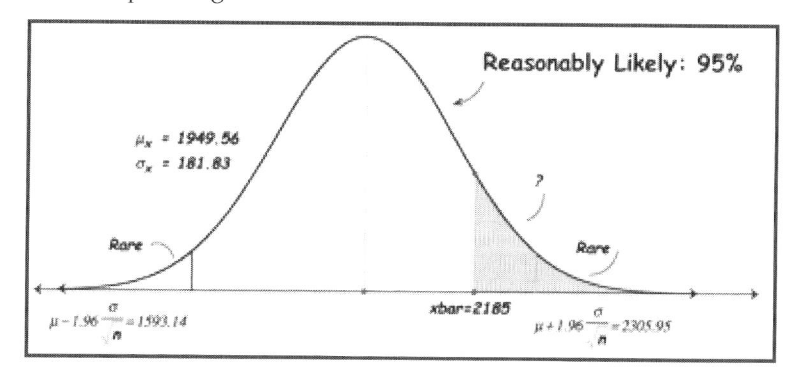

Summary: Sampling Distributions for Quantitative Variables

The ideas in this section about sampling distributions are at the heart of how statisticians think about *inferring* from a sample to a population or from a specific random allocation in an experiment to all possible random allocations. These ideas are abstract; they are not "easy." Expect to be confused for a time—but hopefully not forever! Expect to have to work through the ideas. Avoid taking shortcuts; avoid learning just enough to answer the questions, parrot-like. Ask questions; draw pictures; get the idea.

- The ***sampling distribution of sample means*** is the distribution of the sample means for a specific variable of *all possible* random samples of a given size n drawn from a population.
  - A *sampling distribution* is *not* the same as a *sample distribution*, which is the distribution of a variable for a single sample drawn from a *population*.
  - A *sampling distribution* is *not* the same as a *population distribution*, which is the distribution of a variable for an entire population and which is usually unknown to researchers.
- A good way to understand *sampling distributions* is to think of how they could be constructed:
  - ⇨ Step 1: Randomly sample a sample of a specific size n.
  - ⇨ Step 2: Calculate the sample mean $\bar{x}$ for the sample just collected; store the result.
  - ⇨ Step 3: Repeat steps 1 and 2 a huge number of times to include every possible different sample.
  - ⇨ Step 4: Collect together all the different sample means into a collection.

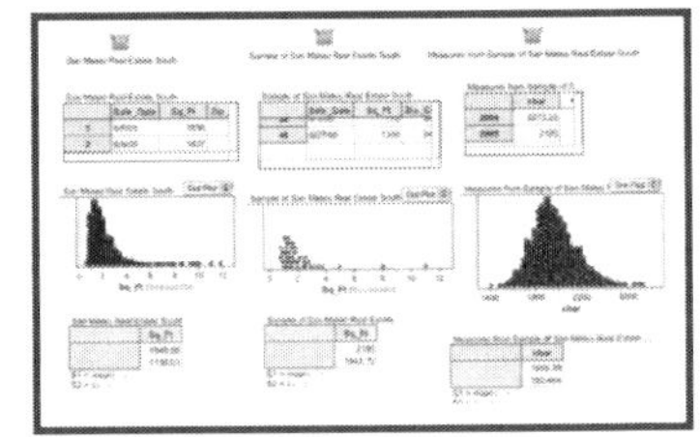

- Another good way to keep straight the differences between *population distributions*, *sample distributions*, and *sampling distributions* is to recall the picture of the simulation done in this section.
- ***Notation: means and standard deviations:***

	Population Distribution	Sample Distribution	Sampling Distribution
Mean	μ	$\bar{x}$	$\mu_{\bar{x}}$
Standard Deviation	σ	s	$\sigma_{\bar{x}}$

- ***Facts about Sampling Distributions of Sample Means Calculated from Random Samples***

  Shape: The shape of the sampling distribution of $\bar{x}$ will be approximately Normal if the sample size n is sufficiently large, even if the population distribution from which the samples were drawn is not Normal.

  Center: The mean of the sampling distribution of $\bar{x}$ is equal to the mean of the population distribution. That is, $\mu_{\bar{x}} = \mu$.

  Spread: The standard deviation of the sampling distribution of $\bar{x}$ is $\sigma_{\bar{x}} = \frac{\sigma}{\sqrt{n}}$, where σ is the population standard deviation.

- ***Central Limit Theorem:*** This theorem says that if a simple random sample of size n is drawn from any shape population with population mean μ and finite standard deviation σ, then if n is sufficiently large, the ***sampling distribution*** of the sample mean $\bar{\boldsymbol{x}}$ is approximately a Normal distribution with mean $\mu_{\bar{x}} = \mu$ and standard deviation $\sigma_{\bar{x}} = \frac{\sigma}{\sqrt{n}}$. The Normal approximation is better the bigger the sample size n and the closer the population distribution is to Normal.

- ***Reasonably Likely and Rare Sample Means***
  - ***Reasonably likely*** sample means $\bar{x}$ are *within* the interval $\mu - 1.96\frac{\sigma}{\sqrt{n}} < \bar{x} < \mu + 1.96\frac{\sigma}{\sqrt{n}}$.
  - ***Rare*** sample means $\bar{x}$ are *outside* the interval $\mu - 1.96\frac{\sigma}{\sqrt{n}} < \bar{x} < \mu + 1.96\frac{\sigma}{\sqrt{n}}$.

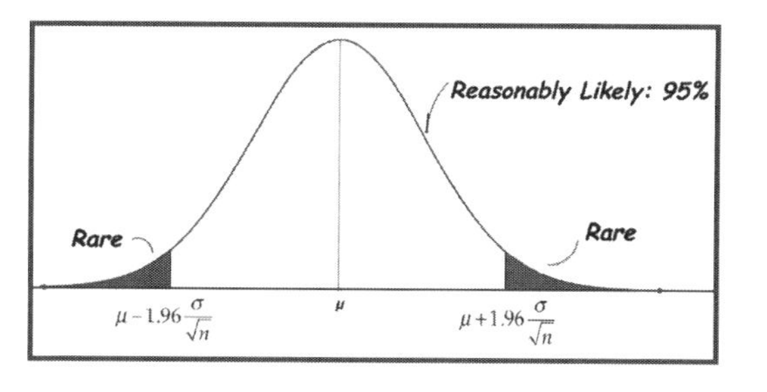

- With a sampling distribution for a specific variable and population, we can answer questions about:
  - The range of *reasonably likely* and *rare* sample means $\bar{x}$ s
  - Determine the likelihood that we see a sample mean $\bar{x}$ that is greater (or lesser) than a specific value
  - Calculate the sample mean that we would expect to see a given percentage of the time if we repeatedly sampled

§3.3 Trusting Data, Part 3: Binomial Distributions

The big ideas of this section

First off, recall from the last section the distinction between population and sample. A population consists of all the things we want to know about—perhaps people, but it does not have to be. A sample consists of the things about which we actually have data, hopefully a representative subset of the population.

- There are plenty of ways to get bad samples, and that even if we do all kinds of analyses on these bad samples, we cannot trust the results. Then, unlikely as it may seem, we saw that the best kinds of samples are those that are randomly chosen. In this section we will begin to see why that is so.
- The question remains, however: can we trust an analysis from a small sample when the population is so large, even if we have a good sample?
- Here is our strategy to answer our question about how to trust random samples—the same strategy as we employed in 3.2. First, we will pretend that we know all about the population. Then we will do two things: we will use the power of computing to get a picture of all the samples that we could possibly get—both truthful and misleading ones. And we will back that up with the power of mathematics to show that our computing picture is correct. In "real statistical life" we never know all about the population, but our pretending here will show us how to cope with our ignorance.

Our statistical question: Obesity amongst young adults

We need an example. Our question is: *What proportion of Americans between the ages of sixteen and thirty-five are obese?* First off, we need a definition of obese. A common definition depends upon the Body Mass Index, which is the mass of a person in kilograms divided by the square of the person's height, measured in meters, so in symbols: $BMI = \frac{mass}{(height)^2}$. Then the definition given by the Centers for Disease Control and Prevention is that adults with a BMI ≥ 30 are considered obese (see: http://www.cdc.gov/obesity/adult/defining.html). So, if we have a sample of adults and have the information to calculate their BMIs, we can classify each as either obese or not obese and then calculate the proportion classified as obese. But can we have to trust our results from a sample?

(What follows can be done without doing Exercise 1 or Exercise 6, but having done one of those exercises will help this section.)

To answer our statistical question, we randomly draw just one sample of adults from the population of all American adults and count the number that is obese in the sample; our idea is that this single sample will tell us about the population. But we know that a *single* sample may mislead us, especially if we choose a very small sample size, such as $n = 5$. Can we do something that will tell us the likelihood that our single sample will show us 20% obese (that is one obese out of five), or 40% (two obese of five), 60%, or perhaps 0%? What we can do, if we have a collection that we can "pretend" is the population, is to sample over and over again—with the same sample size, $n = 5$, every time we sample—and then we see the number of times our sample of $n = 5$ gives us 0 obese, 1 obese, etc. That is what is done in Exercises 1 and 6 where one thousand repeated small samples are used.

Here we show the results of sampling $n = 5$ repeatedly ten thousand times from a pretend population. The "population" that we are drawing from has a proportion obese of about 20% (it is actually just under that figure), and the Summary Table shows the results of this repeated sampling.

Measures from Sample of ...		
xObese	0	3540
	1	4132
	2	1855
	3	421
	4	51
	5	1
Column Summary		10000

S1 = count ( )

The Summary Table shows that 4,132 of the ten thousand repeated samples of $n = 5$ (or 41.32%) gave us just one obese out of the five chosen people, and that 3,540 of the ten thousand repeated samples of $n = 5$ gave us no obese people. You may object that this shows that we need a bigger sample size to more accurately identify the proportion obese, but we can learn something about samples by using this small sample size. Later, to answer our statistical question, we will increase the sample size.

A mathematical model that fits what we see and why: the Binomial Model

Our statistical question and our strategy of repeatedly sampling with the same sample size fit what is called a Binomial Distribution Model. The Binomial Distribution Model is a mathematical model based on four conditions. Here in the box are the four essential conditions for the model.

Conditions for the Binomial Distribution Model

We can fit a Binomial Distribution Model if all of the following conditions are met:

B The outcomes are ***binomial***; that is, there are just two outcomes. Often, the outcomes are called "success" and "failure."

I Each case is ***independent*** of the others. (The cases are often also called "trials.")

N There is a fixed total ***number, n,*** of cases (or "trials").

E We use a probability p of success that is the same ("***equal***") for each case, or "trial."

If the conditions are met, the Binomial Distribution Model will give us the probability that we see 0, 1,...n successes out of n cases or trials. See the next box for the formula that calculates the probabilities.

We can use a Binomial Model because in our situation all of the conditions are met; here is how:

B: We have just two outcomes: obese and not obese. We focused on "obese," which means that "obese" was our "success." (A "success" is just a choice of one of the two outcomes; it need not be something "desirable.") The CDC has four more categories besides "obese." Using these five categories would not meet the condition since we need just two ("bi"). Also, if our outcomes were waist circumference or BMI, our condition would not be met because these variables are quantitative variables. The ***B*** condition demands a categorical variable with just two outcomes.

I: Each case is independent of the others because our samples were randomly drawn. When you roll a die, the next roll is completely independent of the previous roll. The die does not say to itself: "Oh, my, I haven't come up '2' for such a long time; the next one better be '2.'" Also, when software generates random numbers, independence is guaranteed. A random process is without memory and gives independent "trials."

N: Since our sample size was $n = 5$ each time, there is a fixed number of cases or "trials."

E: In our calculations we will fix the p at a specific number. In the example shown just below, we use $p = 0.20$ or 20%. The number that we use depends on out idea for the probability of a "success."

Using the model: P(X = k) and P(X > 1) If the conditions are met then the Binomial Model gives the probabilities of 0, 1, 2,..., up to n successes, using a mathematical formula based on the conditions. Here is an example connected with our statistical question. Here, we want the probabilities of the number X of successes (i.e., obese people) out of $n = 5$. Here is what the Binomial Model gives us. We have seen probabilities like this before. $k = 1$, which can be read "the probability that in a sample of $n = 5$, we see exactly one obese person," is 0.4096. In symbols, this is $P(X = 1) = 0.4096$.

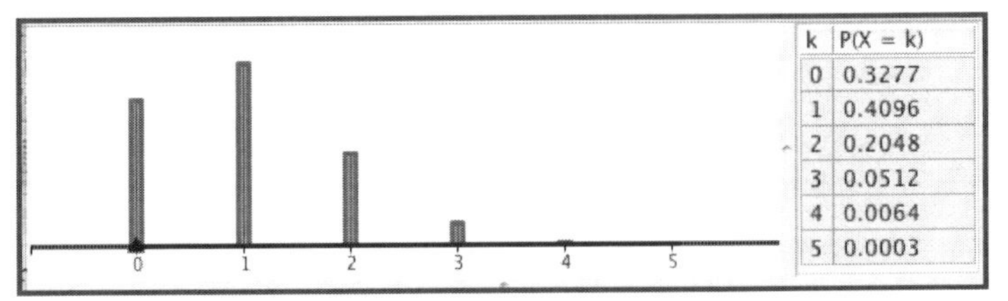

k	P(X = k)
0	0.3277
1	0.4096
2	0.2048
3	0.0512
4	0.0064
5	0.0003

The height of the "spike" for $X = 1$ is the proportion 0.4096. The hand calculation of this number is shown below; here we are content to use the already calculated values.

P(X > 1): By adding probabilities, we can also calculate the probability that we will see *more than one* obese person in a sample of five. That probability will be calculated as:

$$P(X > 1) = P(X = 2) + P(X = 3) + P(X = 4) + P(X = 5)$$
$$= 0.2048 + 0.0512 + 0.0064 + 0.0003$$
$$= 0.2627$$

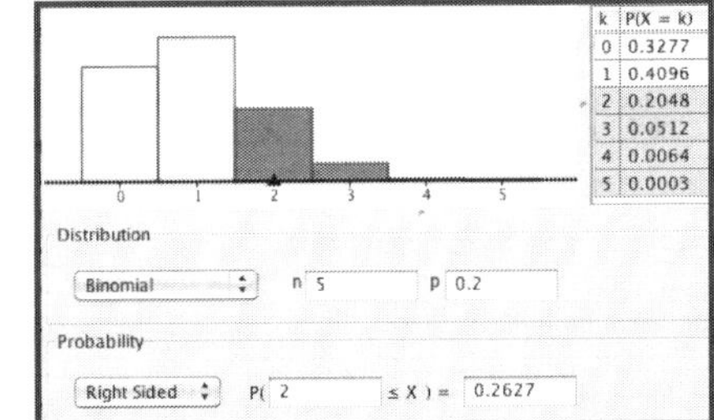

k	P(X = k)
0	0.3277
1	0.4096
2	0.2048
3	0.0512
4	0.0064
5	0.0003

We can also see this graphically with by turning the spikes into a histogram, where the shaded in bars (each having $binwidth = 1$) show $P(X > 1) = 0.2627$. (This graphic comes from ***DistributionCalculator.ggb***.)

How the P(X = k) calculations are done. The formula (which actually has two parts) to calculate these probabilities is given in the box below.

Formula for the Binomial Model and the names of the parts

To calculate the probability of X = k successes out of n cases (or "trials"), use:

$$P(X = k) = \binom{n}{k} p^k (1 - p)^{n-k} \qquad \text{where } \binom{n}{k} = \frac{n!}{k!(n-k)!}$$

and where $n! = n \cdot (n-1) \cdot (n-2) \cdots 2 \cdot 1$ and also $0! = 1$ by definition.

- The symbol $n!$ is named "n factorial".
- $\binom{n}{k}$ tells us the number of ways of placing k things into n boxes (also written ${}_nC_k$) and is named either "n choose k" or "the number of combinations of k things amongst n."

To see how the Binomial Model formula works, see the examples just below this box.

Example 1: P(X = 0) and the meanings of $\binom{n}{k}$ and n! We start with the easiest one; that is, we start with calculating the probability of getting no obese people in the sample of $n = 5$. Here, the value of $k = 0$, so plugging in the numbers, we have:

$$P(X = 0) = \binom{5}{0} p^0 (1 - p)^{5-0} = \left(\frac{5!}{0!5!}\right)(0.20)^0 (0.80)^5 \approx \frac{5 \cdot 4 \cdot 3 \cdot 2 \cdot 1}{1 \cdot 5 \cdot 4 \cdot 3 \cdot 2 \cdot 1}(1)(0.32768) = 0.32768$$

The most mysterious part of this for someone who has never seen it before is probably $\binom{n}{k}$. The formula says that it is $\binom{n}{k}=\frac{n!}{k!(n-k)!}$ and so here $\binom{5}{0}=\frac{5!}{0!(5-0)!}=\frac{5\cdot4\cdot3\cdot2\cdot1}{1\cdot5\cdot4\cdot3\cdot2\cdot1}=1$. First of all, the symbol $n!$ is pronounced "n factorial" (not n spoken with a loud voice!), and means the number n multiplied by one less multiplied by one less than the next one, and so on until you get to 1. There is a special case for 0! Which is defined to be one, so 0! = 1.

n! is pronounced "n factorial" and means $n!=n\cdot(n-1)\cdot(n-2)\cdots2\cdot1$ and 0! =1

The calculation $\binom{n}{k}=\frac{n!}{k!(n-k)!}$ counts the number of ways that we can have k things if we have n to choose from, so, in our example $\binom{n}{k}=\binom{5}{0}=1$ counts the number of ways we can have zero obese people in five choices. The answer is that there is just one way to have zero obese people in five people: all of the boxes (people) must be "not obese." In Exercise 1, the worksheet will look like the one shown here.

1	2	3	4	5		Number Obese
Not	Not	Not	Not	Not		0

Example 2: P(X = 1) and the meaning of $\binom{n}{k}$ ***continued*** Now, what about the calculation for the probability of getting exactly one obese person in the sample of n = 5? Here, $k=1$, so plugging in the numbers, we have:

$$P(X=1)=\binom{5}{1}p^1(1-p)^{5-1}=\frac{5!}{1!(5-1)!}(0.20)^1(0.80)^{5-1}\approx\frac{5\cdot4\cdot3\cdot2\cdot1}{1\cdot(4\cdot3\cdot2\cdot1)}(0.20)(0.4096)\approx5\cdot(0.08192)=0.4096$$

The part of the formula that shows $(0.20)^1(0.80)^{5-1}=(0.20)^1(0.80)^4=(0.20)(0.4096)=0.08192$ calculates the probability that the first of the five people we sample is obese and the rest are not. But there are five different places the one obese person could come, and that is why we multiply by $\binom{n}{k}=\binom{5}{1}=\frac{5\cdot4!}{1!4!}=5$. The one obese person could be the first of the five, the second, the third, the fourth, or the fifth of the five. So the answer of "five different places" makes sense.

Example 3: P(X = 2) and the meaning of $\binom{n}{k}$ ***continued.*** The probability of getting exactly two obese people out of five sampled will have k =2, so we have:

$$P(X=2)=\binom{5}{2}p^2(1-p)^{5-2}=\frac{5!}{2!(5-2)!}(0.20)^2(0.80)^{5-2}\approx\frac{5\cdot4\cdot3\cdot2\cdot1}{2\cdot1\cdot(3\cdot2\cdot1)}(0.04)(0.512)\approx0.2048$$

We calculate: $\binom{n}{k}=\binom{5}{2}=\frac{5\cdot4\cdot3!}{2\cdot1\cdot3!}=\frac{5\cdot4}{2\cdot1}=10$. This calculation means that there are ten different ways that we could place the two obese people in the five boxes. The best way to see that there are ten different ways is to tediously list them, and this is shown on the next page.

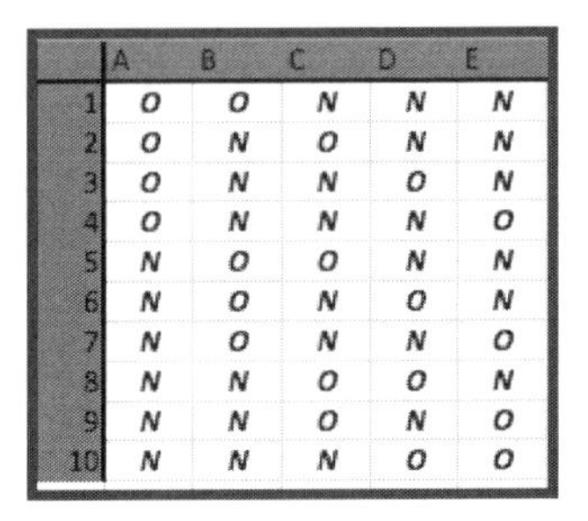

	A	B	C	D	E
1	O	O	N	N	N
2	O	N	O	N	N
3	O	N	N	O	N
4	O	N	N	N	O
5	N	O	O	N	N
6	N	O	N	O	N
7	N	O	N	N	O
8	N	N	O	O	N
9	N	N	O	N	O
10	N	N	N	O	O

With a much bigger sample size n and a much bigger number of successes k, this kind of listing obviously becomes very tiresome. Fortunately, all we need to know is how many different ways there are, and the formula $\binom{n}{k}$ gives that.

The Binomial Model with bigger sample sizes.

Here is what we have seen so far. If we specify a population proportion p (we used $p = 0.20$) and specify an n, which we are taking to be a sample size, then the Binomial Model will give us the probabilities of seeing all the possible outcomes of success, from 0, 1, 2,..., up to n if the conditions for using the model are met. By doing this, we should be convinced that $n = 5$ is too small a sample, since a wrong answer (number obese not 1) is more probable that a correct answer, if the correct answer is 20%. What happens is we increase the sample size?

Binomial Model with n = 20, and p = 0.20. We are using the same value for p, but we have increased the n. Compare this graph with the graph for the n = 5 Binomial Model. You will probably notice that the $n = 5$ graph appears to be much more right skewed than this graph, which seems almost symmetrical. In fact, as we increase the sample size n, the shape of the binomial becomes more like a Normal distribution. Since the Binomial Model is a distribution, we should be able to calculate the mean and standard deviation. The next box shows the mean and the standard deviation of a Binomial Model with specific n and p.

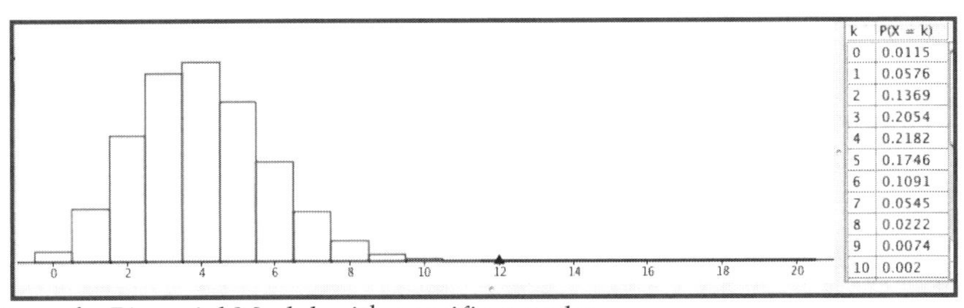

k	P(X = k)
0	0.0115
1	0.0576
2	0.1369
3	0.2054
4	0.2182
5	0.1746
6	0.1091
7	0.0545
8	0.0222
9	0.0074
10	0.002

The calculations become slightly more complicated with a bigger sample size. Look at the bar in the graph for $X = 4$ and also in the table. The table gives $P(X = 4) = 0.2182$, and this is the height of the bar in the graph. The meaning of this number is: "the probability that in a sample of $n = 20$, we get exactly four obese people is 0.2182". The calculation of this number is:

$$\begin{aligned} P(X=4) &= \binom{20}{4} p^4 (1-p)^{20-4} \\ &= \frac{20!}{4!(20-4)!}(0.20)^4 (0.80)^{20-4} \\ &\approx \frac{20\cdot 19\cdot 18\cdot 17\cdot 16!}{4\cdot 3\cdot 2\cdot 1\cdot 16!}(0.0016)(0.028147) \\ &\approx 5\cdot 19\cdot 3\cdot 17\cdot (0.000045036) \\ &\approx (4845)\cdot(0.000045036) \\ &\approx 0.2182 \end{aligned}$$

Calculating the $\binom{20}{4}$ takes a bit of arithmetic; the fraction $\frac{20\cdot 19\cdot 18\cdot 17\cdot 16!}{4\cdot 3\cdot 2\cdot 1\cdot 16!}$ reduces to $5\cdot 19\cdot 3\cdot 17 = 4845$ by seeing that the 16! in the numerator cancels with the 16! in the denominator, and 20 divided by 4 is 5, etc. For large values of n, one must resort to a calculator or an online calculator.

Mean, Standard Deviation, and Shape of the Binomial Model

A Binomial Model with n and p is a distribution that has mean: np

standard deviation: $\sqrt{np(1-p)}$

and a shape that becomes more nearly Normal as the sample size n increases.

If we apply these formulas to our example, we see that for the distribution with $n = 20$ and $p = 0.20$, we get the mean to be $np = 20 \cdot (0.20) = 4.0$. The mean does *not* necessary have to be a whole number, even though the number of successes can only be whole numbers. The standard deviation is:

$\sqrt{np(1-p)} = \sqrt{20 \cdot (0.20)(1-0.20)} = \sqrt{20 \cdot (0.20)(0.80)} = \sqrt{3.2} \approx 1.789$

See the graph above of this distribution and notice that the shape is more "Normal" and less skewed than the distribution for $n = 5$.

Binomial Model with n = 200, and p = 0.20. Let us see what happens when we increase the sample size to $n = 200$ but using the same value for p.

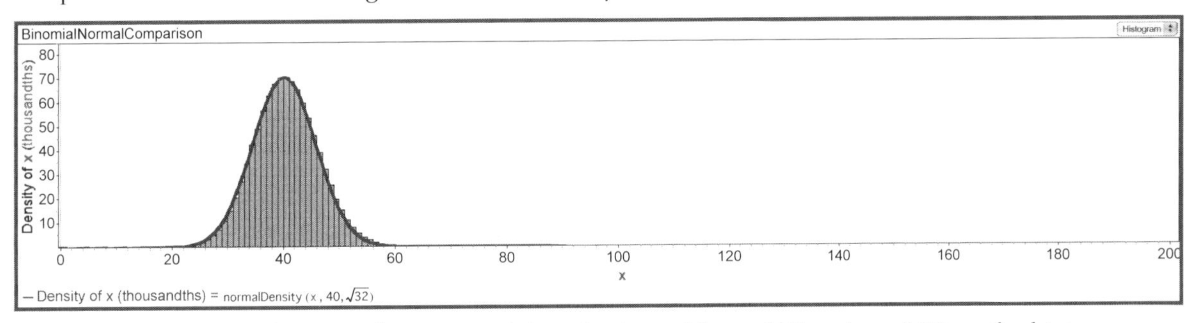

Here is a graphic showing the binomial distribution with $n = 200$ and $p = 0.20$, as the histogram, and the Normal distribution with mean $np = 200 \cdot (0.20) = 40$ and standard deviation

$\sqrt{np(1-p)} = \sqrt{200 \cdot 0.20(1-0.020)} = \sqrt{32} \approx 5.657$

Notice the following about the graphic: the binomial and the Normal are very close but not exactly the same. We should be able to use either the binomial or the Normal for calculations. The entire binomial distribution, although it has a Normal shape, is concentrated on the "left side" of the scale of possible outcomes of successes 0, 1, 2, 3,..., 200.

The meaning of the bars (again) and what we can do. The height of each bar represents $P(X = k)$, which is the probability of getting k successes out of the $n = 200$.

- Using our example, $P(X = 30) = 0.0147$ means that the probability is 0.0147 that we will have *exactly* thirty obese people in a random sample of $n = 200$.
- By adding the probabilities (or getting software to do it), we can calculate $P(X \leq 30) = 0.043$, or the probability that we get thirty or fewer obese in a random sample of $n = 200$. The shaded-in area in the graphic is the number 0.043, or 4.3%.

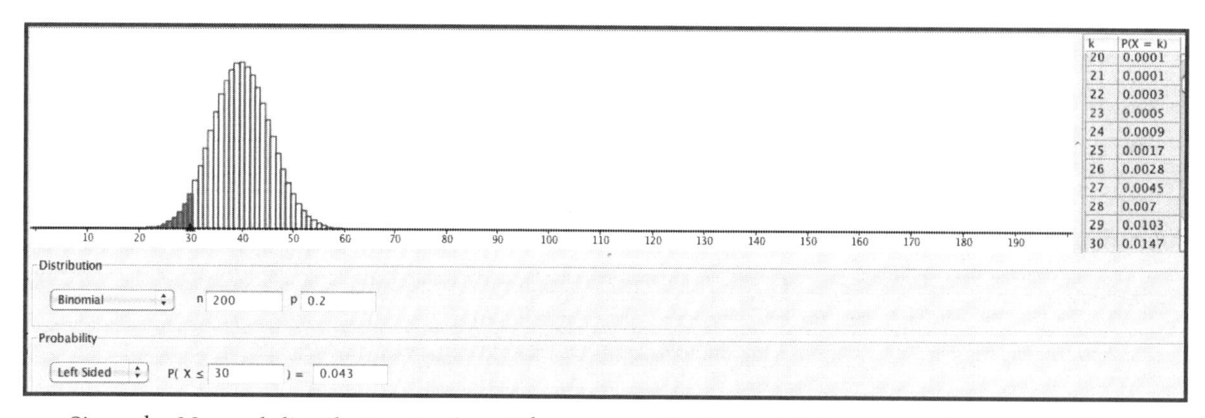

Since the Normal distribution is almost the same as the binomial, it appears that we could get a similar estimate using the Normal distribution, with mean 40 and standard deviation $\sqrt{32} \approx 5.657$. This is a "given a value, find a probability" problem, so we get the z score and consult the Normal Chart:

$$P(X \le 30) \approx P\left(z \le \frac{30-40}{5.657}\right) \approx P(z = -1.77) \approx 0.039$$

This is slightly different from the probability calculated from the binomial but close. The reason that it is different is that the binomial distribution only considers whole numbers (such as 29, 30, 32, etc.), but the Normal distribution is "smooth" and considers all the values between, say, 30 and 31 (such as 30.587675..., for example.) There are ways to correct for this difference (which are used in software), but the important thing is that the two results are close.

How accurate can we get? It depends... Introducing proportions and p-hat.

If the proportion obese is $p = 0.20$, and using the Binomial Model, we can see what happens with various sample sizes. If we add all the heights of the bars from $X = 29$ to $X = 51$, we get the probability that our sample has between twenty-nine and fifty-one obese people.

$$P(29 \le X \le 51) = P(X = 29) + P(X = 30) + \cdots + P(X = 50) + P(X = 51) = 0.9585$$

What this means is that we have more than a 95% chance of getting between 29 and 51 obese people if we have a sample of $n = 200$. The graph of this calculation is shown just below.

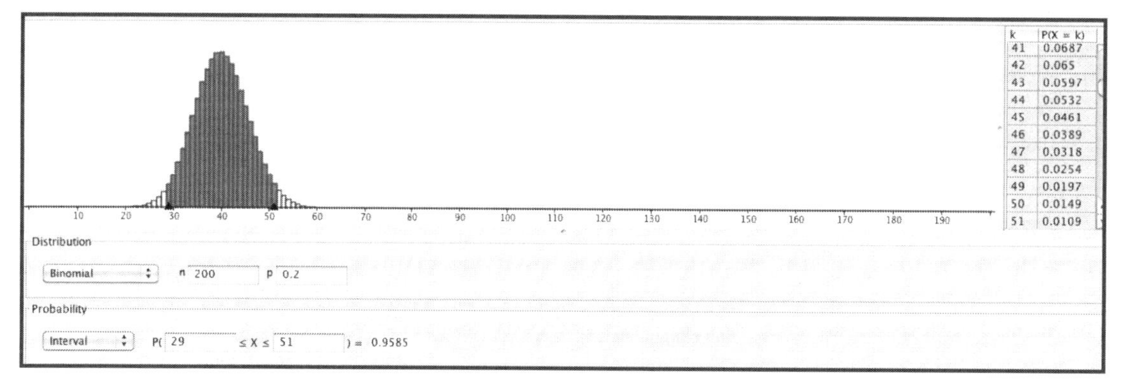

And so? Proportions usually tell more than "counts," so let us calculate proportions and even give them a new name. We will call the proportion $\hat{p}$, pronounced, "p-hat". We use the "hat" to distinguish a proportion calculated from a sample from the proportion for a population.

If we have twenty-nine obese people out of two hundred people then $\hat{p} = \frac{29}{200} \approx 0.145$, and if we have fifty-one then $\hat{p} = \frac{51}{200} \approx 0.255$. (Notice that $\frac{1}{200} = 0.005$, and hence the third decimal.)

Definition of p-hat and p

- For a proportion that refers to—or is calculated from—a sample, we use the symbol $\hat{p}$.
- For a proportion that refers to a population, we use the symbol p.

Generally, it is not possible to actually calculate a population proportion p, but it deserves to have a symbol.

Our calculation $P(29 \le X \le 51) = P(X=29) + P(X=30) + \cdots + P(X=50) + P(X=51) = 0.9585$ can be written as: $P(0.145 \le \hat{p} \le 0.255) = P(\hat{p}=0.145) + P(\hat{p}=0.150) + \cdots + P(\hat{p}=0.250) + P(\hat{p}=0.255) = 0.9585$

We have a 95% chance of getting a sample proportion $\hat{p}$ between 0.145 and 0.255, if the population proportion is 0.20. Can we do better? Yes, by increasing the sample size. Here is the picture for n = 1200. Notice that the Binomial Distribution looks narrower and that $P(213 \le X \le 267) = 0.9529$. This translates into greater accuracy, so we can say $P(0.1775 \le \hat{p} \le 0.2225) = 0.9529$. That is we are 95% certain of getting a sample that gives us a proportion within 2.25% (0.2225 – 0.2000 and 0.2000 – 0.1775) from the "target" $p = 0.2000$.

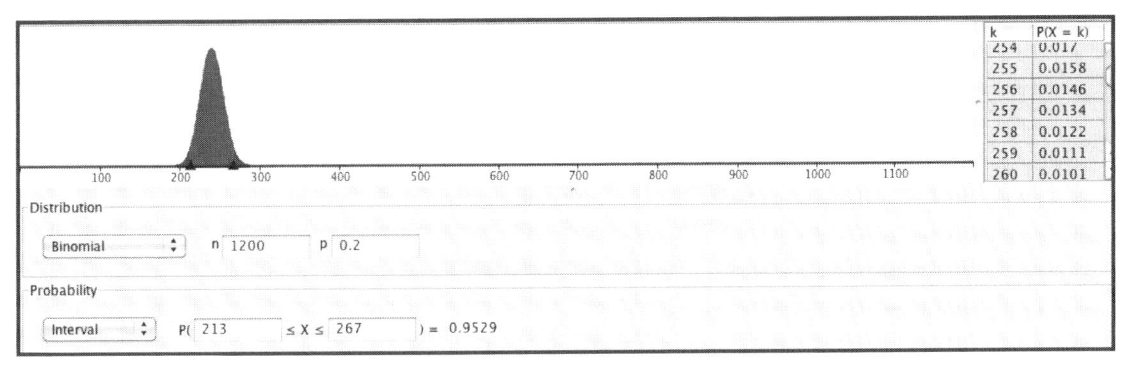

k	P(X = k)
254	0.017
255	0.0158
256	0.0146
257	0.0134
258	0.0122
259	0.0111
260	0.0101

The lesson:

- The Binomial Model can tell us what to expect from our sample. For example, if the sample size is as large as n = 1200, it will say that the proportion of "successes" we will see in a sample $\hat{p}$ will be within 0.0225 of the population p.
- The bigger the sample size, the more accurately the sample will represent the population and therefore the more that we can trust our results.

However, notice that in all of this discussion, we never mentioned the size of the population! What we see is that we need a big sample size (n = 1200) to be accurate, but we need that sample size whether the size of our population is 100,000 or 1,000,000 or 10,000,000, or 100,000,000. What matters is the sample size. The Binomial Distribution Model has a life of its own as a theoretical (and important) distribution. We have been using it as a ***sampling distribution,*** which is the story of sections 3.2 and 3.4.

A mathematical joy ride (if you wish)

This sub-section is for those who are a little dissatisfied with just plugging in the numbers and who want to know why it is that the formula gives us what we want. The formula that we are riding for the joy ride is:

$$P(X=k)=\binom{n}{k}p^k(1-p)^{n-k}$$

This formula gives us the probability of getting k successes in n "trials" or cases. How does it manage to do that? It will help to have an example and one that is simple but not too simple. So, suppose that we have samples of size $n=10$ and $p=0.20$. In the context of out example of obese people, we are choosing ten people at random in a population in which 20% are obese. Suppose we want the probability that exactly three of these in the sample of ten are obese. That will work out to:

$$P(X=3)=\binom{10}{3}0.2^3(1-0.2)^7=\frac{10!}{3!(7)!}(0.20)^3(0.80)^7\approx\frac{10\cdot9\cdot8\cdot7!}{3\cdot2\cdot1(7!)}(0.008)(0.2097152)\approx120\cdot0.0016778\approx0.2013$$

We start with: $p^k(1-p)^{n-k}=(0.2)^3(0.8)^7\approx0.0016778$. Think of the kind of thing that can be true if we have a sample of $n=10$ in which there are exactly three successes—that is, three obese people. Think of the three successes as being placed in three positions in a sample worksheet, something like this:

N	O	N	O	O	N	N	N	N	N

What is the probability of this happening? This is an "and" probability and can be written:

$P(N\ and\ O\ and\ N\ and\ O\ and\ O\ and\ N\ and\ N\ and\ N\ and\ N\ and\ N)$

It is here that the independence condition is used. Recall that two events A and B are independent if $P(A|B)=P(A)$. But since $P(A\ and\ B)=P(A|B)P(B)$, the formula for an "and" probability has an especially simple form for independent events: $P(A\ and\ B)=P(A)P(B)$. Now, since the trials must be independent, we can say that

$P(NONOONNNNN)=(0.8)(0.2)(0.8)(0.2)(0.2)(0.8)(0.8)(0.8)(0.8)(0.8)=(0.2)^3(0.8)^7$.

Also smuggled in here is the requirement that there are just two outcomes and that the probability remains the same. Now think: no matter where the three obese successes go in the ten boxes, the probability of that combination of three successes out of ten will work out the same: $(0.2)^3(0.8)^7$.

But how many of these combinations are there when there are just three successes and seven failures? The answer to that question is in the $\binom{10}{3}=120$, which says that there are 120 different combinations with this same probability. The formula for this, in general, is $\binom{n}{k}=\frac{n!}{k!(n-k)!}$.

Let us see how this works, although in our explanation, the formula will start out looking a bit different. Once again, think of the worksheet:

Where can we put the first success—the first obese? The answer is that there are ten different places; there are ten different choices for it to be. Suppose it lands in the second place, like this. Let us name this first success "A."

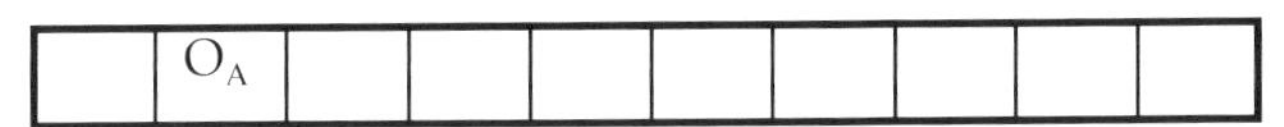

Now for the second success, there are only nine places left; one has been taken by the first success. Let us say that the second one (whose name happens to be "B") lands in the fourth position, like this:

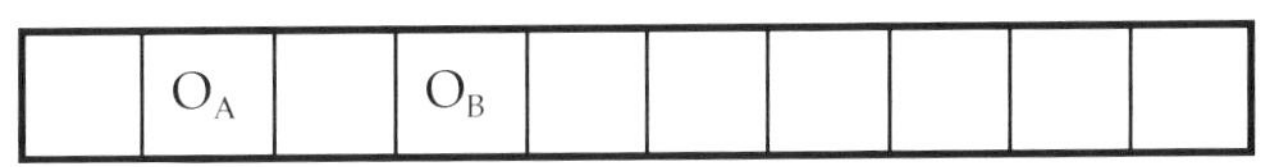

But this is just one possible placement for the two together; there are actually $10 \times 9 = 90$ different possible placements for the two together.. Now for the third (and last) success, there are just eight places left. Let us say the third success ("C") is found in the fifth position:

So, there are $10 \times 9 \times 8 = 720$ different ways we could have done this. However, this is actually too many, and the reason that it is too many is that many of these "landings" are exactly the same combination for us. The same three successes, A and B and C, could have come up like this:

where "B" is the success in the second position and "A" in the fourth position. Or:

	O_B		O_C	O_A					

These two choices (the technical name is permutations) are both the same combination as far as we are concerned; having chosen A, B, and C, we are only concerned that they ended up in the second, fourth, and fifth place, not that specifically A got the second place, B got the fourth place, etc. In fact, with 720 we have listed six *times* too many because for the second place there are three of the A, B, and C available, for the fourth place two, and for the fifth one. Hence $\binom{10}{3} = \frac{10 \cdot 9 \cdot 8}{3 \cdot 2 \cdot 1} = \frac{720}{6} = 120$. The formula:

$\binom{n}{k} = \frac{n!}{k!(n-k)!} = \frac{10!}{3!7!} = \frac{10 \cdot 9 \cdot 8 \cdot 7!}{3 \cdot 2 \cdot 1 \cdot 7!} = \frac{720}{6} = 120$ is just a more convenient way of writing the same thing.

You should be able to work out that they are always equivalent.

One last thing: in the Software simulations, we sampled *with* replacement. That is, the sampled element was returned once chosen, able to be chosen again, and this was done to preserve independence. In real sampling, we usually sample *without* replacement. As long as the population size is more than ten times the size of the sample, independence is not compromised by using sampling *without* replacement.

Summary for the Binomial Distribution Model

The Basic Ideas The ***Binomial Distribution Model*** is a mathematical model that can be applied to many scenarios. In this section we have applied it to look at what we can expect from samples chosen randomly from a population. Specifically,

- If we choose samples of size n and look at a categorical variable with just two values or outcomes,
- Then the ***Binomial Distribution Model*** gives us the probabilities of getting 0, 1, 2,..., up to n "successes" in the sample,
- Where a "success" is one of the two values of the categorical variable.

Conditions for the Binomial Distribution Model We can fit a Binomial Model if *all* of the following conditions are met:

B The outcomes are binomial; that is, there are just *two* outcomes. Often, the outcomes are called "success" and "failure."

I Each case is independent of the others. (The cases are often also called "trials.")

N There is a fixed total number, n, of cases (or "trials").

E We use a probability p of success that is the same ("equal") for each case, or "trial."

Formula for the Binomial Model

To calculate the probability of $X = k$ successes out of n cases (or "trials"), use:

$$P(X=k)=\binom{n}{k}p^k(1-p)^{n-k} \text{ where } \binom{n}{k}=\frac{n!}{k!(n-k)!}$$

and where $n! = n\cdot(n-1)\cdot(n-2)\cdots 2\cdot 1$ and also $0! = 1$ by definition.

$\binom{n}{k}$ tells us the number of ways of placing k things into n boxes (also written ${}_nC_k$) and is "pronounced" "n choose k" or "the number of combinations of k things amongst n."

n! is pronounced "n factorial" and means $n! = n\cdot(n-1)\cdot(n-2)\cdots 2\cdot 1$ and $0! = 1$

Mean, Standard Deviation, and Shape of the Binomial Model

The ***Binomial Distribution Model*** with n and p is a distribution that has mean: np

standard deviation: $\sqrt{np(1-p)}$

and a shape that becomes more nearly Normal as the sample size n increases.

Looking Ahead: Definition of p-hat and p so that we can analyze proportions

- For a proportion that refers to—or is calculated from—a sample, we use the symbol $\hat{p}$.
- For a proportion that refers to a population, we use the symbol p.

§3.4 Trusting Data, Part 4: *Sampling distributions for proportions*

Introduction: Quantitative and categorical variables

Sections 3.2 and 3.3 introduced the idea of a sampling distribution. Our example in §3.2 was about the average area of a house, measured by the quantitative variable *Sq_Ft*. Since *Sq_Ft* is a quantitative variable, we used the sample mean $\bar{x}$ and we developed the idea of the sampling distribution of the sample mean.

In §3.3, we looked at a categorical variable—whether people are obese or not. There we measured the *number* X of obese people in a sample and saw that the Binomial Model served well to predict the number of "successes" in a sample with a given size. Then, toward the end of that section, we transitioned to thinking about proportions, which are simply the number of successes X, divided by the sample size n. We also saw that the Binomial Model started to look very much like the Normal distribution under some circumstances. The story was: the sampling distribution for proportions is essentially binomial, but sometimes the binomial looks a lot like a Normal distribution.

In this section, we look at the Normal sampling distribution as it is used in actual statistical work.

Example: More San Mateo Real Estate

When the San Mateo real estate data were first collected for 2005–2006, property values were high; there was a boom in the housing market. When there is a real estate boom, sellers sometimes find that they actually get more for their house than the listed price; the buyers actually compete with other buyers for the house. Here are data from a random sample of n =100 of the real estate data for 2005–2006. Since the sample size is n = 100, it is easy to calculate the proportion (even in your head) of sellers who got more than they asked; using the notation that we saw in §1.2, we can write that $P(OLP) = 0.53$, if *OLP* stands for the event that the house sold for "Over List Price." We would interpret this by either saying that "53% of the houses in our sample sold for over the listed price" or by saying that "the probability that a house for sale in 2005–2006 sold for over its listed price was 0.53, at least according to our sample data."

Sample of San Mateo Real Estate Y0506

SoldOverList	Not Over List Price	47
	Over List Price	53
	Column Summary	100

S1 = count ( )

New notation for a new sampling distribution. Notice that our data come from a sample and not from a population, and that immediately raises our first question: "Can we generalize this 53% for 'Over List Price' to apply to the population from which the sample was drawn, since it is just from a sample of n = 100?" If we knew about the sampling distribution for a sample proportion, we might be able to answer the question of whether we have a sample result that is *rare*—far from the population value for the proportion of houses that were sold "Over List Price"—or a sample result that is *reasonably likely*—not far from the population value for $P(OLP)$, the proportion of houses that sold for over their listed price.

We need some new notation; we need to distinguish a proportion calculated from sample data from the corresponding proportion in the population. In §3.2, we used the symbol $\bar{x}$ for a sample mean, whose value we can calculate since we have the sample data, and μ for the population mean, whose value we generally do not know.

We have used $P(OLP)$ to stand "the proportion of houses that sold for 'Over List Price.'" However, now we are aware that our calculation is for a sample that comes from a population. To distinguish between a sample proportion and the corresponding population proportion, we use the symbol $\hat{p}$ (pronounced "*p*-hat") for the proportion calculated from a *sample*. For our example, $\hat{p} = 0.53$; that is numerical value that we calculated as $P(OLP) = 0.53$ above. The population proportion for our example would be the proportion of *all* houses in the population that sold over list price, and for that we use the symbol p (p "without the hat.") Usually we are unable to calculate this population proportion; nevertheless, we need a symbol to refer to the population proportion even if we cannot calculate it. We use the $\hat{p}$ and p notation to keep it clear what we are referring to: the sample proportion that we can calculate and therefore know (the $\hat{p}$) or the population proportion that we cannot calculate and do not know but we really want to know (that is the p).

Notation

Sample proportion: $\hat{p}$ ***Population proportion:*** p

Building a Sampling Distribution Using Software

We can apply the idea of a sampling distribution to the sample proportion $\hat{p}$. We will build one using Software and then discuss the properties of the sampling distribution. Instead of the population mean μ and the sample mean $\bar{x}$ we will be interested in the population proportion p and sample proportion $\hat{p}$.

Idea of a Sampling Distribution of a Sample Proportions $\hat{p}$

The ***sampling distribution of sample proportions*** is the distribution of the sample proportions for a specific variable of *all possible* random samples of a given size n drawn from a population.

Simulating a sampling distribution of sample a proportion $\hat{p}$ **.** Once again, we will use a specific example. The population is all of the houses sold in San Mateo County in 2005–2006 (not just from the South Region this time), and we will draw samples of size $n = 100$ (instead of $n = 40$ as we did in the sample mean example.) However, the four steps of the construction (or simulation) will be the same as before:

- ⇨ Step 1: Randomly sample a sample of $n = 100$ from the population of 5,486 houses.
- ⇨ Step 2: Calculate the sample proportion $\hat{p}$ for the sample just collected; store the result.
- ⇨ Step 3: Repeat steps 1 and 2 until 3.641×10^{216} unique samples are found.
- ⇨ Step 4: Collect together all the different sample proportions $\hat{p}$ into a collection.

To see the construction (or simulation) happen with software, follow the bulleted instructions.

- Open the file ***SanMateoRealEstateY0506.*** This file contains data on *all* the houses that were sold in San Mateo County, California, from June 2005 to June 2006. These data are the *population* of houses sold. Consult the ***Software Supplement*** about ***Simulating a Sampling Distribution***.

- For the entire population, have software calculate the proportions of houses in the two categories of the variable ***SoldOverList.*** These are the *population* proportions. The output may resemble what is shown here, although he graphic may well be different.

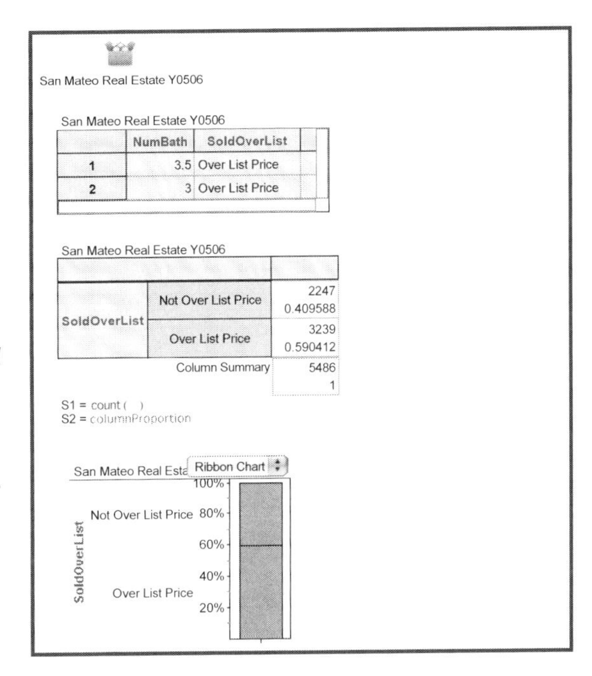

  What you see is the ***population distribution*** of the variable *SoldOverList* for the population of *all* the houses that were sold in 2005–2006 in San Mateo County. Notice that since our variable is categorical, we do not have a dot plot but rather just the ribbon chart. The summary table shows that the population proportion of houses that sold "Over List Price" is $P(OLP) = 0.590$. Using our notation, and knowing that in this instance the proportion is a population proportion, we will write $p = 0.590$. In the population of houses sold in 2005–2006, the proportion of houses that sold over list was 0.59. (Usually, we as researchers will not know the population proportion p; we have chosen an instance where we *do* know the population proportion to show how a sampling distribution is constructed.)
- Have software draw a single sample of size $n = 100$ that includes the variable *SoldOverList* from the collection ***SanMateoRealEstateY0506.*** Arrange the output so that there is a clear distinction between the population and the sample.

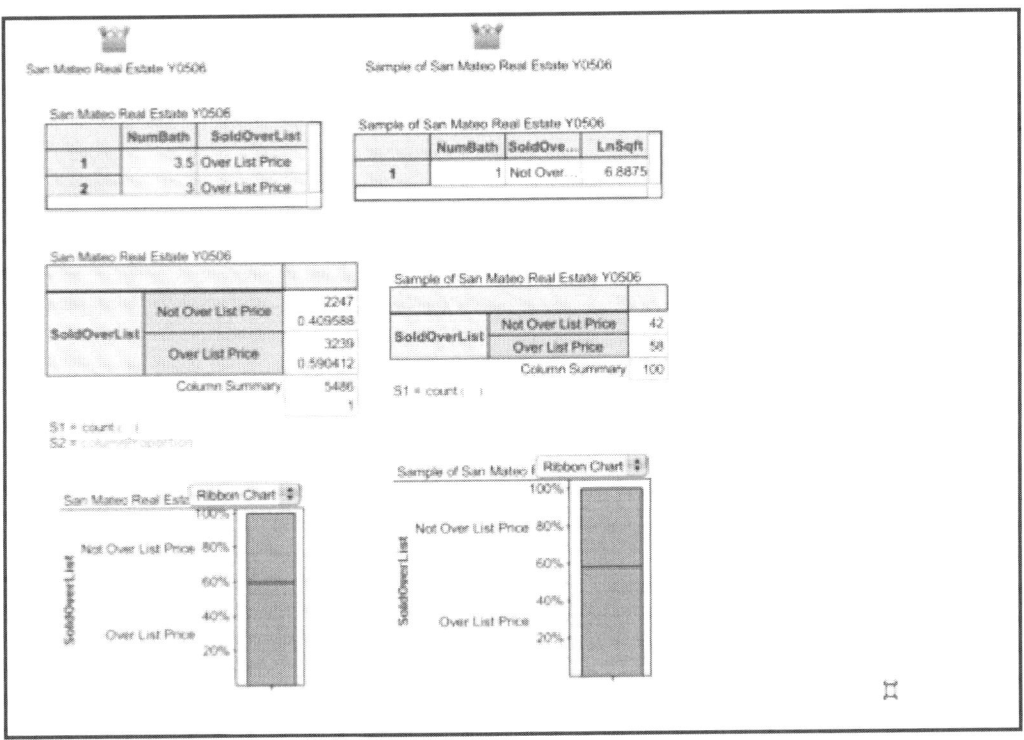

Notice that the sample proportion shown in the example below is $P(OLP)=\frac{58}{100}=0.58$; using our new notation, we would write $\hat{p}=0.58$. This sample proportion is close to the population proportion.

- Use software to continue collecting and accumulating many many samples from the collection ***SanMateoRealEstateY0506*** that include the variable *SoldOverList,* recording the sample proportion for each sample. This is the essence of our simulation of a sampling distribution. Get software to arrange the result so that there is a clear distinction between the population, the sample and the sampling distribution.

For the sampling distribution, we have a dot plot and a mean and standard deviation. Why is this? It is because for the sampling distribution the cases are *phats* ($\hat{p}$ s) and these are quantitative *phats*; the variable for our population and for our sample is categorical, but for each sample we calculated a quantitative measure, the sample proportion, $\hat{p}$. Our collection of $\hat{p}$ s is a distribution of quantitative things.

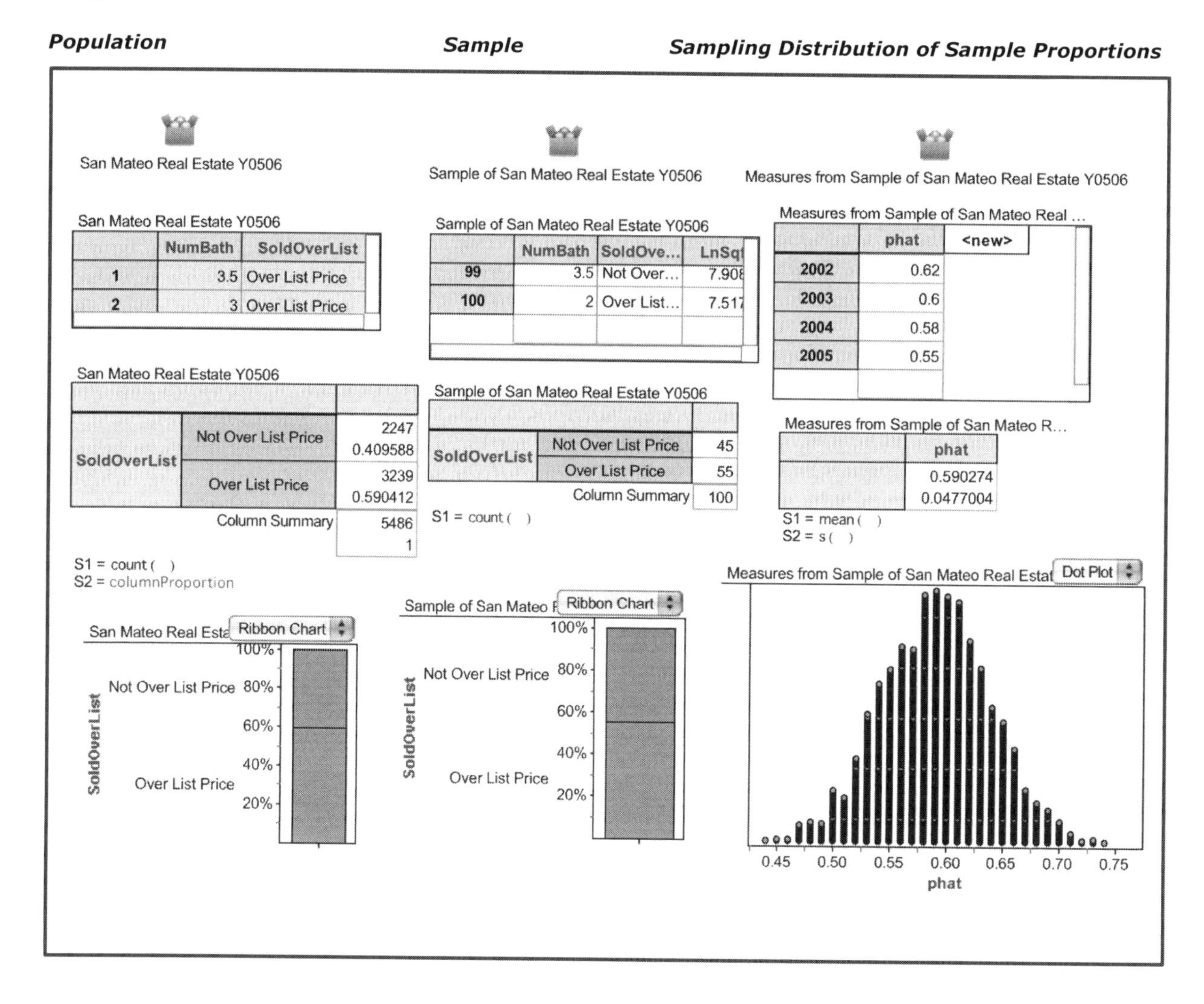

Why there are "spikes" in our sampling distribution. If you compare the plot on the right of the Software output to the similar output in §3.2 for the sampling distribution for quantitative, you see that the values of the $\hat{p}$ here are stacked up in "spikes" (or poles) whereas with the $\bar{x}$ s the values were not in neat vertical stacks. The picture for the sample means $\bar{x}$ s looked like a "normally shaped" collection of grapes rather than spikes. We see "spikes" for our sampling distribution of $\hat{p}$ because in a sample of $n = 100$, there can only be 0, 1, 2, 3,..., 96, 97, 98, 99, 100 houses that sold over list price. The number of houses that sold over list price conceivably could be any one of those numbers but cannot be something like 41.275 houses that sold "Over List Price." Therefore, our *phats* can only take on certain values that come from dividing integers 0, 1, 2, 3,...96, 97, 98, 99, 100 by 100. So we only get certain values, and hence we get the appearance of stacked dots, or "spikes."

The shape of the sampling distribution looks Normal. In our calculations, we will usually use a Normal distribution to approximate the shape of a sampling distribution of a sample proportion. However, the actual sampling distribution is another family of theoretical (model) distributions called the ***binomial distributions,*** which we looked at in §3.3. Under certain conditions, the shapes of these binomial distributions are very close to the shape of a Normal distribution. For example, here is the plot for samples of size $n = 100$ drawn from a population in which the population proportion $p = 0.59$. The graphic is from ***DistributionCalculator.ggb***.

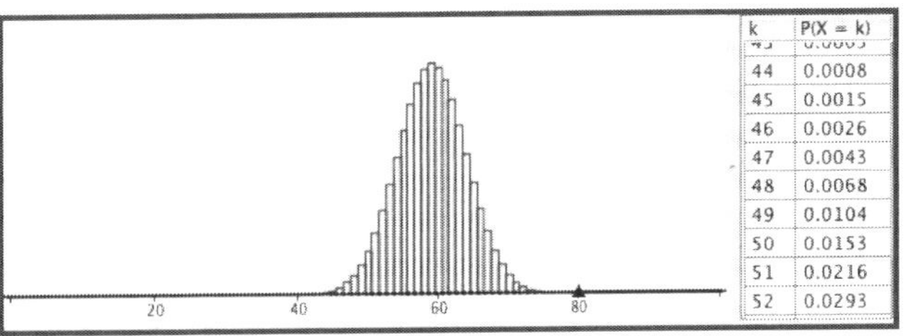

Having a sampling distribution that is Normal was good news because we can calculate what results from random sampling as *reasonably likely* or *rare.* The story is the same here: that the shape of the sampling distribution is approximately Normal allows us to work with the sampling distribution but only (as we shall see) under certain conditions. As with the sampling distribution for sample means $\bar{x}$, we need notation so that we will not lose our way. Here is a guide to the notation.

Notation for mean and standard deviation of the sampling distribution for $\hat{p}$			
Mean:	$\mu_{\hat{p}}$	***Standard Deviation:***	$\sigma_{\hat{p}}$

Now that we have some notation, we can compare the mean of the sampling distribution $\mu_{\hat{p}}$ with the population proportion p. From our simulation of 2005 different samples of $n = 100$, the two numbers look very, very close and exhibit what is actually a fact about the sampling distribution of sample proportions $\hat{p}$; the mean of the sampling distribution is equal to the population proportion:

$\mu_{\hat{p}} = p$.

The standard deviation of the sampling distribution of sample proportions $\hat{p}$ is $\sigma_{\hat{p}} = \sqrt{\frac{p(1-p)}{n}}$. (This formula actually comes from the binomial distributions that we said are the actual forms the sampling distributions for $\hat{p}$ take). For our example where $p = 0.590$ and $n = 100$, this formula works out to be $\sigma_{\hat{p}} = \sqrt{\frac{p(1-p)}{n}} = \sqrt{\frac{(0.59)\bullet(1-0.59)}{100}} = \sqrt{\frac{0.59\bullet 0.41}{100}} = \sqrt{\frac{0.2419}{100}} = \sqrt{0.002419} \approx 0.0492$. When we compare this with our simulation, we see that the standard deviation of the $\hat{p}$ s that we collected is 0.0477, which is close to 0.0492. We can summarize these facts as follows.

Facts about Sampling Distributions of Sample Proportions $\hat{p}$ from Random Samples

Shape:	The shape of the sampling distribution of $\hat{p}$ will be approximately Normal.
Center:	The mean of the sampling distribution of $\hat{p}$ is equal to the population proportion p. That is, $\mu_{\hat{p}} = p$.
Spread:	The standard deviation of the sampling distribution of $\hat{p}$ is $\sigma_{\hat{p}} = \sqrt{\frac{p(1-p)}{n}}$.
Conditions:	These facts hold only if *both* $np \geq 10$ and also $n(1-p) \geq 10$ are met and *also* the population is at least ten times the size of the sample being used.

The conditions $np \geq 10$ and also $n(1-p) \geq 10$ come from the fact that we are using a Normal distribution as an approximation to the binomial distributions. Binomial distributions can be quite skewed to the right or to the left if the p is small (near 0) or big (near 1) and the n also small. If the n is big then the Normal distribution is still a good approximation of the binomial distribution, even if the p is extreme; our guide as to whether the combination of sample size n and the population proportion will work are the rules: it must be true that $np \geq 10$ and also $n(1-p) \geq 10$.

Using Sampling Distributions for Sample Proportions $\hat{p}$ from Random Samples

Checking the conditions The first thing that we always need to do is to check the conditions $np \geq 10$ and also $n(1-p) \geq 10$. With our example of the San Mateo real estate, where $p = 0.59$, we can proceed, since we see that $np = 100 \cdot 0.59 = 59 > 10$ and $n(1-p) = 100 \cdot (1-0.59) = 100 \cdot 0.41 = 41 > 10$. However, if we had a sample size of only $n = 20$, we would *not* certain that the Normal approximation to the binomial distribution would be close enough; when we calculate $np = 20 \cdot 0.59 = 11.8 > 10$ and this "passes the test" (just barely), but when we calculate $n(1-p) = 20 \cdot (1-0.59) = 20 \cdot 0.41 = 8.2 < 10$, this does not pass the test. Both parts of the conditions must be true for us to proceed with confidence.

Reasonably Likely and Rare As we did with sample means $\bar{x}$, we can calculate an interval for what sample proportions $\hat{p}$ we can reasonably expect to see (the ***reasonably likely*** $\hat{p}$ s) and therefore what sample proportions would be ***rare*** if we have some knowledge of the population proportion p. The box below gives the facts, and then just below it there is a worked example.

Reasonably Likely and Rare

Sample proportions $\hat{p}$ that are *within* the interval $p-1.96\sqrt{\frac{p\bullet(1-p)}{n}} < \hat{p} < p+1.96\sqrt{\frac{p\bullet(1-p)}{n}}$ are called

reasonably likely.

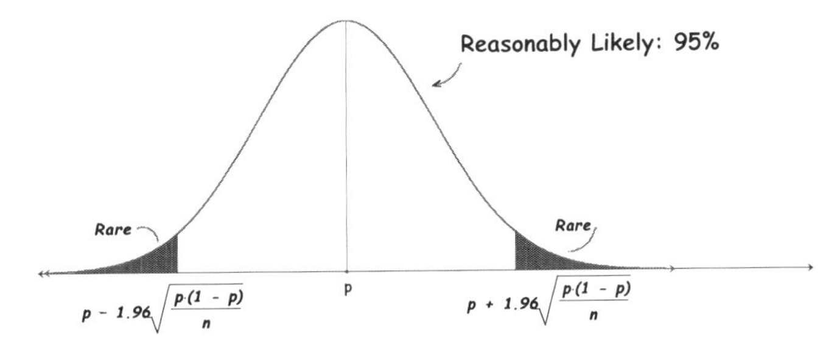

Sample proportions $\hat{p}$ that are *outside* the interval $p-1.96\sqrt{\frac{p\bullet(1-p)}{n}} < \hat{p} < p+1.96\sqrt{\frac{p\bullet(1-p)}{n}}$ are called

rare.

In probability notation, $P\left(p-1.96\sqrt{\frac{p\bullet(1-p)}{n}} < \hat{p} < p+1.96\sqrt{\frac{p\bullet(1-p)}{n}}\right)=0.95$

Reasonably Likely and Rare: San Mateo Real Estate Houses Over List Price To follow the argument, recall that the population proportion of houses that sold for more than the list price was p $=0.59$. Therefore we know that the mean of the sampling distribution of $\hat{p}$ s calculated from samples of $n = 100$ is $\mu_{\hat{p}} = p = 0.59$, and the standard deviation of the sampling distribution is

$\sigma_{\hat{p}} = \sqrt{\frac{0.59\bullet(1-0.59)}{100}} = \sqrt{\frac{0.59\bullet 0.41}{100}} \approx 0.0492$. We can then calculate the *reasonably likely* interval:

Lower limit: $p-1.96\sqrt{\frac{p(1-p)}{n}} = 0.59-1.96\sqrt{\frac{0.59\bullet 0.41}{100}} \approx 0.59-1.96(0.0492) = 0.59-0.096 = 0.494$ and

Upper limit: $p+1.96\sqrt{\frac{p(1-p)}{n}} = 0.59+1.96\sqrt{\frac{0.59\bullet 0.41}{100}} \approx 0.59+1.96(0.0492) = 0.59+0.096 = 0.686$.

So for this example, the interval of *reasonably likely* $\hat{p}$ s is the interval $0.494 < \hat{p} < 0.686$. What does this mean? We can give two interpretations. We can say that if we sampled samples of $n = 100$ repeatedly (and randomly) from our population of all houses, 95% of those samples would have $\hat{p}$ s in the interval $0.494 < \hat{p} < 0.686$. Alternately, we can say that if we draw a random sample of size $n = 100$ houses from our population of houses in San Mateo real estate, we are *95% certain* that our sample proportion $\hat{p}$ of houses that sold over their list price will be in the interval 49.4% to 68.6%. We might be "unlucky" with our random sample and get a $\hat{p}$ outside this interval, but the probability that we will get such a "wild" $\hat{p}$ is only 5%.

How likely is it that we get a sample with "this $\hat{p}$ ***" or one even more extreme?*** What do we mean by this? The full context of exactly *why* we want to know this will be clearer in the next unit; but for now, know that we will be interested in knowing how likely it is that we actually see a $\hat{p}$ as far away from a population p as we have actually seen. We have everything we need to calculate this probability.

Here is an example. The last sample that we collected in our simulation had a $\hat{p}$ = 0.55; here $\hat{p}$ = 0.55 is the "this $\hat{p}$". This *sample* of just n = 100 houses told us that 55% of the houses have sold for over list price. We can ask the question: "If the population proportion is p = 0.59, what is the probability that we get a sample proportion this small or one even smaller?" We are asking the question: "What is $P(\hat{p} < 0.55)$?" This is a "given a value (the value $\hat{p}$ = 0.55)— find a proportion" type of problem, and since we have an approximately Normal sampling distribution, we can solve this problem.

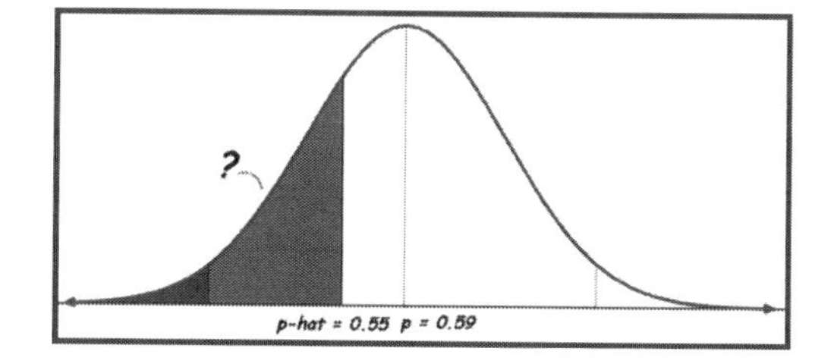

The first step is to draw a picture of the situation showing what we have and what we want—what we want by an arrow; we want a probability. Next calculate the z score so that we get

$$z = \frac{\hat{p} - \mu_{\hat{p}}}{\sigma_{\hat{p}}} = \frac{0.55 - 0.59}{\sqrt{\frac{0.59 \cdot (1 - 0.59)}{100}}} \approx \frac{-0.04}{0.049} \approx -0.813 \approx -0.81 .$$

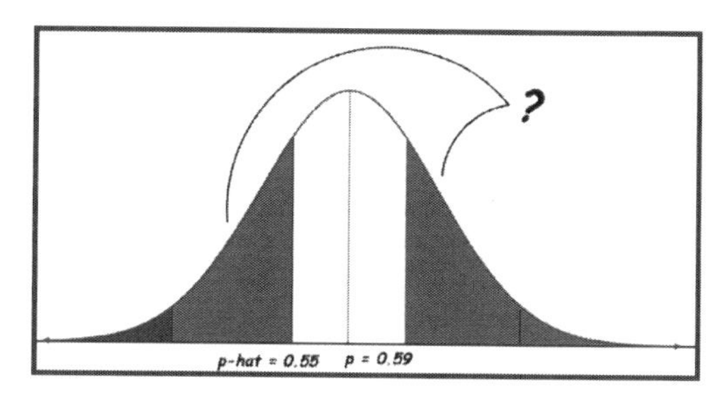

With the z = -0.81, we consult the Normal Probability Chart and write (using the correct notation) $P(\hat{p} < 0.55) = P(z < 0.81) = 0.2090$. The probability of getting a *p-hat* of 0.55 or smaller is about 21%.

The answer $P(\hat{p} < 0.55) = P(z < 0.81) = 0.2090$ is correct for the question "What is $P(\hat{p} < 0.55)$?" That calculation is correct to answer the question: "What is the probability that we get a $\hat{p}$ less than 0.55 if the population proportion is 0.59?" However, it does not do justice to the question: "What is the probability of getting a $\hat{p}$ that is *more extreme?"* Why? How far away from 0.59 is 0.55? Answer: –.04. So anything 0.04 farther away from 0.59 is more extreme; but this can happen (randomly) on the upper side of 0.59 as well as the lower side. "More extreme" could be a $\hat{p}$ greater than 0.59 + 0.04 = 0.63 as well as a $\hat{p}$ less than 0.55. Our picture of "more extreme" should have two tails, rather than one tail like the picture shown.

To actually calculate this "more extreme" probability is very easy because of the symmetry of the Normal distribution. The probability for the two tails is simply twice the probability of one tail. The mathematical way to express this is to use the absolute value notation, since we want to embody both the left and the right tail in one expression. Therefore, we would write, as an expression of "more extreme than $\hat{p}$ = 0.55," $P(\hat{p} > |0.55|) = 2P(z < 0.81) = 2(0.2090) = 0.4180 \approx 0.42$.

Summary: Normal Sampling Distributions for Categorical Variables

- The relevant measure for categorical variables is a ***proportion,*** so we examine the sampling distribution for sample proportions.
- ***Notation:*** ***Sample proportion:*** $\hat{p}$ ***Population proportion:*** p
- ***Notation for mean and standard deviation of the sampling distribution for*** $\hat{p}$

  Mean: $\mu_{\hat{p}}$ ***Standard Deviation:*** $\sigma_{\hat{p}}$

- ***Facts about Sampling Distributions of Sample Proportions*** $\hat{p}$
  - ***Shape:*** The shape of the sampling distribution of $\hat{p}$ will be approximately Normal.
  - ***Center:*** The mean of the sampling distribution of $\hat{p}$ is equal to the population proportion p. That is, $\mu_{\hat{p}} = p$.
  - ***Spread:*** The standard deviation of the sampling distribution of $\hat{p}$ is $\sigma_{\hat{p}} = \sqrt{\frac{p(1-p)}{n}}$.
  - ***Conditions:*** These facts hold only if *both* $np \geq 10$ and also $n(1-p) \geq 10$ are met and *also* the population is at least ten times the size of the sample being used.
- ***Reasonably Likely and Rare***
  - ***Reasonably likely*** sample proportions $\hat{p}$ are those *within* the interval

    $$p - 1.96\sqrt{\frac{p \cdot (1-p)}{n}} < \hat{p} < p + 1.96\sqrt{\frac{p \cdot (1-p)}{n}}.$$

  - ***Rare*** sample proportions $\hat{p}$ are *outside* the interval $p - 1.96\sqrt{\frac{p \cdot (1-p)}{n}} < \hat{p} < p + 1.96\sqrt{\frac{p \cdot (1-p)}{n}}$.

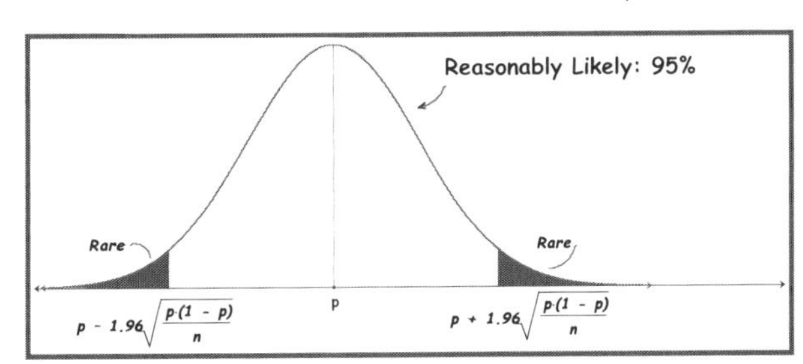

- With a sampling distribution for a specific categorical variable and population, we can:
  - Find the range of *reasonably likely* and *rare* sample proportions $\hat{p}$ s.
  - Determine the likelihood of a sample proportion $\hat{p}$ that is greater (or lesser) than a specific value.
  - Calculate the sample proportion $\hat{p}$ that we would expect to see a given percentage of the time if we repeatedly sampled.

    See the ***Notes*** above for examples of these kinds of calculations.

§4.1 Politics and Confidence: Estimating a Proportion

Approving the President

For decades the Gallup® organization has been asking people about whether they "approve the job the president is doing." There was an interesting website (http://www.usatoday.com/news/washington/presidential-approval-tracker.htm) that that had tracked these approval ratings over time. Here is a graph showing the comparison of the approval ratings for Clinton, Bush (the younger), and Obama (up to July 2012). For Bush and Clinton, the rating is tracked to the end of their second terms.

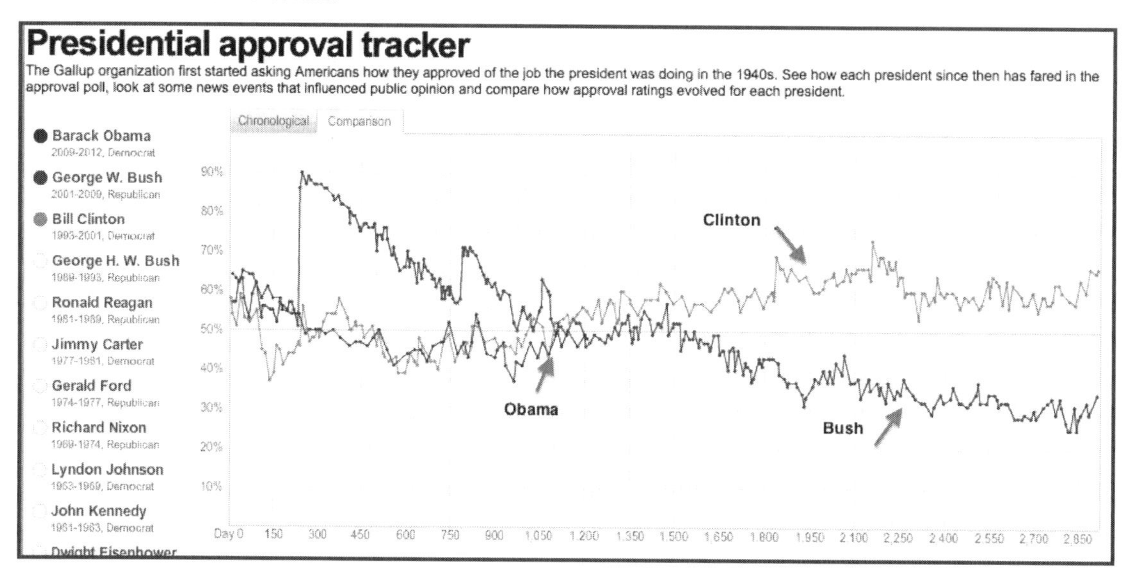

This approval rating is an interesting population proportion p because the population is well defined as "all Americans," but the actual population proportion p is always changing, even by the minute, and no one can know what it "actually" is. The best that can be done is to estimate the p with results from a sample of people. So how is this done? Just after Barack Obama received the Nobel Prize for Peace, the Gallup® organization reported Obama's approval rating was 56% (see http://www.gallup.com/poll/123629/Obama-Job-Approval-56-After-Nobel-Win.aspx). At the foot of that press release, there was the paragraph below that informs the reader how they did the survey. (Gallup® press releases typically give this kind of information, although the details vary.)

Survey Methods

Results are based on telephone interviews with 1,532 national adults, aged 18 and older, conducted Oct. 9-11, 2009, as part of Gallup Daily tracking. For results based on the total sample of national adults, one can say with 95% confidence that the maximum margin of sampling error is ±3 percentage points.

Interviews are conducted with respondents on land-line telephones and cellular phones.

In addition to sampling error, question wording and practical difficulties in conducting surveys can introduce error or bias into the findings of public opinion polls.

This tells us that the sample size was $n = 1532$, that the population included adults eighteen years and older, and that they interviewed the respondents on either land line or cell telephones. Note the sentence: "For results based on the total sample of national adults, one can say with 95% confidence that the maximum margin of sampling error is ±3 percentage points." What Gallup® has calculated is an ***interval estimate*** of the population proportion p of Americans who approve of Obama's job performance. The technical name for the *interval estimate* that they have calculated is a ***confidence interval.*** In this section, we will see:

- How ***confidence intervals*** are calculated
- The meaning of the phrase: "One can say with 95% confidence…"
- The meaning of the term: "margin of sampling error"
- How it is that Gallup® can use a sample of only $n = 1532$ to make their estimate for the entire population
- How what Gallup® has done is connected with the *reasonably likely interval* calculated from a sampling distribution of $\hat{p}$ s, that we looked at in §3.4

What Gallup® has done is to calculate a ***95% confidence interval*** for the population proportion p of Americans who approve of the president's job performance. They can say "with 95% confidence" that the percent of Americans who approved of Obama's performance (for the time they took their survey) was 56%±3%. The 3% is called the ***margin of sampling error*** (or just ***margin of error),*** and by subtracting and adding this *margin of sampling error* we get an interval for the population proportion p: the *interval* in this situation is $53\% \le p \le 59\%$. That is, it is possible that only 53% of Americans approve of Obama's performance or it is possible that the proportion is as high as 59% or any percentage between 53% and 59%. Gallup® can say this with 95% confidence.

How 95% Confidence Intervals Come from Reasonably Likely Intervals

p	$p-1.96\sqrt{\frac{p(1-p)}{n}}$	$p+1.96\sqrt{\frac{p(1-p)}{n}}$
0.950	0.907	0.993
0.900	0.841	0.959
0.850	0.780	0.920
0.800	0.722	0.878
0.750	0.665	0.835
0.700	0.610	0.790
0.650	0.557	0.743
0.600	0.504	0.696
0.550	0.452	0.648
0.500	0.402	0.598
0.450	0.352	0.548
0.400	0.304	0.496
0.350	0.257	0.443
0.300	0.210	0.390
0.250	0.165	0.335
0.200	0.122	0.278
0.150	0.080	0.220
0.100	0.041	0.159
0.050	0.007	0.093

Imagine that we try to repeat what Gallup® did on a college campus. Instead of a sample size of $n = 1532$, we will manage (for simplicity) with a smaller sample size of $n = 100$. However, our sample *must* still be a random sample; everything depends upon the sample being random. One thing we can do even before we collect the data is to calculate intervals of *reasonably likely* $\hat{p}$ s for various *population* p s. Since a President's Approval Rating could be any p between 0.00 and 1.00, we have calculated many reasonably likely intervals by steps of 0.05. For example, if the population proportion $p = 0.60$ then the lower end of the reasonably likely interval is calculated by

$p-1.96\sqrt{\frac{p\cdot(1-p)}{n}}$, so $0.60-1.96\sqrt{\frac{0.60\cdot(1-0.60)}{100}} \approx 0.60-1.96(0.049) = 0.60-0.096 = 0.504$.

The upper end of the interval is $0.60 + 1.96\sqrt{\frac{0.60 \cdot (1-0.60)}{100}} \approx 0.60 + 1.96(0.049) = 0.60 + 0.096 = 0.696$.

That is, the interval $0.504 \le \hat{p} \le 0.696$ contains the reasonably likely sample proportions $\hat{p}$ for samples of size $n = 100$ if the population proportion $p = 0.60$.

Now suppose from our *sample* of $n = 100$, we get fifty-six people who approve of the job the president is doing. This means that we have a $\hat{p} = 0.56$, which falls in the interval $0.504 \le \hat{p} \le 0.696$. Here is the question: how can we use this sample proportion $\hat{p} = 0.56$ to get an *estimate* of the population proportion? In other words, how can we *infer* (or generalize to) the population proportion p using our sample proportion $\hat{p} = 0.56$? To see the logic for this inference, we will use a **Swift Diagram**.

Showing the link between reasonably likely intervals and confidence intervals: Swift Diagrams. The horizontal lines on the **Swift Diagram** show the intervals of reasonably likely $\hat{p}$ s shown in the table on the previous page. You should look at the reasonably likely interval for $p = 0.60$ and see that it does appear to reach from $\hat{p} = 0.504$ to $\hat{p} = 0.696$. Actually, although we have only put the horizontal reasonably likely intervals for $p = 0.05$, $p = 0.10$, $p = 0.15$… up to $p = 0.95$, you should imagine that the spaces between these horizontal lines are filled with other reasonably likely intervals or the values of p between.

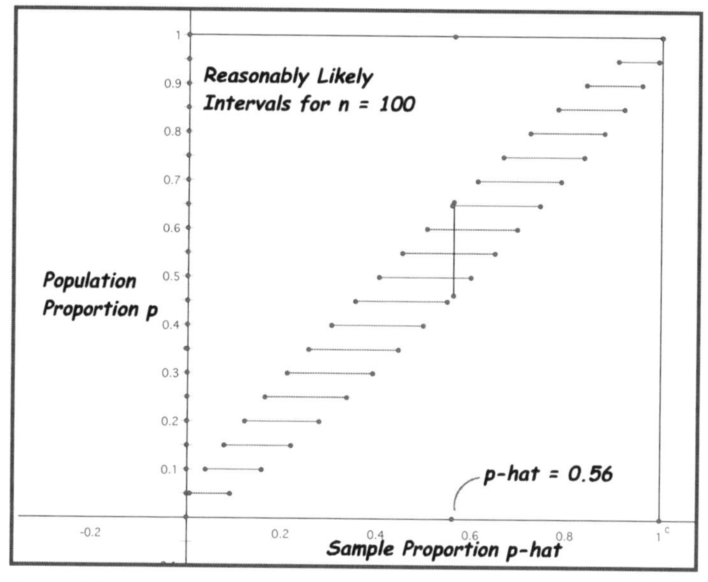

In the Swift Diagram, the *vertical line* shows the $\hat{p} = 0.56$, which is the *phat* that we actually got.

Using the Swift Diagram, here is how we reason. If the population proportion of people approving the job the president is doing were (only) $p = 0.40$ then would we be likely to see the $\hat{p} = 0.56$ we actually got? No, we would not; a $\hat{p} = 0.56$ is far *outside* the reasonably likely interval for $p = 0.40$. If the population proportion were $p = 0.70$, would we be likely to see the $\hat{p} = 0.56$ that we actually got? Again, no, for the same reason; the $\hat{p} = 0.56$ is far *outside* the reasonably likely interval for the population proportion $p = 0.70$. However, the population proportion *could be* $p = 0.50$, or $p = 0.55$, or $p = 0.60$ or $p = 0.65$ because for all of *these* values for p, our value of $\hat{p} = 0.56$ is *inside* the reasonably likely intervals.

We can be even more precise. By looking at the bolded part of the vertical-line Swift Diagram, we can see that the population proportion should be somewhere between $p = 0.46$ and $p = 0.66$. We can say that we are **95% confident** that the population proportion is in the interval $0.46 < p < 0.66$ because those are the values of the population proportion p that have $\hat{p} = 0.56$ inside their 95% reasonably likely intervals for $\hat{p}$.

The values of p *outside* this interval $0.46 < p < 0.66$ have $\hat{p} = 0.56$ *outside* their reasonably likely intervals. The bolded part of the vertical line in the Swift Diagram is the ***95% confidence interval for a population proportion p.***

If you measure the length of the bolded part of the vertical line, you will find that its length was exactly the same as the length of the reasonably likely interval for $p = 0.56$. This gives a way of being very precise in our calculation of the bolded part of the vertical line in the Swift Diagram. The calculation for our example will be

$\hat{p} \pm 1.96\sqrt{\dfrac{\hat{p}\cdot(1-\hat{p})}{n}} = 0.56 \pm 1.96\sqrt{\dfrac{0.56\cdot(1-0.56)}{100}} \approx 0.56 \pm 1.96(0.0496) \approx 0.56 \pm 0.097$, and this results in an interval whose lower end is $0.56 - 0.097 = 0.463$ and upper end is $0.56 + 0.097 = 0.657$. Our 95% confidence interval is $0.463 < p < 0.657$. With our small random sample of $n = 100$, we can be 95% confident that the proportion of people who approve of the president's job performance is between 46.3% and 65.7%.

Calculation of a Confidence Interval for a Population Proportion p

$$\hat{p} \pm z^*\sqrt{\frac{\hat{p}\cdot(1-\hat{p})}{n}}\text{, where}$$

$z^* = 1.645$ for a 90% confidence interval

$z^* = 1.96$ for a 95% confidence interval

$z^* = 2.576$ for a 99% confidence interval

This formula works under the ***conditions*** that:

1. The sample of size n from which $\hat{p}$ was calculated is a simple random sample.
2. Both $n\hat{p} \geq 10$ and $n(1-\hat{p}) \geq 10$.
3. The size of the population is at least ten times the size of the sample.

In the box above, there is a z^* in the formula in the position where we used 1.96 in our calculations. We have z^* instead of 1.96 because we can have confidence intervals that estimate the population proportion p with 90% or 99% confidence, and then the value for z^* will be different, although the basic reasoning is the same. (Actually, we could calculate confidence intervals with any level of confidence we want, but 90%, 95%, and 99% are the most common.)

The Margin of Error for a Confidence Interval

The formula $\hat{p} \pm z^*\sqrt{\dfrac{\hat{p}\cdot(1-\hat{p})}{n}}$ can be thought of as having two parts: the part on the left side of the "±", that is, our sample proportion $\hat{p}$, and the part on the right-hand side of the "±". The part on the left of the ± is sometimes called a ***point estimate*** or just ***estimate.*** This estimate is the number that we derive from calculating a proportion in our sample. The part of the right-hand side is called the ***margin of error*** (or ***margin of sampling error***). The *margin of error* reflects what we know from the sampling distribution theory about how much variability there can be in sample proportions when we use randomization. Putting these together, we can say that a confidence interval is constructed by an ***estimate ± margin of error.***

Margin of Sampling Error (or Margin of Error) for a Confidence Interval
$$ME = z^* \sqrt{\frac{\hat{p}\cdot(1-\hat{p})}{n}}$$

For the sample proportion $\hat{p} = 0.56$ for the president's approval proportion, we can calculate

$$ME = z^* \sqrt{\frac{\hat{p}\cdot(1-\hat{p})}{n}} = 1.96\sqrt{\frac{0.56\cdot(1-0.56)}{100}} \approx 1.96(0.0496) \approx 0.097.$$

When we calculated the confidence interval, we actually calculated the margin of error; it was the part that we subtracted from and added to our $\hat{p} = 0.56$. The ***margin of error*** for our 95% confidence interval is 0.097 or about 9.7%. If we interpret the confidence interval in the way the Gallup® organization has, we would say: "We estimate Obama's approval rating to be 56% with a margin of sampling error of 9.7%." It may be that a 9.7% margin of error is not that impressive; recall that the Gallup® organization said that their margin of error was at most 3%. Why? The answer is that they had a much bigger sample size. We can calculate the margin of error with their sample size and the same $\hat{p} = 0.56$ as:

$$ME = z^* \sqrt{\frac{\hat{p}\cdot(1-\hat{p})}{n}} = 1.96\sqrt{\frac{0.56\cdot(1-0.56)}{1532}} \approx 1.96(0.0127) \approx 0.025$$

The margin of error is 2.5 (less than what Gallup® said; this is because they advertised a "maximum" margin of error of 3%). If we look at the formula for the margin of error, we can easily see that as the sample size increases, the margin of error decreases (if the other numbers are the same) since we are dividing by a larger number for n. The bigger the sample size, the smaller the margin of error. If we decide that we want a specific margin of error (say, 3%), we can do the algebra and calculate what sample size we need for that margin of error.

Sample Size for a Given Margin of Error We can calculate the sample size that is necessary for a 3% (or 0.03) margin of error. We set $0.03 \geq ME$ so we have $0.03 \geq z^* \sqrt{\frac{\hat{p}\cdot(1-\hat{p})}{n}}$ and solve for n.

$ME = z^* \sqrt{\frac{\hat{p}(1-\hat{p})}{n}}$	
$0.03 \geq 1.96\sqrt{\frac{0.56\cdot(1-0.56)}{n}}$	The " $\geq$ " translates ME $\leq$ 0.03 as 0.03 $\geq$ ME
$\frac{0.03}{1.96} \geq \sqrt{\frac{0.56\cdot(1-0.56)}{n}}$	Divide both sides by 1.96
$\left(\frac{0.03}{1.96}\right)^2 \geq \frac{0.56\cdot(1-0.56)}{n}$	Square both sides
$n \geq \frac{0.56\cdot(1-0.56)}{\left(\frac{0.03}{1.96}\right)^2}$	Multiply both sides by n and divide by $\left(\frac{0.03}{1.96}\right)^2$
$n \geq \left(\frac{1.96}{0.03}\right)^2 [0.56\cdot(1-0.56)]$	Dividing by $\left(\frac{0.03}{1.96}\right)^2$ is the same as multiplying by $\left(\frac{1.96}{0.03}\right)^2$
$n \geq (4268.444)\cdot(0.2464)$	
$n \geq 1051.74$	
$n \geq 1052$	Sample sizes must be whole numbers, so round up.

For a 3% margin of error, the sample size that is needed is just over $n = 1000$. This means that we can estimate a population proportion p, whatever the size of the population, if we have a sample size of about $n = 1000$. The margin of error does *not* depend on the size of the population; the margin of error depends on the level of confidence (so the z^*), the $\hat{p}$, and the sample size n. This answers the question of how organizations such as Gallup® can estimate population proportions with what appear to be very small sample sizes, such as $n = 1532$.

It is sometimes simpler to have done the algebra in general and to have the formula already solved for n.

Calculation of the Sample Size for a Given Margin of Error

$$n \geq \left(\frac{z^*}{ME}\right)^2 \left[\hat{p} \cdot (1 - \hat{p})\right]$$

Note: If the $\hat{p}$ is not known (as, for example, before data are collected), use $\hat{p} = 0.50$ since this gives the biggest possible value for $\hat{p} \cdot (1 - \hat{p})$ and gives a slightly larger sample size than is actually needed.

Interpretation of a Confidence Interval

There are a number of ways to correctly interpret a confidence interval. There are also a number of ways to go wrong and give an interpretation that makes no sense. Gallup® and other opinion polls report the estimate—the $\hat{p}$—along with the margin of error, with a note stating the confidence level. For our presidential approval example, Gallup® would report that the rating was 56% and then note that "one can say with confidence that the maximum margin of error is ± 3 percentage points," as in the quote at the beginning of this section. For our example of the students on the campus and the proportion who approve of the president's job performance, we could say (following this pattern): "With 95% confidence, we can say that 56% of students on the campus approve of how the president is doing his job, with a margin of sampling error of 9.7%"

A standard interpretation in textbooks reports the confidence interval, the confidence level, and the population proportion one is estimating. So, for the example we have been using, where the population is the collection of all students on one campus, we would say: "We can be 95% confident that the proportion of all students on the campus who approve of the president's job performance is between 46.3% and 65.7%"

Here is a kind of template for these two ways of interpreting a confidence interval.

Interpretation of a Confidence Interval I

"With 95% (or 90%, or 99%, whatever the level is) confidence, we can say that the proportion for [variable] for *[the population from which the sample was drawn]* is *[the sample proportion]* with a margin of error of *[margin of error]."*

Interpretation of a Confidence Interval II

"We can be 95% (or 90%, or 99%, whatever the level is) confident that the proportion for *[variable]* for *[the population from which the sample was drawn]* is between *[the lower limit]* and *[the upper limit]*. The margin of error is *[margin of error]."*

In both of these interpretations, what is made clear is the (i) confidence level, (ii) the variable being measured, (iii) the population to which we are inferring, and (iv) the results.

Reading technical reports In technical research reports, the interpretation is typically left to the reader, who is assumed to have taken a course in statistics and to understand what these intervals are. Often, the confidence intervals are given (without interpretation), or sample proportions $\hat{p}$ and the margins of error are given, especially when there are many estimates to be reported. However, in this situation, somewhere in the report, researchers make clear what confidence level (90%, 95%, 99%, or some other) is being used and what the population is. In the exercises, some of the common mistakes that are made in interpretation will be pointed out.

Interpretation in terms of a Repeated sampling process There is still another way to interpret confidence intervals.

Here is the Fathom output for a confidence interval for our Presidential Approval Rating example. The last sentence states that if the process were performed repeatedly then 95% of the confidence intervals that would be generated would capture the population proportion. This sounds a bit like the process for making a sampling distribution and is in fact related to it. What is meant by "95% of the confidence Intervals would capture the population proportion" can be illustrated by using a Swift Diagram.

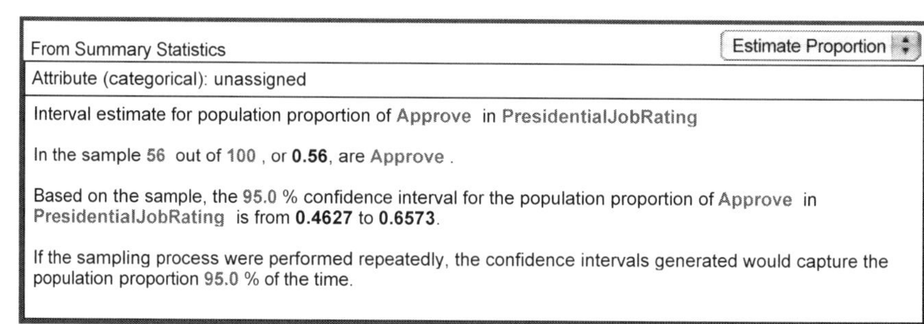

From Summary Statistics | Estimate Proportion

Attribute (categorical): unassigned

Interval estimate for population proportion of Approve in PresidentialJobRating

In the sample 56 out of 100 , or 0.56, are Approve .

Based on the sample, the 95.0 % confidence interval for the population proportion of Approve in PresidentialJobRating is from 0.4627 to 0.6573.

If the sampling process were performed repeatedly, the confidence intervals generated would capture the population proportion 95.0 % of the time.

How the Swift Diagram works; many lines that we usually do not draw. In this diagram, we have specified the "true population proportion" to be 0.54, and that is shown by the thick horizontal line. The thin, solid, vertical lines are possible confidence intervals that we could calculate depending on what our $\hat{p}$ happened to be. We cannot show all of them, of course, but we can show what would happen with a few; 95% of the confidence intervals will be based upon the $\hat{p}$ s that are reasonably likely. Confidence intervals calculated from any of these reasonably likely $\hat{p}$ s will "capture" or cross the true population proportion. The only confidence intervals that will miss will be those based on the 5% *rare* $\hat{p}$ s. This diagram shows two of them, shown as the thicker vertical confidence interval lines. For those rare 5% of the confidence intervals, the confidence interval will *not* cross the true population proportion. We can be 95% confident that our confidence interval is one of those based on a reasonably likely $\hat{p}$ and captures the p.

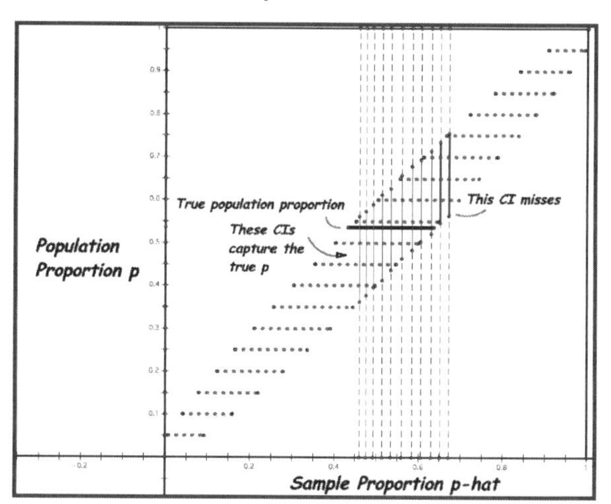

Summary: Confidence Intervals for Proportions

- One formal type of statistical inference is to calculate an ***interval estimate*** of the population proportion p based upon a sample proportion $\hat{p}$, where the sample is randomly drawn from the population.
  - This *interval estimate* is called a **confidence interval.**
  - The connection between the calculation of a *confidence interval* and the *reasonably likely* intervals based upon the sampling distribution of $\hat{p}$ can be seen in a Swift Diagram, in which the reasonably likely intervals are the horizontal lines, and the confidence intervals are the vertical lines. (See the example above.)

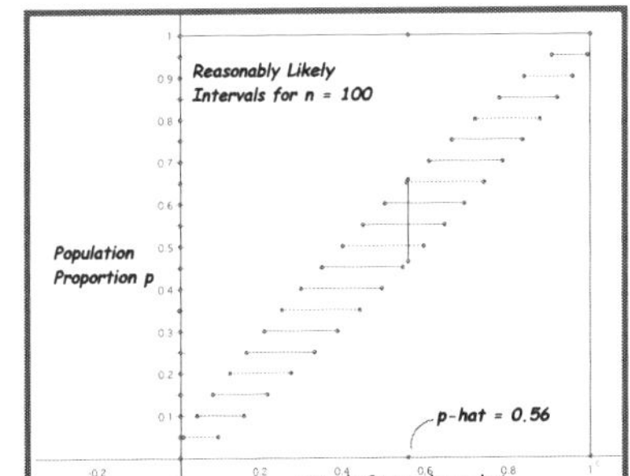

- ***Formula for a Confidence Interval for a Population Proportion p:*** $\hat{p} \pm z^* \sqrt{\frac{\hat{p}\cdot(1-\hat{p})}{n}}$, where

  $z^* = 1.645$ for a 90%, $z^* = 1.96$ for a 95% and $z^* = 2.576$ for a 99% confidence interval,
- The formula given above for a confidence interval works under the ***conditions*** that:
  - The sample of size n from which $\hat{p}$ was calculated is a simple random sample.
  - Both $n\hat{p} \geq 10$ and $n(1-\hat{p}) \geq 10$.
  - The size of the population is at least ten times the size of the sample.
- Another way of expressing the general form of a confidence interval is: ***estimate ± margin of error.***
- ***Margin of Error for a Confidence Interval:*** $ME = z^* \sqrt{\frac{\hat{p}\cdot(1-\hat{p})}{n}}$; the margin of error is the part of the formula that is subtracted from and added to the point estimate.
  - The larger the sample size, the smaller the margin of error, and hence the narrower the confidence interval.
  - It is possible, using algebra, to calculate the sample size needed to achieve a specific margin of error; the following formula may also be used: $n \geq \left(\frac{z^*}{ME}\right)^2 \left[\hat{p}\cdot(1-\hat{p})\right]$.
- ***Interpretation of a Confidence Interval*** There is a number of correct interpretations of a confidence interval and also many incorrect interpretations. Here are two templates for correct interpretations:
  - "With 95% (or 90%, or 99%, whatever the level is*)* confidence, we can say that the proportion for |*variable*| for |*the population from which the sample was drawn*| is |*the sample proportion*| with a margin of error of |*margin of error*|."
  - "We can be 95% (or 90%, or 99%, whatever the level is) confident that the proportion for |*variable*| for |*the population from which the sample was drawn*| is between |*the lower limit*| and |*the upper limit]*. The margin of error is |*margin of error*|."

§4.2 Hypothesis Testing: Handedness

Background: What is the percentage of left-handed people?

The usual answer to this question is that about 10% to 12% of the population is left-handed. However, the proportion has been found to vary across cultures. A web page by the Australian Broadcasting Corporation "News in Science" (www.abc.net.au/science/news/stories/s1196384.htm, Sept. 13, 2004) reports on research done by a team of Australian and American researchers.[5]

> There may be more left-handed people than we realise, an international study has found. If we include the number of people who throw a ball, strike a match or use a pair of scissors with their left hand, the researchers say the world looks more of a left-handed place.
>
> Australian researcher Sarah Medland of the Queensland Institute of Medical Research in Brisbane and team publish the research in the current issue of the journal *Laterality*.
>
> Left-handed people face problems in a world where most things, from scissors to can-openers and computers to power tools, are designed for right-handers.
>
> In the past left-handers faced even greater problems. Some schoolchildren were forced, under threat of the strap, to write with their right hand, regardless of their natural tendency. Medland and team hoped to shed light on the contribution of cultural factors like this on the distribution of handedness.

The researchers studied a sample of $n = 8528$ people from various parts of the world and measured "handedness" in various ways. A previous larger study had found that the proportion of left-handers varied in different countries and in different cultures. For example, they found that the proportion was 2.5% for Mexico and 12.8% for Canada.[6] Other researchers have found very low proportions of left-handedness in Chinese (3.5%) and in Japanese (0.7%) schoolchildren.

These researchers think that the differences in the proportion left-handed are probably cultural rather than genetic. That is, in many cultures there is great pressure to conform to the "norm" of right-handedness, but in some cultures left-handed or ambidextrous people are "allowed" to be left-handed, although they still have to live in a right-handed world and have to make accommodations for computer mice, keyboards, and can openers.

Our statistical question is:

Does the percentage of left-handers in the population of all students at one college differ from the "standard" 11 percent that is found generally?

This college has a very diverse student body, and so our reasoning is that the proportion left-handed may be lower or perhaps higher than the accepted 11%. We are unwilling to say which way, but we want to know if we have evidence that the proportion is different.

Language and Logic of Hypothesis Tests

Estimation and Testing: the Difference We have been doing *inference* (or generalization) from a sample $\hat{p}$ to a population p by calculating a *confidence interval estimate*.

---

5 Medland, S. E., Perelle, I, De Monte, V., and Ehrman, L. (2004) "Effects of culture, sex, and age on the distribution of handedness: An evaluation of three measures of handedness." *Laterality: Asymmetries of Body, Brain and Cognition,* **9**(3): 287–297

6 Medland, et. al (2004), page 288

We can calculate a confidence interval even when we have no idea at all about what the population proportion should be; all we need is a sample proportion $\hat{p}$ drawn from a sample of size n, and we can calculate a confidence interval to get an estimate of the population proportion p.

Hypothesis testing is different: it starts with a preconceived *idea* about the population proportion p and uses the sample proportion $\hat{p}$ to test whether our sample information (that is, the $\hat{p}$) is in agreement with our preconceived idea or (on the other hand) the $\hat{p}$ disagrees with our preconceived idea. In our example, our preconceived idea is that the proportion of left-handers in the population is $p = 0.11$. Our sample data is for the ***CombinedClassData*** for 2009. The proportion of students in our sample of who reported being left-handed was $\hat{p} = \frac{27}{317} = 0.0852$, about 8.5%. This 8.5% is somewhat lower than the 11% that we were expecting. Does this mean, necessarily, that the entire population of CSM has a lower percentage of left-handers than the 11% that is reported generally? No, it does not; remember that our sample proportion $\hat{p}$ varies from the population proportion p just because it is a calculated from a sample and not from the entire population. We need to use the idea of a *sampling distribution*. So how do we do this? There is some standard terminology and structure to hypothesis tests.

Combined Class Data 09

	DominantHand			Row Summary
	Right	Left	Ambidextrous	
	274 0.864353	27 0.0851735	16 0.0504732	317 1

S1 = count ( )
S2 = rowProportion

Null and Alternate Hypotheses. The first thing we do is to formalize our statistical question by expressing our preconceived idea in the form of two competing hypotheses about the population proportion p. One hypothesis is called the ***null hypothesis*** and usually represents what is commonly accepted; in our example, the null hypothesis is that the "population proportion of left-handers among CSM students is $p = 0.11$." A second hypothesis, which is counter (or contrary) to the *null hypothesis* and can be thought of as a *challenge* or a *claim against* the null hypothesis, is called the ***alternate hypothesis.*** For our statistical question, the *alternate hypothesis* is that the "population proportion of left-handers among CSM students *differs* from $p = 0.11$, that is $p \neq 0.11$." There is notation for these two hypotheses, and that notation is shown in the box on the next page. In that box, the symbol p_0 (pronounced "p naught" or "p zero") denotes the basic value that we have for the null hypothesis. In our example, p_0 has the value 0.11.

Notice that both the *null hypothesis* and the *alternate hypothesis* use the same value for p_0, but say different things about the p_0; the null hypothesis says that the population proportion p is equal to the p_0, and the alternate hypothesis says that the population proportion p does *not* equal p_0.

Notation for Null and Alternate Hypotheses for Population Proportions

The *null hypothesis* that the population proportion p is equal to p_0 ,written: $H_0: p = p_0$

The *alternate hypothesis* that the population proportion p is not equal to p_0 ,written: $H_a: p \neq p_0$

[***Note:*** This alternate hypothesis is known as ***two-sided*** because we do not specify in which way the population proportion differs from the p_0—whether p is larger or smaller than p_0. ***One-sided*** tests are discussed later.]

For our example, we would write: $\begin{matrix} H_0: p = 0.11 \\ H_a: p \neq 0.11 \end{matrix}$.

Set-up and Calculations for a Hypothesis Test: First Four Steps

We can think of a conducting a hypothesis test as a having ***five steps***; the fifth and last step is interpretation, and in these **Notes** that step is so important that it has its own sub-section.

The ***first step*** in any hypothesis test is to specify clearly and correctly the null and alternate hypotheses. Then where do we go from there? Checking the conditions is the ***second step***. We check the conditions so that we can be sure that we can use the *sampling distribution implied by the null hypothesis*. For our example, the sampling distribution is the sampling distribution for $\hat{p}$ s for random samples of $n = 317$ drawn from a population in which $p_0 = 0.11$. What is that sampling distribution? We can describe it:

Shape: The shape is approximately Normal because the ***conditions*** for Normality $np_0 = 317(0.11) = 34.87 > 10$ and $n(1-p_0) = 317(1-0.11) = 317(0.89) = 282.13 > 10$ are met.

Center: $\mu_{\hat{p}} = p_0 = 0.11$

Spread: $\sigma_{\hat{p}} = \sqrt{\dfrac{p_0(1-p_0)}{n}} = \sqrt{\dfrac{0.11\cdot(1-0.11)}{317}} \approx 0.01757$

From this we can work out the interval of reasonably likely $\hat{p}$ s using

$p_0 \pm 1.96\sqrt{\dfrac{p_0(1-p_0)}{n}} = 0.11 \pm 1.96(0.01757) = 0.11 \pm 0.03444$, which implies an interval:

$0.0756 < \hat{p} < 0.1444$. The picture of the sampling distribution (shown below) for our example looks like this. There is a general principle involved here; that general principle is that a hypothesis test is *always* based upon the sampling distribution implied by the null hypothesis.

Logic of a Hypothesis Test: This picture gives us an idea of how the testing will proceed. If the $\hat{p}$ that we get from our sample is one of the *rare* $\hat{p}$ s according to the sampling distribution, we will consider getting this rare $\hat{p}$ as evidence *against the null hypothesis.* But if (on the other hand) the $\hat{p}$ that we get from our sample is one of the ***reasonably likely*** $\hat{p}$ s according to the sampling distribution then we will consider that our test has *not* provided us with evidence against the null hypothesis. The logic of the testing is that the sampling distribution of $\hat{p}$ s shows us which $\hat{p}$ s are reasonably likely *if the null hypothesis is true* (that is, if $p = p_0$) and also what $\hat{p}$ s are rare *if the null hypothesis is true* (that is, if $p = p_0$). So, if we get one of the *rare* $\hat{p}$ s, we count that as evidence against the truth of the null hypothesis. If we get a reasonably likely $\hat{p}$ then we cannot count that $\hat{p}$ as evidence against the H_0. If we get a *reasonably likely* $\hat{p}$, it does not necessarily mean that the null hypothesis is true; it only means that we have no evidence that it is *not* true.

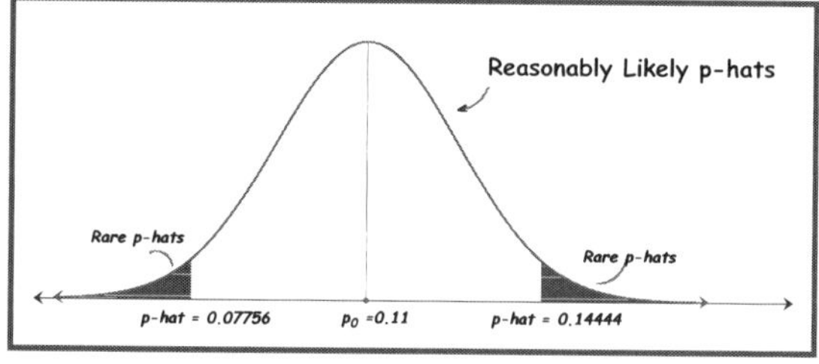

It is the *null hypothesis* that is the basis of the test; the *alternate hypothesis* serves as a challenge to the null hypothesis. Our data shows us whether or not the challenge will be successful. For our example, we got $\hat{p} = 0.0852$. This value happens to be within the interval of reasonably likely $\hat{p}$ s, which is $0.0756 < \hat{p} < 0.1444$. Therefore, for our example, we do *not* have sufficient evidence from our sample against the null hypothesis. Since our $\hat{p} = 0.0852$ is a reasonably likely $\hat{p}$, we cannot challenge the accepted standard of $p_0 = 0.11$ for the proportion of left-handers in the population of students at CSM.

Calculating the Test Statistic. We have come to our conclusion by noting that our $\hat{p} = 0.0852$ is in the interval of reasonably likely $\hat{p}$ s. However, hypothesis testing has a certain structure to it, and the ***third step*** in carrying out a hypothesis test is to calculate the ***test statistic.*** A test statistic is something that is *always* calculated for any hypothesis test, but the nature of what is calculated differs depending on what is being tested. For tests that involve one $\hat{p}$, the test statistic is the z score based upon the sampling distribution for the null hypothesis. The general form of a z score is

$z = \dfrac{\text{observed sample value} - \text{mean}}{\text{standard deviation}}$; we have used it with the Normal model to tell us how far and in what direction a value is—in standard deviation units—from the mean. Now, we are using sampling distributions, so the values are $\hat{p}$ s and the mean of our sampling distribution is $\mu_{\hat{p}} = p_0$ and the standard deviation is $\sigma_{\hat{p}} = \sqrt{\dfrac{p_0(1-p_0)}{n}}$. Our z score takes the form shown just below.

Test Statistic for Testing a Single Population Proportion p

$$z = \frac{\hat{p} - \mu_{\hat{p}}}{\sigma_{\hat{p}}} = \frac{\hat{p} - p_0}{\sqrt{\dfrac{p_0(1-p_0)}{n}}}$$ where p_0 is the value used in the null hypothesis H_0: $p = p_0$.

Below, the *test statistic* for our example is worked out and a graph shows what we have calculated and how it relates to the reasonably likely and rare $\hat{p}$ s.

For our example, where $\hat{p} = 0.0852$ and $p_0 = 0.11$, and $n = 317$, we would calculate:

$$z = \frac{\hat{p} - p_0}{\sqrt{\dfrac{p_0(1-p_0)}{n}}} = \frac{0.0852 - 0.11}{\sqrt{\dfrac{0.11 \cdot (1-0.11)}{317}}} = \frac{-0.0248}{0.01757} \approx -1.41$$

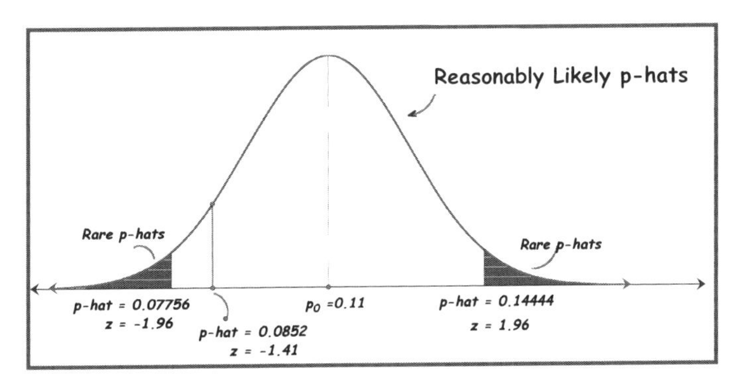

For $\hat{p} = 0.0852$ (or 8.52% left-handers) we get $z = -1.41$; this tells us that our $\hat{p} = 0.0852$ is to the left of the mean of the sampling distribution of $p_0 = 0.11$, and it is 1.41 standard deviations to the left. This also tells us (just from the z score) that our $\hat{p} = 0.0852$ is within the interval of reasonably likely $\hat{p}$ s, since our $z = -1.41$ is between $z = -1.96$ and $z = 1.96$. We knew that our $\hat{p}$ was reasonably likely by seeing that our $\hat{p} = 0.0852$ was inside the reasonably likely interval.

The calculation of the test statistic puts our conclusion in the world of the Normal Distribution Chart. A test statistic that is greater than $z = 1.96$ or less than $z = -1.96$ would be rare, but a test statistic between $z = -1.96$ and $z = 1.96$ will be reasonably likely.

Reasonably Likely p-hats and the Value of the Test Statistic

For a given $\hat{p}$, if the *test statistic* calculated by $z = \dfrac{\hat{p} - p_0}{\sqrt{\dfrac{p_0(1-p_0)}{n}}}$ is within the interval $-1.96 \le z \le 1.96$

then the $\hat{p}$ is *reasonably likely* for the p_0 and the n used in the calculation. If the test statistic z is outside the interval $-1.96 \le z \le 1.96$ then the $\hat{p}$ is *rare.*

Warnings. In Unit 3, there were many things that either had the symbol r or names that began with r, and you needed to keep them straight. Now we are about to have the same problem with the letter p. We have encountered $P(\ )$, $\hat{p}$, p, and p_0—all of them referring to some kind of probability or proportion (hence the letter p) but used in slightly different ways. Now comes still another one; we are about to discuss the ***p-value,*** which is easily confused with all the other "p"s. All of these are used commonly; again, your job is to keep them straight in your thinking. Second warning: the next paragraphs will probably have to be read more than once! These paragraphs concern the calculation of a **p-value** and its meaning. It is not simple.

Calculation of a p-value. This is the ***fourth step.*** We have calculated similar probabilities before. The *p-value* is the probability of getting a random sample with a $\hat{p}$ *as extreme or more extreme than the one that we actually got.* The calculation is another "given a value—find a proportion" type of calculation that uses the Normal distribution, but we will have to see how the "*as extreme or more extreme* "comes in. Using our example of the percentage of left-handed students in our sample 8.52% or $\hat{p} = 0.0852$, we note that this $\hat{p}$ is less than the $p_0 = 0.11$ and ask (first of all): what is the probability that, if the null hypothesis H_0 is true so $p_0 = 0.11$, we end up seeing a $\hat{p} = 0.0852$ *or less*? Anything less than $\hat{p} = 0.0852$ would be "*more extreme*" or farther away from $p_0 = 0.11$ than what we got.

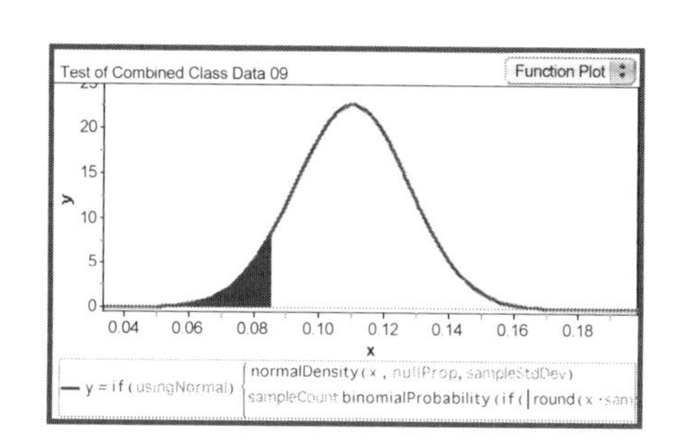

- We should draw a sketch; here is the Software drawing for this part of the problem.
- Then we should calculate a z score remembering that we are dealing with a sampling distribution, so we get

$$z = \frac{\hat{p} - p_0}{\sqrt{\dfrac{p_0(1-p_0)}{n}}} = \frac{0.0852 - 0.11}{\sqrt{\dfrac{0.11 \cdot (1 - 0.11)}{317}}} = \frac{-0.0248}{0.01757} \approx -1.41,$$

  which we recognize as our *test statistic.*
- Then we consult the Normal Distribution Chart and find that $P(z < -1.41) = 0.0793$. That is the area shown shaded here.

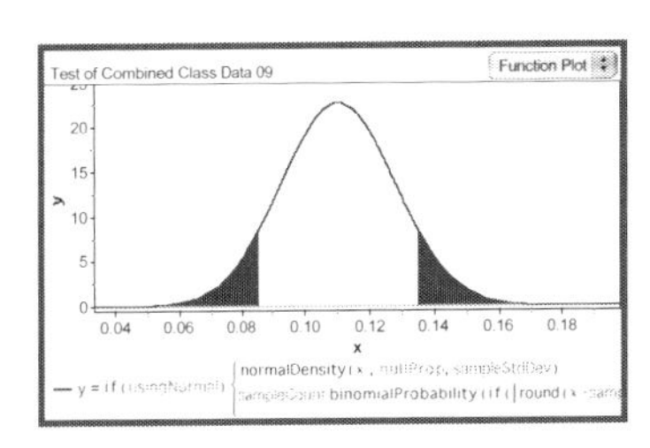

More extreme and its meaning. However, since we are using random sampling, "*more extreme*" could also mean that we get a $\hat{p}$ that is more extreme on the positive side of $p_0 = 0.11$ than the one that we got on the negative side. So our picture for "*more extreme*" should look like the picture on the right. This is not hard to calculate because of the symmetry of the Normal distribution. We multiply the area in the left shaded area by two. In probability notation, we have

$P(\hat{p} \text{ that is more extreme}) = P(Z > |-1.41|) = 2P(Z > 1.41) = 2 \times 0.0793 = 0.1586$. The *capital* Z in this notation refers to the horizontal scale for the standard Normal distribution and not our z score; *our z* score in this example is – 1.41. The ***p-value*** for our example is 0.159, that is: p-value = 0.159, or approximately 0.16.

Many calculations but the same result. It appears that we are calculating several things and that they are all related. This is true. We can see whether our $\hat{p}$ is reasonably likely or rare; we can calculate the test statistic (getting the $\hat{p}$ in terms of standard deviation units); we can calculate the p-value to see the probability that we get this $\hat{p}$ or one more extreme. All of these calculations are related, and the relationships are spelled out in the arrowed bullets below.

- A sample proportion $\hat{p}$ that is ***rare*** will have a ***test statistic*** whose absolute value is relatively ***large*** and therefore a ***p-value*** that is ***small***. (The p-value will be *smaller* than 0.05.)
- A sample proportion $\hat{p}$ that is ***reasonably likely*** will have a ***test statistic*** whose absolute value is relatively ***small*** and therefore a ***p-value*** that is ***large***. (The p-value will be *larger* than 0.05.)

In any particular hypothesis test, all these calculations should agree: either they should show evidence consistent (or in accord) with the null hypothesis or evidence against (or contrary to) the null hypothesis.

Here is the reasoning behind the relationships shown above: we will think about how the *p-value* is going to work; if we get a sample proportion $\hat{p}$ that is *rare* then the test statistic will either be bigger than $z = 1.96$ or smaller than $z = -1.96$. Then the picture like the one just above will have very small shaded areas on either side, and the p-value will be very small. In fact, if we have $\hat{p}$ that is rare, we know that the p-value must be less than 0.05, since the tails amount to just 5%. On the other hand, if we have (as we do here in our example) a $\hat{p}$ that is *reasonably likely* then we know that the area in the shaded part of the picture must be bigger than 0.05, and so the p-value must bigger than 0.05. There are some relationships between p-values, test statistics, and reasonably likely or rare $\hat{p}$ s.

Since we will use a *p-value* in most of our calculations with hypothesis tests, , and because the idea of a p-value is extremely important, we give a general definition, which is found on the next page. It will be a good thing to think carefully about this idea; it is one of the most difficult concepts in the course.

Definition of p-value for a hypothesis test

The p-value for a hypothesis test is the probability of getting a *test statistic as extreme as or more extreme than* the one observed from the null hypothesized value for the population when randomization is used.

This definition is constructed so that it will apply to all the hypothesis testing we will do. In our example here, the test statistic is a z score, and we are thinking of a sample of students as being a simple random sample. We want our definition to apply to data that come from an experiment, where randomization would mean that the experimental units are randomly allocated to the treatments.

The Fifth Step: Interpreting Hypothesis Tests

Logic of the hypothesis test, again. It may be wise to reread the section above with this same title because the way we interpret a hypothesis test depends on this logic. Recall that the data in the form of a sample proportion $\hat{p}$ and the alternate hypothesis form a kind of challenge to the null hypothesis; the challenge can be successful—we have evidence against the null hypothesized p_0—or it can fail: we have evidence, but it is not strong enough, considering sampling variation. If the $\hat{p}$ is rare, and *therefore* the test statistic is relatively large in absolute value, the p-value will be small: *then* we have sufficient evidence *against* the null hypothesis, and the challenge is successful. On the other hand, if the sample proportion $\hat{p}$ is reasonably likely, the test statistic will be relatively small in absolute value, and the p-value is relatively large: *then* we do *not* have sufficient evidence against the null hypothesis over and above sampling variation, and the challenge to the null hypothesis fails. We live with the null hypothesis.

An interpretation of our example. We found that our sample proportion $\hat{p} = 0.0852$ was reasonably likely if $p_0 = 0.11$ is true, and our test statistic $z = -1.41$ is not large in absolute value (it was *not* beyond $z = -1.96$) and the p-value = 0.16 was relatively large (that is, it was larger than 0.05). Therefore, our evidence does *not* successfully challenge the null hypothesized value of $p_0 = 0.11$. We should add that our challenge with the sample size we used (n = 317) was not successful.

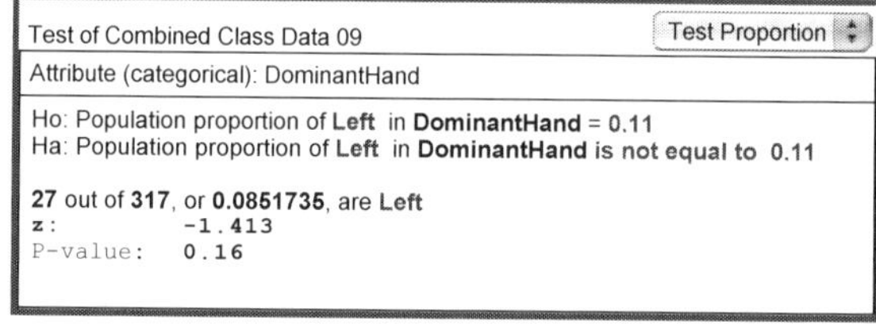

In the context of our data, with a $\hat{p} = 0.0852$, which is reasonably likely if $p_0 = 0.11$, a test statistic of $z = -1.41$, and a p-value of 0.16, we have *insufficient* evidence to say that the percentage of CSM students who are left-handed is different from the usual standard of 11%. This is a good interpretation because it makes clear the hypothesis ("the percentage of CSM students who are left-handed is different from the usual standard of 11%"), the evidence ("a $\hat{p} = 0.0852$, which is reasonably likely if $p_0 = 0.11$, a test statistic of $z = -1.41$, and a p-value of 0.16"), and our conclusion ("we have insufficient evidence to say…").

However, there are certain conventions of language that are used with hypothesis tests. One has to do with "rejecting" the null hypothesis, and the other has to do with the term "statistical significance."

Rejecting the Null Hypothesis—Or Not A very common way of speaking about the conclusion to a hypothesis test speaks about "rejecting the null hypothesis." If we have evidence against the null hypothesis—a very small *p-value*, for example—then we could say that we *reject* the null hypothesis H_0. If, instead of finding twenty-seven left-handers in our sample, we had found just twenty-one then our results would be as the Software output shown here. In this scenario, we *would* have evidence *against* the null hypothesis that the population proportion of left-handers is 11%. Notice how all the evidence is consistent: a test statistic whose absolute value $|-2.49| = 2.49$ is larger than 1.96 and a p-value that is smaller than 0.05. The $\hat{p}$ we got cannot be the result of sampling variation.

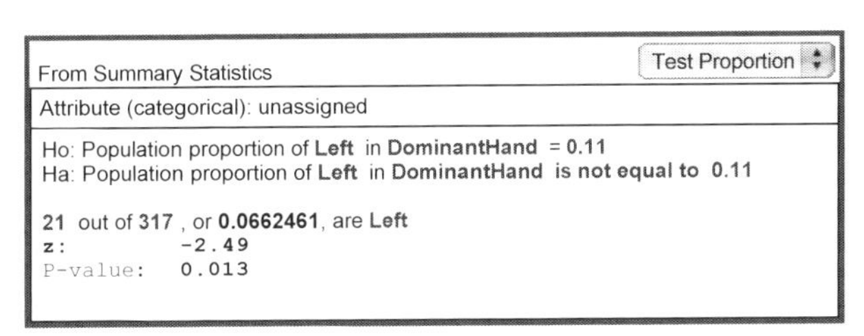

Our example did not in fact look like this. We did not have sufficient evidence against the null hypothesis. We would say that we *fail* to reject the null hypothesis. The percentage of left-handers that we saw can easily be explained as a reasonably likely result when we sample randomly from a population. Notice that we do not say that we "accept" the null hypothesis; the accepted terminology is that we "fail to reject H_0." We brought a challenge to the H_0 that did not work.

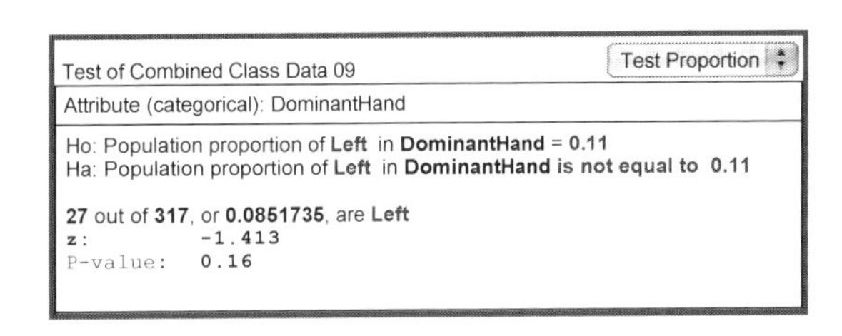

Statistically Significant—Or Not. The term ***statistically significant*** has a very specific meaning. Once again, we think of the null hypothesis: if the null hypothesis is ($p = 0.11$), there is a reasonably likely range of sample proportions $\hat{p}$ that we expect just by random variation. However, if our $\hat{p}$ is outside that range then we believe that something more than sampling variation going on. So, in the scenario above, if we had $\hat{p} = \frac{21}{317} \approx 0.066$, we would say that this test was "statistically significant." In our actual test with the sample proportion of $\hat{p} = 0.0852$ we would say that the test was *not* statistically significant. The word "significant" should be modified with "statistically"; moreover, the term does *not* mean necessarily that a result is important or profound. It merely means that what we saw could not have been explained by sampling variation from random sampling or allocation. Some publications try not to use the word "significant" and instead use the term ***statistically detectable.***

How small is small? How big is big? We have said that a small p-value and a large test statistic means we have a *rare* sample proportion $\hat{p}$; these three go together and are evidence against the null hypothesis. However, this raises the question: how small does a *p-value* have to be, and how large does the test statistic need to be? We have already hinted at the standard above. Notice that in the box below, both the test statistic and the *p-value* are mentioned. These two go together; if you get a small z and a big p-value, you have made some mistake. The box below gives some standards that depend upon taking 5% as rare.

Standards: How small is small, how big is big? If *rare* means the 5% most extreme $\hat{p}$ s, then:

- If the p-value < 0.05 and the test statistic z is *outside* the interval $-1.96 < z < 1.96$ then we *have* evidence against the null hypothesis, and
  - the test *is statistically significant* and
  - we can *reject* the null hypothesized p_0.
- If the p-value > 0.05 and the test statistic z is *inside* the interval $-1.96 < z < 1.96$ then we do *not* have sufficient evidence against the null hypothesis, and
  - the test is *not statistically significant* and
  - we *fail* to *reject* the null hypothesized p_0.

Interpreting our example, again, and with caution. We must be clear on four points:

- What the population is
- The variable in the null hypothesis and the hypothesized value
- The evidence: the p-value can be mentioned, or the value of the test statistic can be mentioned, or both
- Whether the evidence against the null hypothesis is sufficient or not, and here we can say:

  Whether the test is statistically significant or not and

  Whether we reject the null hypothesis or not

Therefore, a good interpretation of our example could go something like this:

"We found that 8.52% of our sample of $n = 317$ CSM students were left-handed. This 8.52% is not significantly different from the hypothesized proportion of 11%. The calculations give a p-value of 0.16, and so 8.52% is a reasonably likely sample result if 11% of all CSM students are left-handed. So we fail to reject the standard that 11% of CSM are left-handed."

There is one more important thing that should be added to this interpretation, however. One of the conditions for doing these tests is that the sample be a simple random sample (a SRS). We know that our sample of students is *not* a simple random sample of CSM students, and so we should add a cautionary note to our interpretation. We should say that our results must be treated with caution. Now it may be that with respect to handedness, our sample is as good as a SRS, but we actually do not know whether it is or not.

Other Standards

We have been taking the definition of ***rare*** to be the most extreme 5% of $\hat{p}$ s, leaving 95% as ***reasonably likely*** as in the picture to the right. However, although this 5% is very common (even something like an "industry standard"), it is not the only definition of *rare* possible. We could define *rare* to be the most extreme 2% or the most extreme 1%. The choice is left to the researcher doing the analysis.

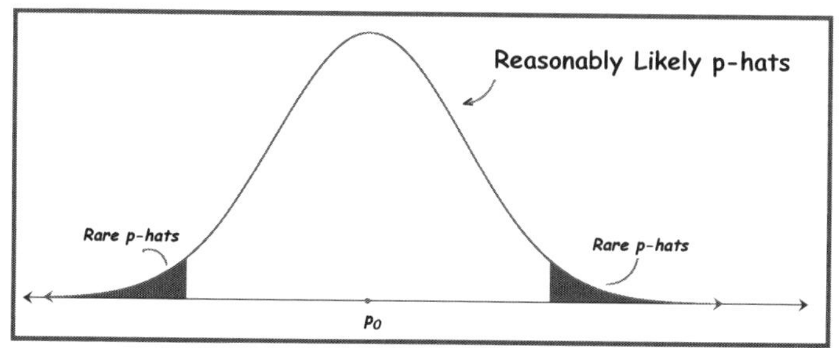

Because that choice is available, this choice has a name and symbol. It is called the ***level of significance,*** and its symbol is the Greek letter α. If *rare* means the most extreme 5% of $\hat{p}$ s then we have chosen $\alpha = 0.05$.

In research publications, the phrase "significant at the 1%"means that the hypothesis test would be statistically significant if α=0.01. Since the choice of standard is left to researchers, it is quite common now to report p-values rather than significance levels, leaving the choice of the level of significance to the reader.

We can put these considerations together and revise the "How small is small" box.

Other standards: If we define *rare* as the 100α percent most extreme $\hat{p}$ s then:

- If the p-value < α and the test statistic z is *outside* the interval $-z^*_{\alpha/2} < z < z^*_{\alpha/2}$ then we *have* evidence against the null hypothesis, and the hypothesis test *is statistically significant* and we can *reject* the null hypothesized p_0.
- If the p-value > α and the test statistic z is *inside* the interval $-z^*_{\alpha/2} < z < z^*_{\alpha/2}$ then we do *not* have sufficient evidence against the null hypothesis, and the hypothesis test is *not statistically significant* and we *fail* to *reject* the null hypothesized p_0.

Where $z^*_{\alpha/2}$ refers to the z^* from the Normal Distribution Chart, that cuts off the lowest $\alpha / 2$ and highest $\alpha / 2$ of the standard Normal distribution. If α=0.01, for example, $z^*_{\alpha/2} = 2.576$.

Summary: Hypothesis Tests

- The basic notion of a hypothesis test is to see if sample data (in this section, a sample proportion $\hat{p}$) agrees with or disagrees with a preconceived idea (in this section, an idea about a population value p), taking account of the fact that the sample values may vary by random sampling variation.
  - The preconceived idea is formalized into the ***null hypothesis*** and the ***alternate hypothesis***.
  - The test is based upon the sampling distribution for the ***test statistic*** assuming the null hypothesis true.
- ***Definition of p-value for a hypothesis test.*** The p-value for a hypothesis test is the probability of getting a *test statistic as extreme as or more extreme than* the one observed from the null hypothesized value for the population when randomization is used.
  - Small p-values indicate that, given the null hypothesis, the sample result is relatively *rare*.
  - Large p-values indicate that, given the null hypothesis, the sample result is reasonably *likely*.
- ***Standards.*** Researchers or readers of research are the ones who decide how small a p-value is necessary to declare that a sample result is *rare* (as opposed to *reasonably likely*). These standards are known as ***levels of significance***, expressed as proportions and denoted by the Greek letter α. The most common standard is $\alpha = 0.05$. This means that the ***test statistic*** will be rare with a probability of 5% when the null hypothesis is true if the hypothesis test were repeatedly done.
- When the evidence from a hypothesis test is *rare* enough according to the significance level chosen, the hypothesis test is deemed ***statistically significant.*** A statistically significant result means that the sample results must have been rare enough to provide evidence against the null hypothesis.

- ***Interpretation of hypothesis tests.*** A good interpretation should include the following:
  - A clear description of the *population*, the *variable* in the null hypothesis, and the hypothesized *values*.
  - The evidence: the p-value can be mentioned, or the value of the test statistic can be mentioned, or both.
  - Whether the evidence against the null hypothesis is sufficient or not, and here we can say: Whether the test is statistically significant or not and whether we reject the null hypothesis or not.

Summary: Five-Step Hypothesis Testing

Step 1:	*Set up the null and alternate hypotheses using an idea for the population proportion* p_0*.* $H0: p = p_0$ $Ha: p \neq p_0$
Step 2:	*Check the conditions for a trustworthy hypothesis test.* *The sample must be a simple random sample from a population ten times n.* *Both* $n\, p_0 \geq 10$ *and also* $n(1 - p_0) \geq 10$*.*
Step 3:	*Calculate the test statistic* $z = \dfrac{\hat{p} - p_0}{\sqrt{\dfrac{p_0(1-p_0)}{n}}}$ *using the* $\hat{p}$ *from the sample and the hypothesized* p_0*.*
Step 4:	*Calculate the p-value, by getting* $P(\hat{p} \text{ more extreme}) = 2P(Z \geq \mid z \mid)$ *where* $\mid z \mid$ *is the absolute value of the test statistic you calculated in Step 3.*
Step 5:	*Evaluate the evidence that the p-value and the test statistic give you to determine whether your test successfully challenges the null hypothesized population proportion p0 or not. Give an interpretation in the context of the data using the terminology of "statistical significance" and "rejecting the null hypothesis."*

§4.3 Comparing Proportions, or What Is the Difference?

Tattoos Again

The first statistical question we asked was: *"Are male or female students more likely to have a tattoo?"* We have looked at various samples (Penn State and from California), but here are the data for the ***CombinedClassDataY09***. To answer our statistical question, one of the first things we did was to compare the proportions of males and females having a tattoo. Using the notation from §1.2, we would write $P(T \mid M) = \frac{23}{142} \approx 0.162$, and $P(T \mid F) = \frac{55}{175} \approx 0.314$. So it appears that females (in a college in California in 2009) are more likely to have a tattoo than males.

Combined Class Data 09

		Tattoo		Row Summary
		N	Y	
Gender	F	120 0.685714	55 0.314286	175 1
	M	119 0.838028	23 0.161972	142 1
Column Summary		239 0.753943	78 0.246057	317 1

S1 = count ( )
S2 = rowProportion

But are we certain? We have since learned that to *generalize* from a sample—to *infer* from a sample to a population—takes some special techniques; we cannot just say that because we saw it in a sample, it *must* be true for a population that we have in mind. We can generalize but only after doing some work and only under certain conditions. Now we come to two very common questions.

Two Questions: The first question is:

Is there any difference between two groups in the population from which the samples were drawn?

For our example, we are asking whether there really is a difference in the percentage of females and the percentage of males who have tattoos among all CSM students. For this question, the "Is there any difference?" question, we will use a hypothesis test.

The second question is:

How big is the difference between two groups and in what direction?

In our sample, it looks like the difference is about 15%, but we would like to have an estimate for the size and direction of the difference in the population. For this question we will *estimate* using a *confidence interval.*

More new notation: Since we have two groups instead of just one, we need some new notation. We need to distinguish between the proportions for each of the groups, both for our sample data and also for the population. For our example, there are different sample sizes for the males and the female (and the sample sizes do *not* have to be the same!), so in our calculations, we want to distinguish between them. It will be convenient to use subscripts and write $n_M = 142$ and $n_F = 175$. For the sample proportions, rather than having just one $\hat{p}$, we now have two, so we would write $\hat{p}_M \approx 0.162$, and $\hat{p}_F \approx 0.314$. And we must not forget the population proportion symbols; remember that even though we never know the exact values for the population proportions, we must still have symbols for them so that we can talk about them. For our example, we would write p_M and p_F—these not wearing hats because they are population proportions.

Advice: you will find it best to assign letters to the subscripts of the two groups to help keep track of which group is which. In the general formulation, since we do not know what we will use them for, we will have to use numerical subscripts. Try to use meaningful letters in any actual application.

Notation for Comparing Groups:

	Sample Size	*Sample Proportion*	*Population Proportion*
Group 1:	n_1	$\hat{p}_1$	p_1
Group 2:	n_2	$\hat{p}_2$	p_2

A sampling distribution for differences of sample proportions. Everything that we did with confidence intervals (in §4.1) and hypothesis testing (in §4.2) depended on having a sampling distribution. Now, we will be interested in the sampling distribution for differences of sample proportions, or $\hat{p}_1 - \hat{p}_2$. What does this mean? In our example above, $\hat{p}_F - \hat{p}_M \approx 0.314 - 0.162 = 0.152$. These kinds of differences are what go into our sampling distribution.

(You may wonder whether it matters which proportion comes first. It does not matter—if you reversed the order you would get a negative number in this case but the same numerical difference. But you *do* have to be consistent; if you start with $\hat{p}_F - \hat{p}_M$ then everything must be in terms of $\hat{p}_F - \hat{p}_M$ and not $\hat{p}_M - \hat{p}_F$.)

Sampling Distribution for the Difference of Sample Proportions

The sampling distribution of $\hat{p}_1 - \hat{p}_2$ for samples of size n_1 and n_2 drawn randomly from a population with population proportions p_1 and p_2 will have shape, center, and spread according to:

Shape: The shape is approximately Normal under the ***conditions*** that n_1p_1, n_2p_2, $n_1(1-p_1)$, $n_2(1-p_2)$ are all 5 or more.

Center: The mean is $\mu_{\hat{p}_1-\hat{p}_2} = p_1 - p_2$.

Spread: The standard deviation is $\sigma_{\hat{p}_1-\hat{p}_2} = \sqrt{\dfrac{p_1(1-p_1)}{n_1} + \dfrac{p_2(1-p_2)}{n_2}}$.

The next thing is to see how we actually use this sampling distribution. We will take up our first question: "Is there actually a difference of proportions in the population?" Perhaps the sample difference we have seen is actually just the result of random sampling variation that we naturally have when we randomly sample. (Notice that, for the moment, we are thinking of our sample as a SRS.) We answer the question with a hypothesis test, and we will use the same five steps but with different formulas.

Example: Is there any difference? Our specific question for the tattoo data is "Is there really a difference in the population proportions of males compared with female students at CSM who have a tattoo?" Realistically, there probably is *some* difference in the tattooed population proportion of males and females. We would be very surprised to find *exactly* the same proportions in any collection. For simplicity in doing the test, however, we will take as our preconceived idea that there is *no difference;* that is our standard. (When we get to confidence intervals then we can estimate the size of any difference.) What does it mean that there is no difference in the population?

"No difference" would mean $p_F = p_M$, that the two population proportions are equal. In fact, if it is true that the proportion of females having is the same as the proportion of males having a tattoo then it must be the same proportion, and we could write $p_F = p_M = p$, where *p* is the common proportion.

The idea that the population proportions of the two groups are the same will form the ***null hypothesis*** for all of the hypothesis tests that we will do with two groups. (It is possible to have other null hypotheses; we could test whether the population proportions differ by 0.12, for example. The most common null hypothesis is one of "no difference.") Having made this decision, we are now ready to go through our ***five steps*** of a hypothesis test with our example, after one preliminary step that we already made.

We have to decide whether what proportions we are comparing. Here we could compare the proportions of males and females that *do* have a tattoo or the proportion of males and females that *do not* have a tattoo. The results would be the same, but we must make a choice. The one that we choose ("Yes"—tattooed, in our example) is called a ***success*** and defines what counts go into the numerators of our proportions.

***Step 1:* Setting Up the Hypotheses**: There are two equivalent ways of writing the null and alternate hypotheses:

$$H_0: p_F = p_M \quad \text{or} \quad H_0: p_F - p_M = 0$$
$$H_a: p_F \neq p_M \quad \quad \quad H_a: p_F - p_M \neq 0$$

(where the proportions refer to the proportions having a tattoo)

***Step 2:* Checking the Conditions:** To check the conditions, we must calculate one more thing. The sampling distribution is based on the null hypothesis, which says that $p_F - p_M = 0$. If it is true that $p_F - p_M = 0$ then $p_1 = p_2 = p$. Of course we do not know this *p*, so our best estimate is the *overall proportion of students who have a tattoo* in our sample. This estimate of *p* is designated $\hat{p}$, written without any subscript, since it refers to both groups. For our example, $\hat{p} = \dfrac{\text{total number of students with a tattoo}}{\text{Total number of students}} = \dfrac{78}{317} \approx 0.246$.

Combined Class Data 09

		Tattoo N	Tattoo Y	Row Summary
Gender	F	120 0.685714	55 0.314286	175 1
	M	119 0.838028	23 0.161972	142 1
Column Summary		239 0.753943	78 0.246057	317 1

S1 = count ( )
S2 = rowProportion

One condition for our test is that $n_1\hat{p}$, $n_2\hat{p}$, $n_1(1-\hat{p})$, $n_2(1-\hat{p})$ are five or greater. We can calculate these: for example: $n_M\hat{p} = (142)\cdot(0.246) \approx 34.9 > 5$. In practice, it suffices to calculate only those that may appear problematic. Here the calculation we have chosen is for the cell with the smallest value. However, notice that we use an "expected" value for *p* of $\hat{p} = 0.246$ for these calculations; the $\hat{p} = 0.246$ is based on our best estimate of the overall proportion of students who have a tattoo.

We can also now calculate the standard deviation of the sampling distribution using the formula $\sqrt{\hat{p}(1-\hat{p})\left[\dfrac{1}{n_1} + \dfrac{1}{n_2}\right]}$; this is really the same formula as in the box above but simplified because $p_1 = p_2 = p$, and *p* is estimated with $\hat{p}$.

For our example, where $\hat{p} \approx 0.246$, we calculate

$$\sqrt{\hat{p}(1-\hat{p})\left[\frac{1}{n_1}+\frac{1}{n_2}\right]} = \sqrt{0.246(1-0.246)\left[\frac{1}{175}+\frac{1}{142}\right]} \approx 0.0486.$$

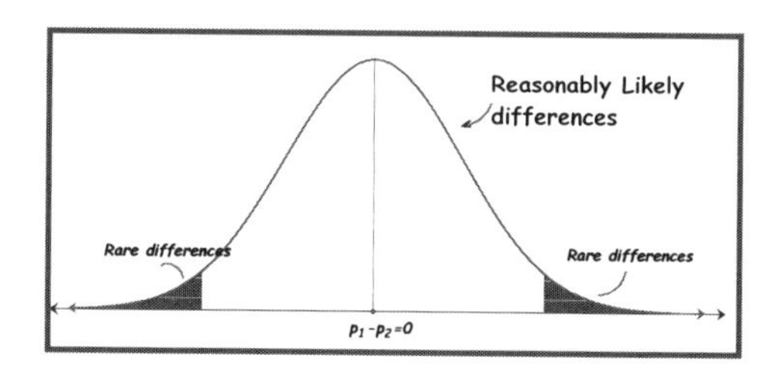

Then the sampling distribution looks like this Normal distribution, where the interval of reasonably likely differences is $0 - 1.96(0.0486) < \hat{p}_F - \hat{p}_M < 0 + 1.96(0.0486)$ or $-0.0953 < \hat{p}_F - \hat{p}_M < 0.0953$. For this example, the shaded (rare) region on the left is to the left of $\hat{p}_F - \hat{p}_M = -0.0953$ and the shaded (rare) region on the right is to the right of $\hat{p}_F - \hat{p}_M = 0.0953$. These are *differences of sample proportions. Our* actual sample difference happened to be $\hat{p}_F - \hat{p}_M \approx 0.314 - 0.162 = 0.152$. You should easily be able to determine whether *our* difference of proportions is reasonably likely or rare. (Is 0.152 inside or outside the reasonably likely interval?)

***Step 3:* Calculating the Test Statistic:** The test statistic is just a z score, and its formula comes from the fact that our null hypothesis is $p_F - p_M = 0$, and that therefore (under the null hypothesis) $p_1 = p_2 = p$, which we are estimating from our sample with $\hat{p} \approx 0.246$. The test statistic for this test is as shown in the box.

Test statistic for a hypothesis test for comparison of proportions.

$$z = \frac{(\hat{p}_1 - \hat{p}_2) - 0}{\sqrt{\hat{p}(1-\hat{p})\left[\frac{1}{n_1}+\frac{1}{n_2}\right]}} \quad \text{where } \hat{p} = \frac{\text{Total number of cases that are "successes"}}{\text{Total number of cases}}$$

Here is the calculation for our example.

$$z = \frac{(\hat{p}_1 - \hat{p}_2) - 0}{\sqrt{\hat{p}(1-\hat{p})\left[\frac{1}{n_1}+\frac{1}{n_2}\right]}} = \frac{0.314 - 0.162}{\sqrt{0.246(1-0.246)\left[\frac{1}{175}+\frac{1}{142}\right]}} \approx \frac{0.152}{0.0486} \approx 3.13$$

Since $z = 3.13 > 1.96$, we can see immediately that our observed sample difference is *rare* and that we have evidence *against* the null hypothesis that $p_F - p_M = 0$. Since the z is big, it is very unlikely that we would see as big a difference (or one bigger) that we have seen in our sample, if it were really true that $p_F - p_M = 0$. We can calculate the probability of getting a difference of proportions this big or more by calculating the *p*-value.

***Step 4:* Calculating the p-value:** We first find the probability of getting $z = 3.13$ or greater by consulting the Normal Distribution Chart and getting $P(z > 3.13) = 1 - P(z < 3.13) = 1 - 0.9991 = 0.0009$. The *p*-value will be double this so that *p*-value = $2P(z > 3.13) = 2(0.0009) = 0.0018$. Since this *p*-value is far *less* than $\alpha = 0.05$, we see again that our result for the difference of proportions is rare if the null hypothesis is true (that is, if $p_F - p_M = 0$).

Step 5: ***Interpretation:*** The test statistic and the p-value point in the same direction (as they must always!), and that is that we have observed a rare difference of proportions if it were true that $p_F = p_M$. Since our result is so rare, we are led to reject the null hypothesis that the proportion of female students (in the population) who have tattoos is the same as the proportion of male students in the population of all CSM students who have tattoos. Our result is *statistically significant.* We think that we have evidence that the proportions of male and females who have tattoos really do differ; we seem to have successfully challenged the null hypothesis.

However, remember that our sample is not random; it is just possible that the tattooed males (as opposed to the tattooed females) avoid taking statistics. It may be that the non-tattooed males have a greater likelihood of taking statistics. For the test to be trustworthy, we need a random sample.

Two-Sided and One-Sided. All of the hypothesis tests that we have done have had null and alternate hypotheses of the form $\begin{matrix} H_0 : p = p_0 \\ H_a : p \neq p_0 \end{matrix}$ (the ones we did in §4.2) or of the form $\begin{matrix} H_0 : p_1 = p_2 \\ H_a : p_1 \neq p_2 \end{matrix}$ (in this section). We have been challenging the null hypothesis with the idea that the proportion *differs* or is *not equal* to the standard or that the proportion for one group *differs* or is *not equal* to the proportion for the second group. All of these tests are ***two-sided tests***. In our example of the proportions of males and females who have tattoos, we tested $\begin{matrix} H_0 : p_F = p_M \\ H_a : p_F \neq p_M \end{matrix}$. However, we may be able to bring a more refined "preconceived idea" to the test before we actually do it. We may have reason to believe (*before* we look at the data) that the proportion of tattooed females is bigger than the proportion of tattooed males. We can make this refined preconceived idea to be the challenge to the null hypothesis that $H_0 : p_F = p_M$. If so, then what we are doing is a ***one-sided test***. In a one-sided test, the null stays the same, but the alternate is different. The hypotheses would be: $\begin{matrix} H_0 : p_F = p_M \\ H_a : p_F > p_M \end{matrix}$.

Differences between two-sided tests and one-sided tests. Some calculations remain the same whether the test is one- or two-sided. The test statistic will be the same. However, since the hypotheses are different, the calculation of the p-value will be different. We are making a hypothesis in only one direction, so in this situation "more extreme" does not take account of random variation in both directions. Hence, the p-value will be calculated on only one shaded area. The test statistic is the same, but the p-value will be different. In our interpretation, we will mention that we hypothesized that one proportion would be greater than the other.

We can see these differences by looking at the Software output for the two-sided and one-sided tests for our example. Inspect the output carefully in the graphics on the next page to see the differences. (Note: the graphs shown are from another test to show the typical differences in the pictures; the shaded areas for *our* test would be so small that you would not be able to see them.)

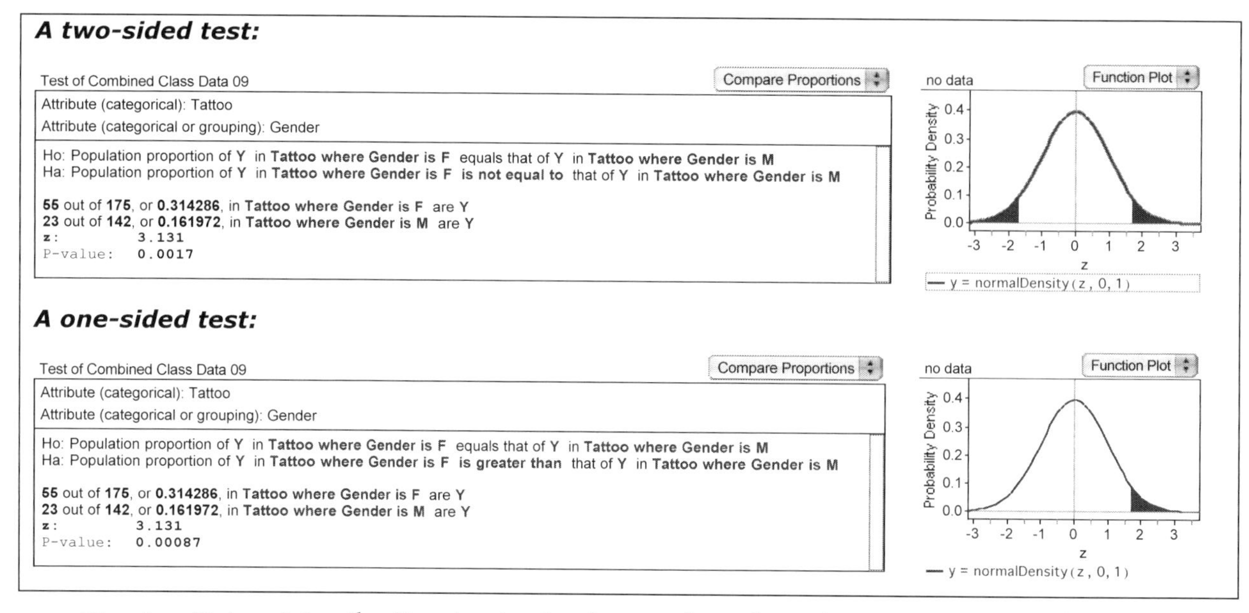

Warning. Determining the direction for the alternate hypothesis for a *one-sided test* should not be based on the data that is used *for the test.* In our example, it would be "cheating" to say $H_a : p_F > p_M$ based on the data that we have. The idea for $H_a : p_F > p_M$ must come from ideas that the tester had before seeing the data.

Example: Estimating how big the difference is—the second question

The second question that we had concerned the size of the difference of the proportions. We want to estimate how big the difference is between the proportions in our population. *Estimation* means using confidence intervals. In this case what we are estimating is the difference of the proportions of tattooed females and tattooed males among *all the students,* our population. The formula is given in the box below.

Formula for the Confidence Interval for the Difference of Two Proportions

$$(\hat{p}_1 - \hat{p}_2) \pm z^* \sqrt{\frac{\hat{p}_1(1-\hat{p}_1)}{n_1} + \frac{\hat{p}_2(1-\hat{p}_2)}{n_2}}$$

where $z^* = 1.645$ for a 90% confidence interval

$z^* = 1.96$ for a 95% confidence interval

$z^* = 2.576$ for a 99% confidence interval

From the summary table shown again below, we see as before that $\hat{p}_F = 0.314$ and $\hat{p}_M = 0.162$, and so we can calculate a 95% confidence interval for the difference in proportions having a tattoo between female and male students using the formula shown in the box above:

$$\begin{aligned}(\hat{p}_F - \hat{p}_M) \pm z^* \sqrt{\frac{\hat{p}_F(1-\hat{p}_F)}{n_F} + \frac{\hat{p}_M(1-\hat{p}_M)}{n_M}} &= (0.314 - 0.162) \pm 1.96\sqrt{\frac{0.314(1-0.314)}{175} + \frac{0.162(1-0.162)}{142}} \\ &= 0.152 \pm 1.96\sqrt{0.001231 + 0.000956} \\ &\approx 0.152 \pm 1.96(0.0468) \\ &= 0.152 \pm 0.09166\end{aligned}$$

and this implies a confidence interval of $0.0603 < p_F - p_M < 0.2437$.

We can interpret the interval by saying:

- "With 95% confidence, 15.2% (with a margin of sampling error of 9.2%) more female students than male students have a tattoo among the entire student body," or
- "We are 95% confident that the difference in the proportions of female and male students having a tattoo among all of the students at the college is between 6.0% and 24.4%," or
- "With 95% confidence, we can say that the likelihood that a female student at the college has a tattoo is 15.2% greater than the likelihood that a male student has a tattoo, with a margin of error of 9.2%."

Combined Class Data 09

		Tattoo		Row Summary
		N	Y	
Gender	F	120 0.685714	55 0.314286	175 1
	M	119 0.838028	23 0.161972	142 1
Column Summary		239 0.753943	78 0.246057	317 1

S1 = count ( )
S2 = rowProportion

Another example, showing software output

We can ask another similar question comparing male and female students. The Summary Table shows the numbers and the proportions of male and female students who have visited "More than Seven States" against those who have visited "Seven or Fewer States." If a student has visited more than seven states then we call that student "Well-Travelled" or WT. So, using our notation from §1.2, we can see that

Combined Class Data 09

		Travelled		Row Summary
		More than seven states	Seven or fewer states	
Gender	F	56 0.325581	116 0.674419	172 1
	M	59 0.430657	78 0.569343	137 1
Column Summary		115 0.372168	194 0.627832	309 1

S1 = count ( )
S2 = rowProportion

$P(WT \mid F) = \frac{56}{172} \approx 0.326$, and $P(WT \mid M) = \frac{59}{137} \approx 0.431$. Using the notation for this section, we would write $\hat{p}_F = \frac{56}{172} \approx 0.326$ and $\hat{p}_M = \frac{59}{137} \approx 0.431$. We have two questions:

- In the college as a whole, is there a difference in the proportions of "well-traveled" students by gender, or is the proportion essentially the same for male and female students?
- If there is a difference, what are the size and the direction of that difference?

The first question will be answered using a hypothesis test, since it is testing an idea: "Is there a difference by gender?" The second question calls for an estimate, so we will use a confidence interval.

For the first question, we can set up the null and alternate hypotheses (*Step 1*) in symbolic form as

$$H_0: p_F - p_M = 0$$
$$H_a: p_F - p_M \neq 0$$

We can do a quick check on the conditions of a Normal sampling distribution by calculating

$n_M \hat{p} \approx (137) \cdot (0.372) \approx 51.0$ where $\hat{p} = \frac{115}{309} \approx 0.372$ (*Step 2*). However, we still have our usual concerns about the sample not being a random sample from all students at the college. The output shows the results of the two steps (***Step 3: Calculating the test statistic*** and ***Step 4: Finding the p-value***).

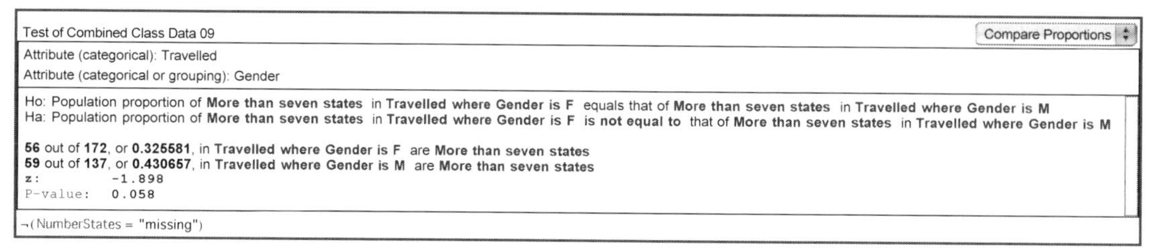
Test of Combined Class Data 09 — Compare Proportions
Attribute (categorical): Travelled
Attribute (categorical or grouping): Gender

Ho: Population proportion of **More than seven states** in **Travelled where Gender is F** equals that of **More than seven states** in **Travelled where Gender is M**
Ha: Population proportion of **More than seven states** in **Travelled where Gender is F** **is not equal to** that of **More than seven states** in **Travelled where Gender is M**

56 out of **172**, or **0.325581**, in **Travelled where Gender is F** are **More than seven states**
59 out of **137**, or **0.430657**, in **Travelled where Gender is M** are **More than seven states**
z: -1.898
P-value: 0.058

¬(NumberStates = "missing")

How do we interpret these calculations? Is the test statistically significant? Can we reject the null hypothesis that the proportions of well-travelled are equal for male and female students? No, and no. The test is *not statistically significant* at $\alpha=0.05$ because the *p*-value of 0.058 is bigger than 0.05. However, notice that it is very close, so we have *some evidence* of a gender difference. Also, the test statistic $z=-1.898$ is *not* outside the interval $-1.96 < z < 1.96$, although, again, it is close. (The number is negative because the calculations are doing the subtraction $\hat{p}_F - \hat{p}_M$, and the proportion for the males is bigger than the proportion for the females.) Both of these facts also say that we can*not reject* the null hypothesis of no difference between genders in proportion of well-travelled. We do not have enough evidence to say that in the college as a whole, the proportion of males who are well-travelled is higher than the proportion of females who are well-travelled.

For the second question, we can calculate a confidence interval. The Software output is shown below. Notice that one end of the confidence is negative and the other end is positive, although it is not very much larger than zero. Intervals that include zero are clumsy to put into elegant English (or perhaps other languages as well), so we have to say something like: "We are 95% confident that the likelihood of being well-travelled if female is anything from about 21% less than the likelihood for a male to just barely more likely (0.3%) of being well-travelled." Alternately, we could say: "We estimate with our data that there is a difference of 10.3% in the likelihood of being well-travelled between males and females for college students, but that comes with a margin of error of 10.8% so that it is possible that there is no difference at all, or that the difference is as much as 21%."

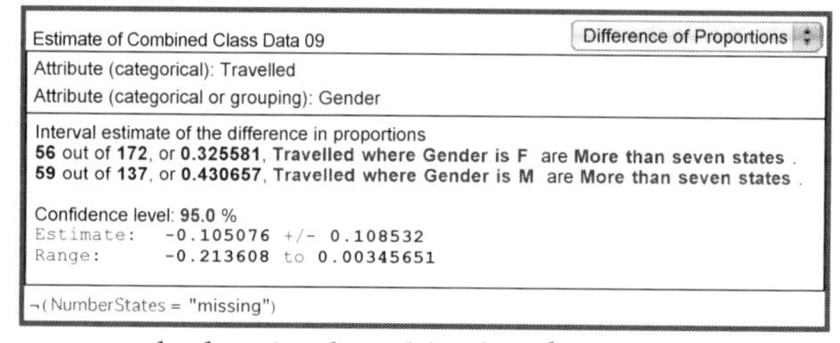
Estimate of Combined Class Data 09 — Difference of Proportions
Attribute (categorical): Travelled
Attribute (categorical or grouping): Gender

Interval estimate of the difference in proportions
56 out of **172**, or **0.325581**, **Travelled where Gender is F** are **More than seven states**.
59 out of **137**, or **0.430657**, **Travelled where Gender is M** are **More than seven states**.

Confidence level: **95.0** %
Estimate: -0.105076 +/- 0.108532
Range: -0.213608 to 0.00345651

¬(NumberStates = "missing")

The fact that the confidence interval includes *zero* is consistent with the conclusion that we did *not* reject the null hypothesis. That is, our sample data is plausible if in the population there is "no difference between males and females."

Summary and warning

The examples that we have looked at in this section were chosen because it was thought they would be fairly easy to understand. They illustrate the calculations nicely. However, they come with a big warning: we cannot be certain that the samples that were used are random samples of any population. The sample may *approach* being a random sample of students at a Northern California community college who are pursuing an academic program. This is because a great many students following an academic (rather than a vocational program) must take statistics. Some vocational programs (such as nursing) also require statistics. However, in the end, we do not know what population is being sampled.

All of the calculations that have been done presume that we have randomization.

The best research practice—such as the weight-loss experiment that was an example in §3.1—makes certain that randomization is used. However, many collections of data, including the data used in these sections, were not collected using randomization explicitly. The data may in effect be a random sample of some population, but we do not know. Proceed with caution.

Summary: Hypothesis Test for Comparing Proportions

Step 0: *Decide which of the two categories of a categorical variable is to be called a "success" so as to define the proportions clearly and decide which group in the second categorical variable is "Group 1" and which is "Group 2."*

Step 1: Set up the null and alternate hypotheses using an idea for the population proportion p_0.

$H_0: p_1 = p_2$ OR $H_0: p_1 - p_2 = 0$

$H_a: p_1 \neq p_2$ $H_a: p_1 - p_2 \neq 0$

Step 2: Check the conditions for a trustworthy hypothesis test.

1. *The sample must be a simple random sample from a population ten times n.*
2. *All* $n_1\hat{p}$, $n_1(1-\hat{p})$, $n_2\hat{p}$, *and* $n_2(1-\hat{p})$ *must be five or greater.*

Step 3: *Calculate the test statistic* $z = \dfrac{(\hat{p}_1 - \hat{p}_2) - 0}{\sqrt{\hat{p}(1-\hat{p})\left[\dfrac{1}{n_1} + \dfrac{1}{n_2}\right]}}$ *where* $\hat{p}$ *is calculated from the sample by*

$$\hat{p} = \frac{\text{Total number of "successes" in the sample}}{\text{Total Number of cases in the sample}}$$

Step 4: *Calculate the p-value by getting* $P(\hat{p} \text{ more extreme}) = 2P(Z \geq |z|)$ *using the absolute value of the test statistic you calculated in Step 3.*

Step 5: *Evaluate the evidence that the p-value and the test statistic give you to determine whether your test successfully challenges the null hypothesized population proportion* p_0 *or not. Give an interpretation in the context of the data using the terminology of "statistical significance" and "rejecting the null hypothesis."*

Summary: Confidence Interval for a Difference of Proportions

$$(\hat{p}_1 - \hat{p}_2) \pm z^* \sqrt{\frac{\hat{p}_1(1-\hat{p}_1)}{n_1} + \frac{\hat{p}_2(1-\hat{p}_2)}{n_2}}$$

where $z^* = 1.645$ *for a 90% confidence interval*
$z^* = 1.96$ *for a 95% confidence interval*
$z^* = 2.576$ *for a 99% confidence interval*

§4.4 Do We Have Independence? Chi-square

Introduction: One, two, three, more... One of the exercises in §4.3 looked at this table relating *PoliticalView* and *Gender*. We compared the proportions of males and females who have "Liberal" political views. We compared $P(L \mid M) = \frac{70}{142} \approx 0.493$ with $P(L \mid F) = \frac{95}{175} \approx 0.543$ (where L stands for "Liberal").

Combined Class Data 09

		PolticalView			Row Summary
		Liberal	Moderate	Conservative	
Gender	F	95	69	11	175
	M	70	61	11	142
Column Summary		165	130	22	317

S1 = count( )

We then did a hypothesis test, so we used the notation $\hat{p}_M \approx 0.493$ and $\hat{p}_F \approx 0.543$. We tested the idea that females are more likely to hold liberal political views than males as against the null hypothesis that there is no difference by gender in political views. (As it happened, our test was not statistically significant.)

However, there are really three gender comparisons that could be made, not just one; we could compare $P(Moderate \mid M)$ with $P(Moderate \mid F)$ and also $P(Conservative \mid M)$ with $P(Conservative \mid F)$ as well as $P(L \mid M)$ with $P(L \mid F)$; it appears that statistical life got still more complicated. However, doing all of these comparisons would actually *not* be a good way to proceed. Instead, we ask a different question about the table *as a whole* and not about the various parts of the table. Our statistical question is:

Is there an association between the variables Gender and PoliticalView?

We will answer this question by building a model of "no association" between the variables and then seeing if the data that we have in the table successfully challenges the model of "no association." This sounds like a hypothesis test, and that is what it is. Where we have a table relating two variables (such as the one above) where the cells show the *counts* or *frequency* of cases in each intersection of categories (Female-Liberal, Male-Liberal, Female-Moderate, etc.), we call such a table a ***r x c contingency table*** because the table has ***r rows*** and ***c columns***. Moreover, the numbers in the table may show how one variable is *contingent* or depends on the other.

Definition of an r x c Contingency Table
An $r \times c$ *contingency table* shows the counts (or frequencies) of cases in the intersections of the categories of two categorical variables.

The Opposite of Association: Independence

Our statistical question begins: "Is there an association ...?" To answer this, we start by looking at what "no association" would look like. If there is *no* association between variables, then we say that the variables are ***independent***. No association between variables means independence between the variables; association means "not independence" or dependence between the variables.

In §1.2, we introduced the notion of ***independence*** for events. We will now expand this definition to apply it to variables. Here is the definition, once again (on the next page).

Independence of events

If $P(A \mid B) = P(A)$ then the events A and B are ***independent.*** In words, if the probability of A *given* B is the same as the probability of A then we say that the events A and B are ***independent***.

So we could begin (tediously!) doing calculations to see if *events* are independent. Using the *Gender* and *PoliticalView* example shown above, we can calculate that $P(L) = \frac{165}{317} \approx 0.521$, but we see that $P(L \mid F) = \frac{95}{175} \approx 0.543$; so the events L and F—the events liberal political view and female gender—are *not* independent, since $P(L \mid F) \neq P(L)$. The events L and M (liberal and male) are also *not* independent because $P(L \mid M) \neq P(L)$. If we do the calculations for $P(Mod)$ and $P(Mod \mid M)$ and $P(Mod \mid F)$ we will see that the events *Moderate* and M are also *not* independent, and the events *Moderate* and F are *not* independent. Very seldom do we find exact independence when we are analyzing real data; we may come close, but, because of variation in samples, events that may seem to be *logically* independent will seldom meet the strict definition of independence. Moreover, things in the real world are simply too interconnected for strict independence; two variables may not be related in any simple fashion, but they may be in some extremely complicated way.

Combined Class Data 09

		PolticalView			Row Summary
		Liberal	Moderate	Conservative	
Gender	F	95	69	11	175
	M	70	61	11	142
Column Summary		165	130	22	317

S1 = count( )

But we want to transcend this tedious calculation and look at variables rather than events. The next section shows how we use the idea of independent events to build a "model of no association" — that is what the data would look like if there were *no* association between the variables.

Building the Model: Independence of Variables and Expected Counts

To build our model, we will do two things. First we will apply the idea of independence to *all* the categories of the categorical variables so that we will speak of variables (as well as the categories within the variables) as being independent. Then we pursue this idea and work backwards to answer the question: *What would the counts in the table be if the variables Gender and PoliticalView were independent?*

Here is how we reason. If the variables *Gender* and *PoliticalView* were independent then we must have $P(L \mid F) = P(L) \approx 0.521$ and also because $P(L) = \frac{165}{317} \approx 0.521$. What would the count for the "liberal females" have to be in order for this to be true? There are $n_F = 175$ female students. For $P(L \mid F) = P(L) \approx 0.521$, we need

$$P(L \mid F) = \frac{x}{175} = \frac{165}{317} = P(L)\text{, where } x \text{ is the number of "liberal females."}$$

If we use algebra, and solve for x in this equation (by "cross multiplying"), we get

$x = 175 \times \frac{165}{317} \approx 91.088$ "liberal females" would be needed in the table to make $P(L \mid F) = P(L) \approx 0.521$.

Instead of 95 female students with liberal political views that we observe, we expect to have to be 91.1 for independence.

Likewise, for the male students, if we multiply $P(L)=\frac{165}{317}\approx 0.521$ by the number of male students $n_M=142$, we get $x=142\times\frac{165}{317}\approx 73.912$. So, instead of 70 males with liberal views that we actually observe in our sample, we would expect to have 73.9 males if the variables *PoliticalView* and *Gender* were independent.

Then we do the same kind of thing for the "Moderate" and "Conservative" categories, but we use $P(Mod)=\frac{130}{317}\approx 0.410$ and $P(C)=\frac{22}{317}\approx 0.069$ and calculate $x=175\times\frac{130}{317}\approx 71.767$ for the moderate females and $x=142\times\frac{130}{317}\approx 58.233$ for the moderate males. The counts necessary for independence for the students with conservative political views are calculated in a similar fashion.

Test of Combined Class Data 09 — Test for Independence

Column attribute (categorical): PolticalView
Row attribute (categorical): Gender

		PolticalView			Row Summary
		Liberal	Moderate	Conservative	
Gender	F	95 (91.1)	69 (71.8)	11 (12.1)	175
	M	70 (73.9)	61 (58.2)	11 (9.9)	142
Column Summary		165	130	22	317

```
Column attribute:          PolticalView
  Number of categories:    3
Row attribute:             Gender
  Number of categories:    2
Ho: PolticalView is independent of Gender
Chi-square:    0.8541
DF:            2
P-value:       0.65
```

The numbers in parentheses in the table are expected counts.

These counts are called ***expected counts*** (the symbol is ***E***) and are shown in parentheses in the Software output here. Notice that even though we cannot have a fractional number of students (e.g. 58.233 students, for example) we do *not* round to the nearest whole number, and in the calculations, Software (and we) will use as many decimal places for the expected counts as we can manage. The expected counts are the *model* based upon the independence of the two variables. They are found by applying the overall proportion of one category of one variable (for example: $P(L)=\frac{165}{317}\approx 0.521$) to the overall numbers in the categories of the second variable.

We can easily calculate that for any particular cell in a table, the expected count works out to be the row total for that cell multiplied by the column total for that cell and then divided by the total number of cases. Even more important is to realize that these counts *are the counts expected if the two variables are independent* of each other using the statistical definition of independence.

Expected Counts E for an r x c Contingency Table for the Model of Independence

The ***expected counts*** *(or frequencies)* for the model of independence are calculated by

$$E=\frac{(Row\ total\ for\ that\ cell)\times(Column\ total\ for\ that\ cell)}{Total\ Number\ of\ cases}$$

and are the counts expected if the two variables in the contingency table are independent.

Observed Counts O for an r x c Contingency Table

The *observed counts* are the actual data observed in the contingency table.

Testing the Model with Data

We raised the question: *Is there an association between the variables Gender and PoliticalView?* If there is no relationship or association between the variables then it appears that the variables are *independent.* If the variables are independent then the probability that a student has (for example) moderate political views is the same whether the student is a male or a female. Our hypothesis test to answer this will have the same *five steps,* but what goes into these five steps will come from the model of independence — or *no association* — that we have been building.

Step 1: Hypotheses The structure of the hypotheses for our test is quite simple:

H_0 : *The two variables are independent*

H_a : *The two variables are not independent*

We could use symbols involving *p,* but the notation would be very complicated. For our example, we would write:

H_0 : *PoliticalView is independent of Gender*

H_a : *PoliticalView is not independent of Gender*

Step 2: Conditions The *expected counts* must not be too small. There are several rules of thumb that are used, but the simplest is that ***all expected counts must be five or more.*** Notice that it is the expected counts that must be five or more and not the ***observed*** counts for the condition. If we wish to infer from sample data to a population then the second and usual condition is that the data must be a ***simple random sample*** from the population of interest.

Step 3: Calculating the test statistic What we want to do is to compare the *observed counts* to the *expected counts.* If the observed counts differ much from the expected counts, the counts that we expect if the null hypothesis is true and the variables are independent, then we will have evidence *against* the null hypothesis of independence. But if the differences between the observed and expected counts are not great, we will have insufficient evidence against the null hypothesis of independence. We need a measure that will involve every single cell and count all the differences between observed and expected. That measure is the ***chi-square goodness of fit statistic.*** (Pronunciation of χ: ***Chi*** stands for the Greek letter $\chi$and is pronounced as the first syllable in "Kaiser." It does not have the sound "chi" as in "chili.")

Chi-Square Goodness of Fit Statistic

$$\chi^2 = \sum_{All\ cells} \frac{(O-E)^2}{E}$$

where *O* = *Observed Counts* and *E* = *Expected Counts*

For our example, refer to the observed counts (95, 69, etc.) and the expected counts in parentheses (91.1, 71.8, etc.) in the Software output and see how they are used, and you should get an idea of how the calculation is done.

$$\chi^2 = \sum_{All\ cells} \frac{(O-E)^2}{E}$$

$$= \frac{(95-91.1)^2}{91.1} + \frac{(69-71.8)^2}{71.8} + \frac{(11-12.1)^2}{12.1} + \frac{(70-73.9)^2}{73.9} + \frac{(61-58.2)^2}{58.2} + \frac{(11-9.9)^2}{9.9}$$

$$\approx 0.1680 + 0.1067 + 0.1080 + 0.2070 + 0.1314 + 0.1331$$

$$\approx 0.8541$$

Test of Combined Class Data 09 — Test for Independence

Column attribute (categorical): PolticalView
Row attribute (categorical): Gender

		PolticalView			Row Summary
		Liberal	Moderate	Conservative	
Gender	F	95 (91.1)	69 (71.8)	11 (12.1)	175
	M	70 (73.9)	61 (58.2)	11 (9.9)	142
Column Summary		165	130	22	317

```
Column attribute:          PolticalView
  Number of categories:    3
Row attribute:             Gender
  Number of categories:    2
Ho: PolticalView is independent of Gender
Chi-square:   0.8541
DF:           2
P-value:      0.65
```

The numbers in parentheses in the table are expected counts.

Think about how this statistic will behave; if all of the observed values were exactly the same as the expected, the value would be zero. The more the observed are *different* from the expected, the bigger will be the value of the χ^2. Notice also that the value cannot be negative, since all the terms in the sum are squared. Our calculation agrees with the value that Software calculated. (Had we used the rounded values, we might have encountered rounding error.)

Sampling Distributions Again: The Chi-Square Distributions

However, is what we have got a big value or a small value? To determine whether we have a big value of χ^2 or a small value of χ^2, we need a sampling distribution of the values of χ^2 that could come from all possible cell values with the same values for the row and column totals. The meaning of a sampling distribution is the same as we have seen before; we worked with the sampling distribution of $\hat{p}$s from all possible samples of size n and found that it was a Normal distribution, under certain conditions. When we worked with the differences of two sample proportions $\hat{p}_1 - \hat{p}_2$ we found that the sampling distribution was also Normal, again under certain conditions. We might expect that all sampling distributions are Normal; not so.

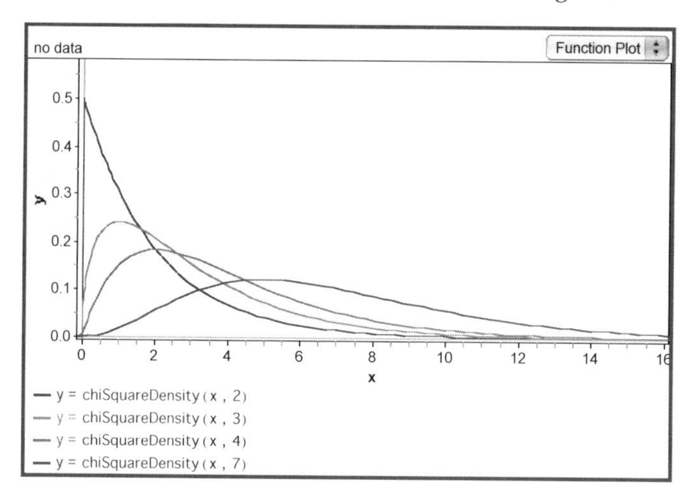

The sampling distributions for the chi-square statistic are called the ***chi-square distributions.*** There is an entire *family* of distributions, and for that reason everything is in the plural (*chi-square distributions*). Here is a graphic showing the shapes of some of the *chi-square distributions.* The different distributions we see in the plot have different ***degrees of freedom***, and ultimately the shapes go back to the mathematical definition of the distributions.

Which degrees of freedom and therefore which distribution we use depends upon the use to which we are putting the distribution, but for our ***Test of Independence for a Contingency Table*** the degrees of freedom is defined by the formula $df = (r-1)(c-1)$, where r means the number of rows in the table, and c means the number of columns in the table. So, for our table relating *Gender* and *PoliticalView* we have $df = (r-1)(c-1) = (2-1)(3-1) = 1 \cdot 2 = 2$ degrees of freedom

Degrees of Freedom for an r x c Contingency Table

$$df = (r-1)(c-1)$$

where r means the number of rows in the table, and c means the number of columns in the table.

We use this distribution in a similar way to the way we used the Normal table; we start by defining *reasonably likely* and *rare*. We have been using $\alpha = 0.05$ (that is 5%) as our definition of rare, which means that 95% or $1 - \alpha$ of the results are reasonably likely. For a chi-square distribution, the rare outcomes will be in the *right tail* because (in our application to contingency tables) these large values of χ^2 are the ones that challenge the null hypothesis of independence. Remember that a small value of χ^2 means that the observed counts are quite close to the expected counts and we have a sample result that could have come from two variables being independent. A large value of χ^2 means that the observed counts are far from the counts expected if the null hypothesis is true and the variables are independent. Work through this interaction to get some notion of the shapes of chi-square distributions and also what *rare* outcomes for the χ^2 look like.

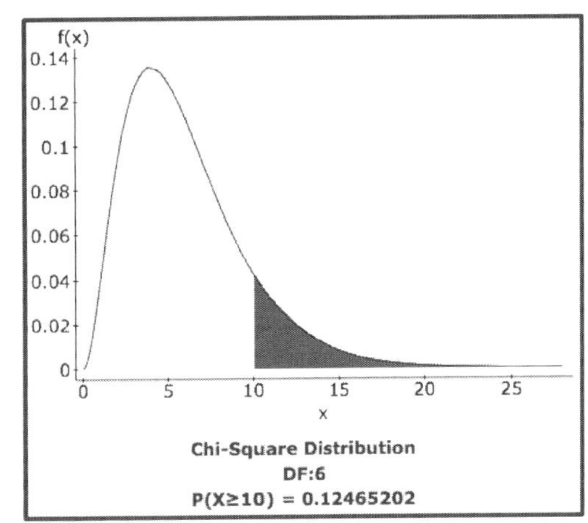

Here is a plot showing the Chi-Square Distribution for $df = 6$, and also showing the probability that the value of the test statistic is greater than $\chi^2 = 10$. The probability is about 0,125. (Will this result be statistically significant?)

Values of the chi-square statistic χ^2 that are greater than what is called a ***critical value*** are rare (if we choose $\alpha = 0.05$), and values of χ^2 that are less than this 5.991are reasonably likely. For the test whether the variables *PoliticalView* and *Gender* are independent or not, the test statistic is $\chi^2 = 0.8541$. Since this number is far less than the critical value 5.991, we certainly *do not* have evidence against our null hypothesis that the variables *Gender* and *PoliticalView* are independent. The test is *not* statistically significant; we do not have evidence that there is a difference in male and female college students' political views overall.

There is a chart, the ***Chi-Square Distribution Chart,*** that records the *critical values* for various df and for various choices of α– choices other than the $\alpha = 0.05$ that we have been using.

Reading the Chi-Square Distribution Chart The chi-square chart is not like the Normal chart. In the Normal chart, the entries in the body of the table were probabilities, and the critical values for the test statistic *z* were on the margins (the left-hand side, with the second decimal place across the top.) The Chi-Square Distribution Chart is organized in exactly the opposite fashion, and the chart is much more "compressed" and gives less information because we have to cope with many different chi-square distributions, one for each value of degrees of freedom, *df*. Here is a picture of a part of the chart. The ***degrees of freedom = df*** are listed on the left-hand side, and the various choices of rare, the α, are listed on the top of the chart. For our example, we had $df = 2$, and we had chosen $\alpha = 0.05$; if you look at the intersection of the row and the column for these $df = 2$ and $\alpha = 0.05$, you will see the number 5.99. If we had $df = 4$ but still wanted $\alpha = 0.05$ then our *critical value* would be 9.49. That is, in that situation, for us to say that our test was significant, we would have to have our test statistic to be bigger than 9.49, or, in other words, for $\chi^2 > 9.49$.

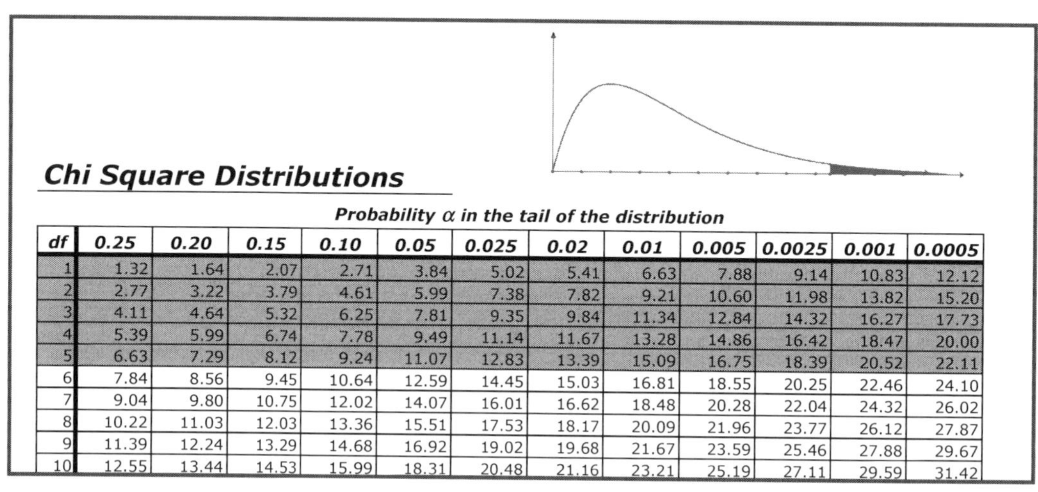

Chi Square Distributions

Probability α in the tail of the distribution

df	0.25	0.20	0.15	0.10	0.05	0.025	0.02	0.01	0.005	0.0025	0.001	0.0005
1	1.32	1.64	2.07	2.71	3.84	5.02	5.41	6.63	7.88	9.14	10.83	12.12
2	2.77	3.22	3.79	4.61	5.99	7.38	7.82	9.21	10.60	11.98	13.82	15.20
3	4.11	4.64	5.32	6.25	7.81	9.35	9.84	11.34	12.84	14.32	16.27	17.73
4	5.39	5.99	6.74	7.78	9.49	11.14	11.67	13.28	14.86	16.42	18.47	20.00
5	6.63	7.29	8.12	9.24	11.07	12.83	13.39	15.09	16.75	18.39	20.52	22.11
6	7.84	8.56	9.45	10.64	12.59	14.45	15.03	16.81	18.55	20.25	22.46	24.10
7	9.04	9.80	10.75	12.02	14.07	16.01	16.62	18.48	20.28	22.04	24.32	26.02
8	10.22	11.03	12.03	13.36	15.51	17.53	18.17	20.09	21.96	23.77	26.12	27.87
9	11.39	12.24	13.29	14.68	16.92	19.02	19.68	21.67	23.59	25.46	27.88	29.67
10	12.55	13.44	14.53	15.99	18.31	20.48	21.16	23.21	25.19	27.11	29.59	31.42

Getting an approximate p-value using the chart: "little boxes" A *p*-value is the probability of getting a result as extreme as or more extreme than the one we got from our sample *if* the null hypothesis is true. More extreme for a chi-square distribution means just one thing: area in the *tail on the right*, so it looks fairly simple. However, since the chi-square chart is compressed, we cannot get an exact *p*-value from the chart the way we were able to from the Normal chart. In fact, most of the time we will depend upon the calculation of a *p*-value from software. We can get an approximate *p* value from the chart by using a "little box." Here is an example. Suppose we had a contingency table with $df = 4$, and in our hypothesis test, our test statistic came out as $\chi^2 = 12.07$. If we look at the chart, we do not see this number in the row for $df = 4$, but we see numbers near to it and their corresponding areas in the tail of the chi-square distribution, and we can put them into a small box:

Area (probability) in the tail:	**0.02**	**0.01**
Critical value:	*11.67*	*13.28*

We chose these values because our sample value χ^2= 12.07 falls between them. Then we can approximate our p value as $0.01 < p - value < 0.02$. We do not know exactly what the p-value is, but we know that it is somewhere in this interval $0.01 < p - value < 0.02$. This information about the p-value is all that we need; however, if we have chosen $\alpha = 0.05$ then this information tells us that the test will be *statistically significant*, since 0.02<0.05.

What can we say about the p-value for our example of the relationship between *Gender* and *PoliticalView*? Our test statistic is $\chi^2 = 0.8541$, which is far less than the critical value of 5.99, so we can say that the result is not rare, and is *not statistically significant*, and we would *not* be able to reject the null hypothesis. Can we approximate the p-value from the chart? Looking at the chart for the row df = 2, we see that the smallest value is 2.77 and that this value corresponds to a possible p-value of 0.25. Since we have $\chi^2 = 0.8541$, which is *smaller* than 2.77, we can say (from the chart) that the p-value is greater than 0.25. Software calculates that the p-value = 0.65, and that is certainly bigger than 0.25, so our approximation of p-value > 0.25 is true, if unimpressive as an approximation. It is helpful to get a graphic of the p-value shown on the chi-square distribution. Notice how much of the curve is shaded in; if the test had been statistically significant then the shaded portion would be very small and would show just a sliver of area in the tail of the distribution. Again, the p-value is shown as an area.

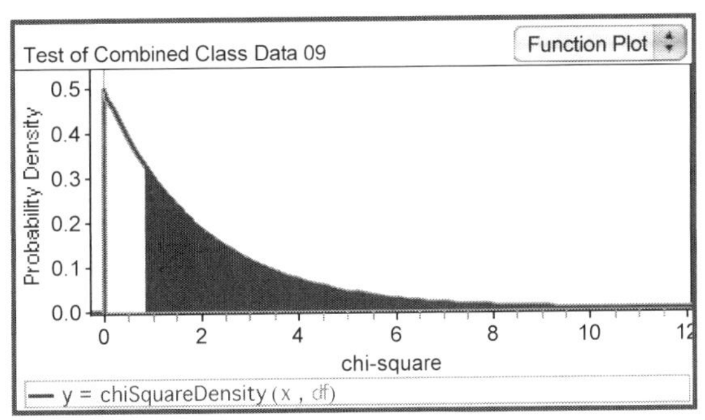

Step 5: Interpretation The ingredients are all here; looking at our example, we have:

- A test statistic $\chi^2 = 0.8541$ that shows a reasonably likely outcome (since $0.8541 < 5.99$) *if* the null hypothesis is true, meaning that the variables *Gender* and *PoliticalView* are independent.
- A p-value = 0.65 that also shows a reasonably likely outcome if the null hypothesis is true.

Therefore, we can say (as we have before) that our test is *not statistically significant* and we cannot reject the null hypothesis of independence between the variables *PoliticalView* and *Gender* for the population of students at the college. We do not sufficient evidence to say that there is any difference between male and female students in their political views (whether liberal, moderate, or conservative). As far as our data informs us, the political views of male and female students are similar. However, once again, our conclusions need to be taken cautiously because we know that our sample is *not* a random sample of the population, and so it is possible that the non-randomness of the sample has affected the results that we have seen in some way.

A Two-by-Two Table: Another (Familiar) Example

It is possible to apply the chi-square analysis to a table that has two rows and two columns. To illustrate this, we will look at the relationship between *Tattoo* and *Gender.* We have already done this with a comparison of proportions. We came to the conclusion that the test *is statistically significant* and that we could reject the null hypothesis of *no difference* in the proportions between male and female students, since the test statistic $z = 3.13 > 1.96$, and the p-value $= 0.0017 < 0.05$. Both the test statistic and the p-value indicate that what we have seen is very rare if the null hypothesis is correct. We conclude that we have evidence that there *is* a difference by gender in the probability of having a tattoo.

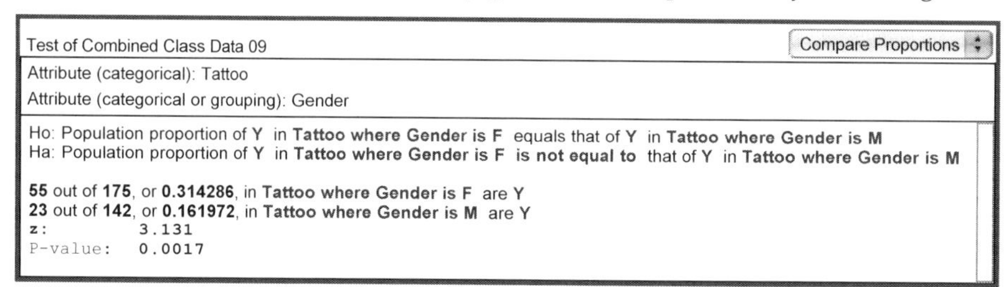

Test of Combined Class Data 09 — Compare Proportions

Attribute (categorical): Tattoo

Attribute (categorical or grouping): Gender

Ho: Population proportion of Y in **Tattoo where Gender is F** equals that of Y in **Tattoo where Gender is M**

Ha: Population proportion of Y in **Tattoo where Gender is F** **is not equal to** that of Y in **Tattoo where Gender is M**

55 out of **175**, or **0.314286**, in **Tattoo where Gender is F** are Y

23 out of **142**, or **0.161972**, in **Tattoo where Gender is M** are Y

z: 3.131

P-value: 0.0017

We can analyze these same data using the chi-square techniques; here is software output showing the results. We come to the same conclusion. Since the test statistic $\chi^2 = 9.803$ is bigger than the *critical value* for $df = 1$ of 3.84, and since the p-value is less than $\alpha = 0.05$, the hypothesis test is *statistically significant,* and we are led to *reject* the null hypothesis that *Tattoo* is independent of *Gender.* In fact, notice that the p-value is exactly the same; that is not a coincidence. There is another connection between the two tests. If we square $z = 3.131$, we will find that we get $\chi^2 = 9.803$. This is also not a coincidence.

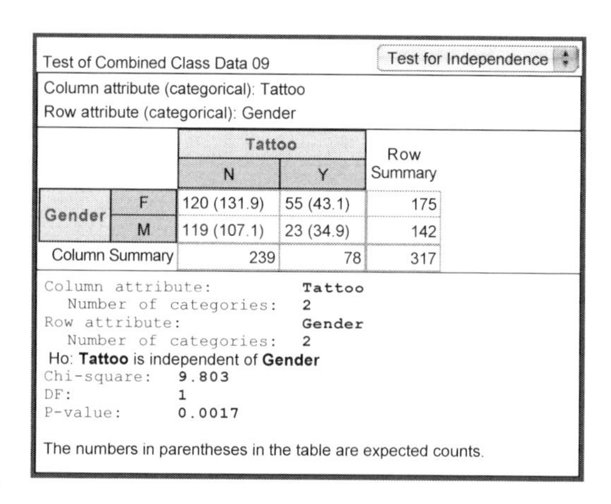

Test of Combined Class Data 09 — Test for Independence

Column attribute (categorical): Tattoo

Row attribute (categorical): Gender

		Tattoo: N	Tattoo: Y	Row Summary
Gender	F	120 (131.9)	55 (43.1)	175
	M	119 (107.1)	23 (34.9)	142
Column Summary		239	78	317

Column attribute: Tattoo

Number of categories: 2

Row attribute: Gender

Number of categories: 2

Ho: **Tattoo** is independent of **Gender**

Chi-square: 9.803

DF: 1

P-value: 0.0017

The numbers in parentheses in the table are expected counts.

Test of Independence for a 2 x 2 Table

- The test of independence for a contingency table with two rows and two columns is equivalent to the *two-sided* hypothesis test where the null hypothesis is $H_0 : p_1 = p_2$ and alternate hypothesis is $H_a : p_1 \neq p_2$.
- The p-values of the two tests are the same and $z^2 = \chi^2$.

Notice that the box above specifies that the tests are equivalent for a two-sided test. The chi-square test of independence does *not* test the same thing as a one-sided hypothesis test; with a one-sided test, we are essentially bringing in another preconceived idea.

Summary: Chi-Square Test for Independence

- The overall relationship between two categorical variables that have two or more categories can be analyzed using an ***r x c contingency table*** and a ***model for independence of variables.***
  - An ***r x c contingency table*** shows the counts (or frequencies) of cases in the intersections of the categories of two categorical variables.
  - The ***model for independence of variables*** is based upon the definition for the independence of events, which says that events A and B are ***independent*** if $P(A \mid B) = P(A)$.
- The ***model for independence of variables*** builds an ***r x c contingency table*** whose counts in the cells of the table are the counts ***expected*** if the summary numbers for the two variables are the same but variables are independent.
  - The ***expected counts*** *(or frequencies)* for the model of independence are calculated by

$$E = \frac{(\textit{Row total for that cell}) \times (\textit{Column total for that cell})}{\textit{Total Number of cases}}$$

  and are the counts expected if the two variables in the contingency table are independent.
  - The ***observed counts*** are the actual data observed in the contingency table.
- The *observed* counts are compared with the *expected* counts of the *model for independence* by the ***chi-square goodness of fit statistic*** $\chi^2 = \sum_{\textit{All cells}} \frac{(O-E)^2}{E}$ where O = *observed* and E = *expected.*
  - The values of the *chi-square goodness of fit statistic* χ^2 are always positive.
  - The larger the value of the *chi-square goodness of fit statistic* χ^2, the more difference there is between the observed and expected.
- The sampling distributions of the *chi-square goodness of fit statistic* χ^2 are the family of ***chi-square distributions.***
  - There is a different ***chi-square distribution*** for different values of ***degrees of freedom, df.***
  - For *r x c contingency tables,* the *degrees of freedom* is calculated by $df = (r-1)(c-1)$.
  - *Chi-square distributions* are typically right-skewed.
  - ***Critical values*** are values for a given *df,* and a given choice of *level of significance* α shows the smallest value for the *chi-square goodness of fit statistic* χ^2 that is considered rare.

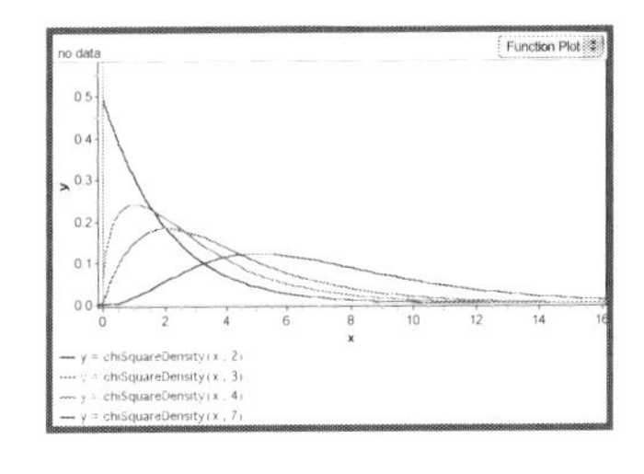

  - ***Critical values*** can also be read off the ***Chi-Square Distribution Chart.***
- ***Test of Independence for a 2 x 2 Table***
  - The test of independence for a contingency table with two rows and two columns is equivalent to the *two-sided* hypothesis test where the null hypothesis is $H_0 : p_1 = p_2$ and alternate hypothesis is $H_a : p_1 \neq p_2$.
  - The p-values of the two tests are the same, and $z^2 = \chi^2$.

Summary: Test for Independence of Variables in an r x c Contingency Table

Step 1: *Set up the null and alternate hypotheses using an idea for the population proportion p_0.*

H_0: *The two variables are independent*

H_a: *The two variables are not independent*

Step 2: *Check the conditions for a trustworthy hypothesis test.*

a. *The sample must be a simple random sample.*

b. *All of the expected counts (see step 3 below) must be five or greater.*

Step 3: *Find the test statistic*

a. *Calculate the expected counts for the r x c table using:*

$$E=\frac{(\text{Row total for that cell})(\text{Column total for that cell})}{\text{Total Number of cases}}$$

b. *Calculate the chi-square goodness of fit statistic:*

$$\chi^2 = \sum_{\text{All cells}} \frac{(O-E)^2}{E}$$

Step 4: *Calculate the p-value, either by using software or by approximating using the Chi-Square Distributions Chart and the "little box" method.*

Step 5: *Evaluate the evidence that the p-value and the test statistic give you to determine whether your test successfully challenges the null hypothesis of independence or not. Give an interpretation in the context of the data using the terminology of "statistical significance" and "rejecting the null hypothesis."*

§4.5 Hypothesis Test, Confidence Interval, Both, or Neither?

Introduction: The Assignment

This section starts with an assignment to work with and make sense of data using most of what we have been looking at. The sections you have studied so far have presented various *descriptive* techniques (Units 1 and 2) and *inferential* techniques (Units 3 and 4), and this section's goal is to pull all of that together. When confronted with a statistical question and some data, what technique should be used? How do you decide? And what can go wrong? What mistakes can be made, and how can you spot them?

Perhaps more importantly (but less enjoyably!), suppose it is not you that had the assignment but someone else. If you are reading the analysis and interpretation of some data that someone else has made, what helps you to know if that person has done the job well or has blundered in some way? One way you can spot others' possible mistakes is to be exposed to the possibility of making them yourself—even actually making the mistakes! So we will analyze some data to answer some statistical questions.

The Assignment You have been working with the Census at School data for Australia. The same kind of data (see http://www.censusatschool.com/en/about) have been collected in Canada, New Zealand, South Africa, and the UK, and some of the same data—the same variables—were collected in all of these places. So our *assignment* is:

What are the similarities and differences in high school students and their lives in these different places?
You can easily imagine that this could be an assignment for a sociology course or an education course or it could be a journalist's assignment: "Write us an article that shows how the lives of high school students are basically the same in different countries" or "write us an article about how the lives of high school students differ in different countries." The idea is to use the data to say something.

We begin with three essential questions.

- What are the cases?
- What are the variables?
- How were the data collected?

The data. The cases for the data are students in schools. All students at participating schools answer the questions online rather than using paper questionnaires. It is called the Census at School because all the students in a school participate. The schools include all grade levels, although the variables measured for the secondary-level students differ from those measured for the elementary-level students. It is important to know that school participation is voluntary and the schools that participate are *not* a random sample of schools in the country. Here is what the ***case table*** looks like for the data we will analyze.

Australia Canada Comp

	Gender	Age...	NumLan...	Hand	Reflex	Height	ArmSpan	RightFoot	NumHH	TimeConc	Transport	NoBreak...	Cereal	Sport	StatePr...	Country
199	Female	14	(a) One	Right han...	0.4	164	2	6	5	33	(a) Walk o...	No	Yes	Running,...	NSW	Australia
200	Male	14	(a) One	Right han...	0.37	173.5	190.5	26.5	5	41	(c) Public ...	No	Yes	Football/...	NSW	Australia
201	Female	16	(a) One	Right han...	0.41	158	165	22	4	25	(b) Car	No	No	Netball	Vic	Australia
202	Female	13	(a) One	Right han...	0.43	160	159	23	3	46	(b) Car	No	Yes	Basketball	Vic	Australia
203	Male	14	(b) Two	Right han...	0.3	167.5	169	25.5	4	63	(b) Car	No	No	Other	Vic	Australia
204	Male	17	(b) Two	Right han...	0.31	175	183	27	4	32	(b) Car	No	Yes	Basketball	Vic	Australia
205	Male	15	(a) One	Left han...	0.45	177	178	27	5	33	(c) Public ...	No	No	Football/...	Vic	Australia
206	Female	15	(b) Two	Right han...	0.34	160	156	24	4	55	(c) Public ...	Yes	No	Dancing	Vic	Australia
207	Female	16	(a) One	Right han...	0.33	155	155	24	2	32	(a) Walk o...	No	Yes	Swimming	Vic	Australia

You will find a complete list of the variables and their definitions at the end of this section of the ***Notes***; you should be familiar with some of the variables in any case, as you have previously worked with them. Some of the variables are ***quantitative*** and some of the variables are ***categorical.*** Recall that there are different techniques for analyzing quantitative and categorical variables.

Statistical questions with no inference: descriptive analyses

The basic question for this assignment, *"What are the similarities and differences in high school students and their lives in these different places?"* is far too broad and must be broken down into specific questions about variables. Statistical questions generally involve comparisons between groups or relationships between variables, or a comparison of a variable to some standard. Here are some examples for the first two or three variables listed in the case table.

Example 1: *Are there differences in the ages of the students in our Australian and Canadian samples?* How do we answer this question? First, we notice that *Age* is a quantitative variable, and, secondly, we notice that our question is only about our samples and not the population from which the samples were drawn. The analysis appropriate for comparing groups on a quantitative variable is to compare ***distributions.*** The graphics we use are dot plots or box plots or histograms. For numerical summaries, we could compare the means, medians, standard deviations, and IQRs for the two samples. The means and the medians tell us that, on average, the Canadian students sampled are slightly older than the Australian students sampled. For our measures of spread, the standard deviation indicates a slightly greater spread for the ages of the Canadian students, but recall that the s can be sensitive to skewness and outliers. The IQR for the two samples shows us the spread for the middle 50%; this spread is the same, and the appearance of the box plots confirms this. Does this indicate that secondary-school students are younger generally (not just in our sample)? That would be an ***inferential*** question because it is a question about Australian students and Canadian students in general and not only about our samples. We cannot answer that question yet; inference for quantitative variables is the topic of Unit 5.

Example 2: *Is there a relationship between* age *and the scores on the concentration game (*TimeConc*)?*

The students were asked to play a "concentration game"; the time to complete the task was measured. Shorter times indicate a deeper level of concentration on the game. Shown below is the "prompt" that was given to the Australian students. As a square is selected, an image appears, which is then hidden until that image's double is located under one of the other squares. Success at this game requires that the player remember (concentrate on) the location of the various images.

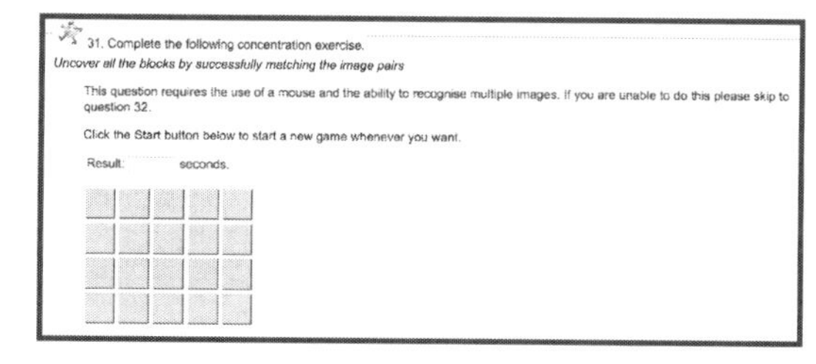

The variable *TimeConc* measures how well students do on this game. It is natural to ask (since we are thinking of age) whether there is a relationship between *Age* and *TimeConc*; do older students do better than younger students? Is their schooling having an effect at all? Or are the scores on this game relatively unrelated to age and dependent on other variables that we have not measured? These questions lead to our general question about the relationship between *TimeConc* and *Age*.

We consider whether the variables are categorical or quantitative. Here we have *two quantitative* variables, and since we are interested in the relationship between two quantitative variables, the appropriate techniques involve linear models, and specifically the *correlation coefficient* r and the fitting of a least squares linear model, evaluated by the *coefficient of determination* R^2, which were studied in Unit 2. Here are the graphics and the results of the calculations.

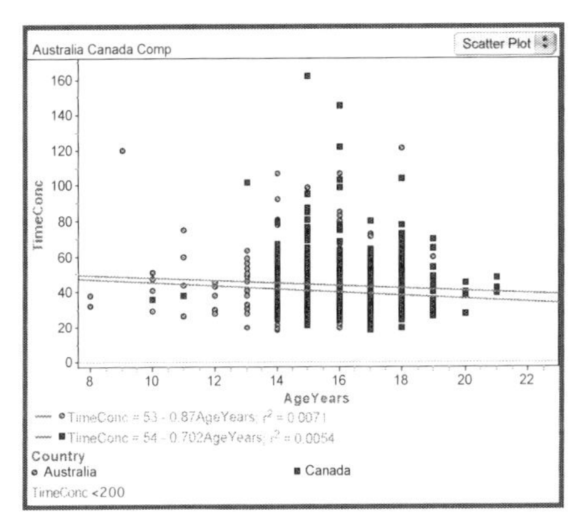

Age, apparently, does *not* affect the scores on the concentration game. The most important numbers here are the *coefficient of determination R2*. The R^2 are nearly zero for the students in both countries, and those small numbers indicate that the variation in the *TimeConc* scores cannot be explained by age at all. We have to look elsewhere for explanations of differences in the scores. We have treated this analysis as a *descriptive* analysis and not an *inferential* analysis concerned about generalization to a population beyond the samples we have. Inference for linear models uses some of the same tools as in Unit 5, but we leave it to the next course in statistics.

Both *Example 1* and *Example 2* require descriptive analysis and interpretation based on that analysis. These questions involve no formal inference to a larger population. The focus is on the samples and not a population beyond the samples. In the next sub-section, we ask questions where formal inferential analysis is implied in the question or the background to the question. Descriptive analysis is still worthwhile; indeed, descriptive analysis should always precede formal inferential analysis.

Statistical Questions with Inference

Here is an interesting summary table comparing students in Australia and in Canada on the number of languages they speak. Both Australia and Canada are countries peopled by immigrants, and so you would expect a sizable proportion of high school students to speak more than one language. Moreover, Canada has two official languages, English and French. In Quebec, French is the predominant as well as the official language. French speakers are found in all parts of Canada, and it is common for students to be bilingual if they are a minority (e.g., English-speaking in a majority French area or French in a majority English area). For that reason alone we would expect that a sizable proportion of students will speak more than one language.

Australia Canada Comp

		Country		Row Summary
		Australia	Canada	
NumLanguages	(a) One	432	335	767
	(b) Two	110	190	300
	(c) Three or More	29	75	104
	Column Summary	571	600	1171

S1 = count ( )

We have also looked at the proportion of students in a California community college who are bilingual and found that proportion to be over 50%. We can write down a number of statistical questions that require formal *inferential* procedures.

Combined Class Data 09

NumLanguages	(a) One	137 0.432177
	(b) Two	143 0.451104
	(c) Three or More	37 0.116719
	Column Summary	317 1

S1 = count ( )
S2 = columnProportion

Example 3: *Determine whether the proportion of monolingual students (those who speak just one language) in Australia differs significantly from the proportion of monolingual students in Canada.*

Example 4: *If there is a difference, how big is the difference between Australia and Canada in the proportion of monolingual students?*

Example 5: *Determine if, in the population of all students in the two countries, there are differences between the countries in the proportions of students who speak one, two, or three or more languages? (Or is NumLanguages independent of Country?)*

Example 6: *Does the proportion of students in Canada who speak two or more languages differ from the standard of 50%, which we think describes the typical percentage among college students in Northern California?*

Example 7: *Can we estimate the proportion of all high school students in Australia who speak two or more languages?*

We have broken down our overall question—*What are the similarities and differences in high school students and their lives in these different places?*—into specific questions about specific variables. For all of these examples we will use *inference.* We will consider each of these examples below.

Conditions and a potential problem One of the first things that we do for any inferential technique is to check the ***conditions.*** One condition that is common to all of the inferential procedures is that the data come from random samples. For these data, we know that they are *not* random samples. First, they are certainly not simple random samples of the student populations; we would not expect them to be, as random cluster samples are much more practical to collect. However, we know that the schools participating in both countries were not randomly selected. The school authorities had to volunteer to participate. This is a big enough problem that we have devoted a special section to it. (See below: ***Problems with non-randomness.)***

The sample size ***condition*** differs among the various procedures, so we will address that as we go. But how do we choose which procedure to use? We have done *hypothesis tests* and *confidence intervals* and we have done these for one proportion and for the difference of proportions. And when we were faced with more than two proportions, we conducted a hypothesis test called a *test of independence.* How do we decide which of these to use?

Choosing between estimating and testing We have done two kinds of procedures—*hypothesis tests* and *confidence intervals*—and although there are connections between these procedures, the way these questions are worded will direct us to either a hypothesis test or calculating a confidence interval. Notice that some of the questions ask to ***estimate*** a population proportion or to find "how big" a difference is. A *confidence interval* is an estimate, and so for these questions we want a *confidence interval.* Other questions have the form: "Are there differences...?"or "Is the proportion different or the same as...?"or "Do the proportions differ...?" *All* of these kinds of questions indicate hypothesis tests. They all ask whether we have evidence that a statement is true or not true. So the first thing to do will be to see what kind of procedure is implied by the language.

Guideline to choosing a procedure

Decide, by reading carefully, whether the question implies a *confidence interval* or a *hypothesis test* by looking at the structure of the question and whether it appears to be asking for an estimate or a test.

Examples: choosing a procedure.

Example 3 : *Determine whether the proportion of monolingual students (those who speak just one language)in Australia differs significantly from the proportion of monolingual students in Canada.* We will of course calculate proportions of monolingual high school students in the two countries. However, the words *whether* and "differs significantly" suggest a hypothesis test. Since we have *two* proportions, we can use a ***compare proportions test.*** "Monolingual" means speaking just one language, and we see from the Software output that the test statistic is large and that the *p*-value is small. This shows that the test is *statistically significant* and that we do have evidence that the proportions of monolingual students are different in the two countries.

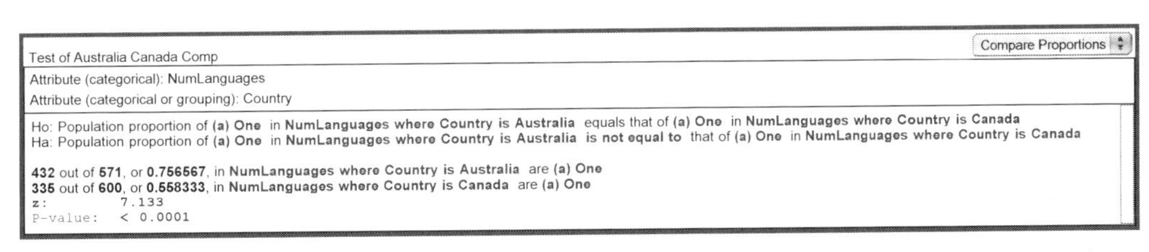

Example 4: *If there is a difference, how big is the difference between Australian and Canada in the proportion of monolingual students?* To infer about "how big" a difference is, the question is asking for an estimate rather than asking *whether* there is a difference. We need a confidence interval for the difference of the proportion of monolingual students in Australia as compared with Canada. Here is the Software output for the difference in proportions.

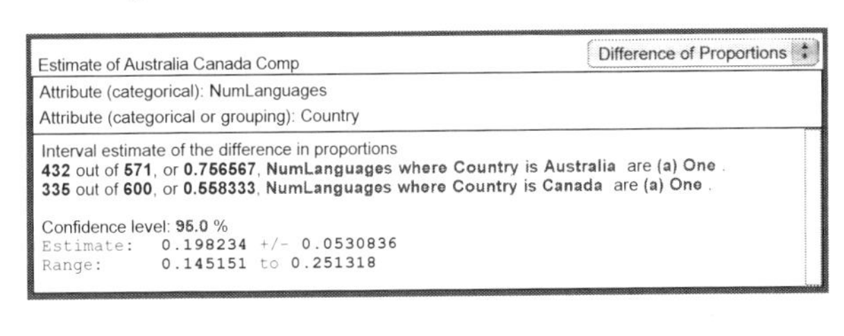

We can interpret this confidence interval by saying: "With 95% confidence, we can say that the proportion of monolingual high school students in Australia is 19.8% higher than the proportion of monolingual high school students in Canada, with a margin of error of 5.31%." Or we could say, "We are 95% confident that the proportion of monolingual high school students in Australia is from 14.5% to 25.1% higher than the proportion of monolingual high school students in Canada."

How a confidence interval can tell you about a hypothesis test Notice that both ends of the confidence interval in the calculation of Example 5 are positive. If one end of the confidence interval were negative and the other end were positive then the interval would include zero percent, or no difference, which would mean that, in the population, it is plausible that there is no difference in the proportions. In that scenario—where the confidence interval of two proportions *includes zero*—the hypothesis test would be *not statistically significant.* In our scenario—where the confidence *excludes zero*—a zero difference in proportions in the population is not plausible, and the hypothesis test will be *statistically significant.* So here is a situation in which getting the confidence interval also informs us about a hypothesis test implied by the confidence interval.

One has to be careful, as the formulas for the estimated sampling distribution standard deviation are slightly different; however, in most cases, a confidence interval for the difference of two proportions that includes zero indicates that the hypothesis test result will be *not significant*, and a confidence interval that *excludes zero* indicates a *significant* result to a hypothesis test. If you draw the sampling distribution for two proportions you will see why this will be so; a significant difference will come when the $\hat{p}_1 - \hat{p}_2$ is rare for the null hypothesized $p_1 - p_2 = 0$. However, a confidence interval based upon a rare $\hat{p}_1 - \hat{p}_2$ should *miss (exclude)* zero, whereas a confidence interval based upon a reasonably likely $\hat{p}_1 - \hat{p}_2$ will "hit" or include the null hypothesized $p_1 - p_2 = 0$.

The box summarizing this phenomenon is below. There are some situations where the confidence interval is more useful than the hypothesis test, since the hypothesis test can be read from the confidence interval.

Confidence Intervals and Hypothesis Tests for Differences of One Proportion:

If a confidence interval for one proportion

- *Excludes* zero then the associated hypothesis test will be *statistically significant.*
- *Includes* zero then the associated hypothesis test will be *not statistically significant.*

This principle works only if

- The level of confidence used is $1 - \alpha$ (for example, 95% with $\alpha = 0.05$), and
- The hypothesis test is a **two-sided test.**

More Examples: choosing a procedure. Example 5: *Determine if, in the population of all students, there are differences between the countries in the proportions of students who speak one, two, or three or more languages? (Or is NumLanguages independent of Country?)*

This particular question asks about all the proportions for the variable *NumLanguages*, and we see that there are three such proportions. The words "determine if" suggest a hypothesis test, but we cannot use a comparison of proportions test because we have three proportions. Actually, the alternate form of the question shows what is needed: *Is NumLanguages independent of Country*? indicates a *test of independence.*

Here is the Software output for the *test of independence*. Notice that the sample size conditions are met since all the *expected counts* are bigger than five. Since the *p*-value is very small, we can say that the differences of proportions that we see in the samples are unlikely to have arisen just from sampling variation from random sampling (if we had random sampling!). There would appear to be a *statistically significant* difference in the proportions of students in Canada and Australia in the proportions of student who speak one, two, or three or more languages. We are led to reject the null hypothesis that Australian and Canadian high school student populations have essentially the same proportions speaking one, two, and three or more languages. The students' language backgrounds differ.

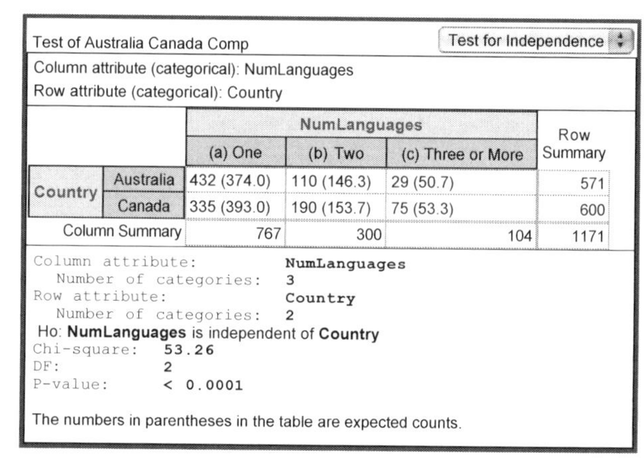

Test of Australia Canada Comp — Test for Independence

Column attribute (categorical): NumLanguages
Row attribute (categorical): Country

		NumLanguages			Row Summary
		(a) One	(b) Two	(c) Three or More	
Country	Australia	432 (374.0)	110 (146.3)	29 (50.7)	571
	Canada	335 (393.0)	190 (153.7)	75 (53.3)	600
Column Summary		767	300	104	1171

```
Column attribute:          NumLanguages
  Number of categories:    3
Row attribute:             Country
  Number of categories:    2
Ho: NumLanguages is independent of Country
Chi-square:    53.26
DF:            2
P-value:       < 0.0001
```

The numbers in parentheses in the table are expected counts.

Example 6: *Does the proportion of students in Canada who speak two or more languages differ from the standard of 50%, which we think describes the typical percentage among college students in Northern California?* Despite the absence of the words "determine whether," the wording "*Does the proportion...*" implies "does" or "does not," and that is the language of a hypothesis test. Moreover, a specific standard (50%), and just one population (Canada), is mentioned. All of this leads us to a two-sided hypothesis test for one proportion. The sample proportion can be calculated from

$P(\textit{Two or More} \mid C) = 1 - P(\textit{One} \mid C) = 1 - \frac{335}{600} \approx 1 - 0.558 = 0.442$. Since we are doing a hypothesis test for one proportion, we will call this $\hat{p} = 0.442$ and compare this to our idea that $p_0 = 0.50$. Here is output for this test. Again the *p*-value is extremely small, indicating statistical significance, so that we have evidence that the proportion of bilingual (or more) high school students in Canada is different from what we are accustomed to seeing in our California community colleges. Note: if the word "differ" in the question had been "greater than," we would be led to a one-sided and not a two-sided hypothesis test.

Example 7: *Can we estimate the proportion of all high school students in Australia who speak two or more languages?* The word "estimate" indicates that we want a confidence interval, and the other specifications indicate that we want a confidence interval for just one proportion. Here is Software output, whose interpretation is that we can be 95% confident that between 40.2% and 48.1% of all Australian students speak two or more languages, if all the conditions are met. We can apply the principle that connects confidence intervals with hypothesis tests for two proportions. The principle is: does the confidence *include* or *exclude* the null hypothesized value? Here the null hypothesized value is $p = 0.50$ (and not zero), so we see if the confidence interval *includes* or *excludes* 0.50. The confidence interval *excludes* $p = 0.50$, so the associated two-sided hypothesis test will be statistically significant.

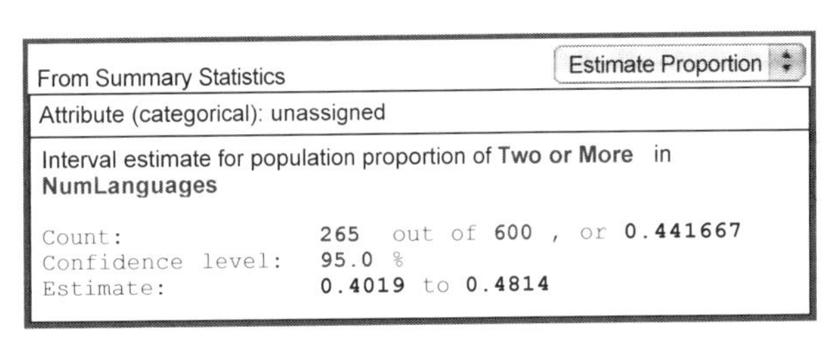
From Summary Statistics   Estimate Proportion
Attribute (categorical): unassigned
Interval estimate for population proportion of Two or More in NumLanguages

Count:	265 out of 600 , or 0.441667
Confidence level:	95.0 %
Estimate:	0.4019 to 0.4814

Confidence Intervals and Hypothesis Tests for One Proportion:

If a confidence interval for one proportion

- *Excludes* the hypothesized p_0 then the associated hypothesis test will be *statistically significant.*
- *Includes* the hypothesized p_0 then the associated hypothesis test will be *not statistically significant.*

This principle works only if

- The level of confidence used is $1 - \alpha$ (for example, 95% with $\alpha = 0.05$), and
- The hypothesis test is a **two-sided test.**

Problems with non-randomness

The most common thing that can go wrong is to use *non-random samples*. Strictly random samples are hard to collect, and it is quite common for all of these formal inferential procedures to be applied to samples that are not random samples, as we have done here—partly so that we can make this point.

(Note, the software cannot tell if our samples are random or not; it is up to the researcher.) We said at the outset that our samples were not strictly random samples of students and not even random cluster samples of schools. Had the schools been randomly selected then the analyses here would be approximately correct, although an organization such as Gallup would probably use slightly different formulas. Our samples are *voluntary response samples* in the sense that the school authorities volunteered to participate. For the Canadian Census at School, we are able to get some idea of how the participation of the schools under- and over-represents various parts of Canada. (The Census at School data in the first column is from http://www19.statcan.gc.ca/04/04_0809/prov/04_0809_prov_001-eng.htm, and the Enrollment by Province is from Statistics Canada, *Education Indicators in Canada: Report of the pan-Canadian Educators Indicators Program, catalogue 81-582-XIE, 2007.*)

We can see by comparing the percentages of enrollments by province to the percentages by province of secondary students who participated in the Census at School program that students in the provinces of Ontario, Nova Scotia, and Manitoba were overrepresented, whereas those from Saskatchewan and British Columbia and to some extent Quebec are underrepresented. How does this affect our conclusions about number of languages spoken by high school students? The answer is that we cannot say. It may be that if students in British Columbia and Quebec were proportionally represented (through random sampling), the percentage of bilingual students would be higher, but we are just guessing here. If that were so, it would make the contrast between Canada and Australia more striking. However, the contrast between Canada and California would not be as striking.

Province	**C@S participation (% by Province)**	**Enrollments (% by Province)**
Newfoundland	0.00	1.56
Prince Edward Island	0.00	0.45
Nova Scotia	6.50	2.95
New Brunswick	2.16	2.38
Quebec	18.90	21.66
Ontario	50.56	40.85
Manitoba	5.95	3.62
Saskatchewan	1.34	3.46
Alberta	9.49	10.80
British Columbia	4.74	11.79
Yukon Terr.	0.35	0.11
North West Territory, Nunavut	0.00	0.37
	100.00	100.00

What does one do in the face of no-random samples? There are several choices.

One choice is to do no inferential analyses at all. Since the conditions are not met, the calculations may be meaningless. Or, if you discover that another researcher has done inferential analyses where the conditions (especially the randomness condition) are not met, you can ignore the inferential part.

A second choice is to try to assess the direction of non-randomness. That is what we have done here (but only in a rudimentary way). Sometimes when this is done, one can argue that the direction of under- or overrepresentation would just make the contrasts sharper, as we have done here with the comparison with Australia. Other times, making a judgment about the direction of under- or overrepresentation will not be possible. In all situations, in reporting results, one has to be forthcoming and transparent about the quality of the data.

A third (and wise) choice is to consult an expert.

However, know that the application of inferential techniques to non-random data is widespread. On the next two pages there is a summary of formulas for inference for categorical variables.

Summary of Inference Procedures for Categorical Variables

Type	Confidence Interval	Test of Significance	Comments (Mostly Conditions)
One Proportion	$\hat{p} \pm z^* \sqrt{\frac{\hat{p}\cdot(1-\hat{p})}{n}}$ where $z^* = 1.645$ for a 90% Confidence Interval, $z^* = 1.96$ for a 95% Confidence Interval, $z^* = 2.576$ for a 99% Confidence Interval, and $\hat{p}$ is the sample proportion, and n is the sample size.	1. *Hypotheses:* $H_0: p = p_0$ $H_a: p \begin{cases} < p_0 \\ > p_0 \\ \neq p_0 \end{cases}$ 2. *Test Statistic* $z = \frac{\hat{p} - p_0}{\sqrt{\frac{p_0(1-p_0)}{n}}}$ 3. Use the Standard Normal Distribution Chart to find *p*-values	*The sample from which $\hat{p}$ is calculated is a simple random sample where the population is at least 10 times the sample size.* *For Confidence Intervals:* *Both* $n\hat{p} \geq 10$ *and* $n(1-\hat{p}) \geq 10$ *For Hypothesis Tests:* *Both* $np_0 \geq 10$ *and* $n(1-p_0) \geq 10$
Comparing Two Proportions	$(\hat{p}_1 - \hat{p}_2) \pm z^* \sqrt{\frac{\hat{p}_1(1-\hat{p}_1)}{n_1} + \frac{\hat{p}_2(1-\hat{p}_2)}{n_2}}$ where z^* as for one proportion, $\hat{p}_1$, $\hat{p}_2$ are the sample proportions, and n_1, n_2 the sample sizes for the two samples.	1. *Hypotheses:* $H_0: p_1 - p_2 = 0$ $H_a: p_1 - p_2 \begin{cases} < 0 \\ > 0 \\ \neq 0 \end{cases}$ 2. *Test Statistic* $z = \frac{(\hat{p}_1 - \hat{p}_2) - 0}{\sqrt{\hat{p}(1-\hat{p})\left[\frac{1}{n_1} + \frac{1}{n_2}\right]}}$ *where* $\hat{p} = \frac{\text{Total count of successes over both samples}}{\text{Total number of cases in both samples}}$ 3. Use the Standard Normal Distribution Chart to find *p*-values	*The samples from which $\hat{p}_1$ and $\hat{p}_2$ are calculated are simple random samples where the populations are at least 10 times the sample size.* *For Confidence Intervals:* $n_1\hat{p}_1 \geq 5$, $n_1(1-\hat{p}_1) \geq 5$, $n_2\hat{p}_2 \geq 5$, $n_2(1-\hat{p}_2) \geq 5$ *For Hypothesis Test,:* $n_1\hat{p} \geq 5$, $n_1(1-\hat{p}) \geq 5$, $n_2\hat{p} \geq 5$, $n_2(1-\hat{p}) \geq 5$

Type	Test of Significance	Comments (Mostly Conditions)
Test for Independence	1. *Hypotheses*: H_0 : *The two variables are independent* H_a : *The two variables are not independent* 2. *Test Statistic:* $\chi^2 = \sum_{All\ cells} \frac{(O-E)^2}{E}$ *where* *O = observed frequencies, E = expected frequencies* *And the expected frequencies are the frequencies expected if the null hypothesis is true. They can be calculated for a specific cell by:* $E = \frac{\text{(Row Total)(Column Total)}}{\text{Grand Total}}$ 3. Use the Chi-Square Distribution Chart to approximate *p*-values	*If inferring to a population, the sample containing the data must be a random sample.* *The expected frequencies must be ≥ 5.*

Summary: Hypothesis Tests and Confidence Intervals

- ***Guidelines to using inferential or descriptive procedures***
  - Descriptive analyses are always appropriate (even if inferential procedures are also used) and serve to illuminate the data at hand.
  - Inferential procedures should be chosen if the conditions are met, and there is an interest in generalizing *beyond* the data at hand.
  - If the conditions for inferential are not met, abandon inferential procedures or proceed knowing that the results may not be able to be generalized.
- ***Guidelines to choosing an inferential procedure*** Decide, by reading carefully, whether the question implies a *confidence interval* or a *hypothesis test* by looking at the structure of the question and whether it appears to be asking for an estimate or a test.
  - Language that implies making a decision ("whether, " "if," "Are the differences or not") are indicators that a hypothesis test is appropriate.
  - Language that implies an estimate (i.e., "estimate," "how big…") are indicators that a confidence interval is required.
- ***Confidence Intervals and Hypothesis Tests for the Difference of Two Proportions:***

  If a confidence interval for one proportion
  - *Excludes* zero then the associated hypothesis test will be *statistically significant.*
  - *Includes* zero then the associated hypothesis test will be *not statistically significant.*

  This principle works only if
  - The level of confidence used is $1 - \alpha$ (for example, 95% with $\alpha = 0.05$), and
  - The hypothesis test is a ***two-sided test.***
- ***Confidence Intervals and Hypothesis Tests for One Proportion:***

  If a confidence interval for one proportion
  - *Excludes* the hypothesized p_0 then the associated hypothesis test will be *statistically significant.*
  - *Includes* the hypothesized p_0 then the associated hypothesis test will be *not statistically significant.*

  This principle works only if
  - The level of confidence used is $1 - \alpha$ (for example, 95% with $\alpha = 0.05$), and
  - The hypothesis test is a ***two-sided test.***

§5.1 Inference for Quantitative Variables

Moving from categorical variables to quantitative, from proportions to means

All of Unit 4 was about confidence intervals and hypothesis tests for categorical variables. We looked in detail at sample proportions $\hat{p}$ and how we can infer beyond sample proportions to a population proportion p. Now, we move on to quantitative variables, and our focus will be on sample means $\bar{x}$ and how we can infer to population means μ. (For these sections, we leave behind proportions—temporarily.)

We looked at inferring from sample means $\bar{x}$ to population means μ in §3.2, when we studied the sampling distribution for a sample mean $\bar{x}$. Here is some useful notation to remember from that section. Again, we are moving from the world of proportions, $\hat{p}$ and p, to the world of quantitative variables and $\bar{x}$, s, and the population mean μ.

Notation for means and standard deviations

	Population Distribution	*Sample Distribution*	*Sampling Distribution*
Mean	μ	$\bar{x}$	$\mu_{\bar{x}}$
Standard Deviation	σ	s	$\sigma_{\bar{x}}$

The summary box about the sampling distribution for sample means $\bar{x}$ is also relevant to this section.

Facts about Sampling Distributions of Sample Means Calculated from Random Samples

Shape: The shape of the sampling distribution of $\bar{x}$ will be approximately Normal if the sample size n is sufficiently large, even if the population distribution from which the samples were drawn is not Normal.

Center: The mean of the sampling distribution of $\bar{x}$ is equal to the mean of the population distribution. That is, $\mu_{\bar{x}} = \mu$.

Spread: The standard deviation of the sampling distribution of $\bar{x}$ is $\sigma_{\bar{x}} = \frac{\sigma}{\sqrt{n}}$.

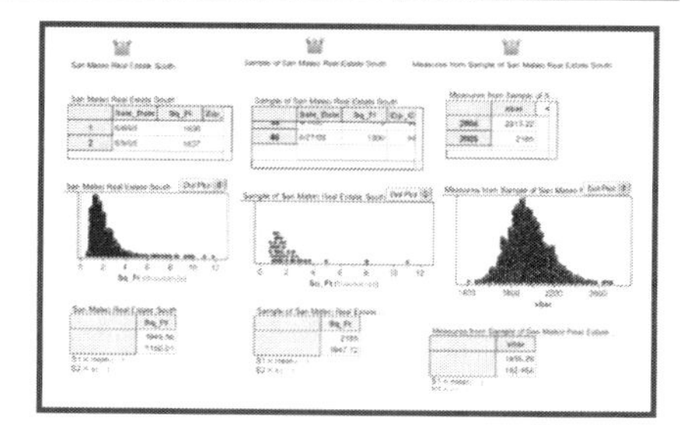

Although the graphic is not big enough to read here, the graphic showing the difference between a population distribution, a sample distribution, and a sampling distribution is a good picture to have in mind. You can refer to the large version in §3.2.

To see how we *infer* from $\bar{x}$ to μ, how the ideas of confidence intervals and hypothesis tests are worked out with quantitative variables, we consider an example.

Example: How long does it take to get to Seattle?

If you log on to https://www.alaskaair.com/shopping/ssl/shoppingstart.aspx for Alaska Airlines and specify Oakland to Seattle (as an example), you will see the screen something like the one displayed here. Besides the times that the flights depart and arrive, they also show the duration of the flight.

From: **Oakland (OAK)** To: **Seattle (SEA)** on Fri, Dec 18

Sort flights by: Prefer Alaska/Horizon

Operated By	Departs	Arrives	Flight	Stops	Hot Deals	Bargain	Value	Full Flex	First Class
Alaska View seat map	7:00 **am**	9:10 **am**	AS353 Total Duration: 2 hr 10 min	0		$97.00	$212.00	$227.00	$257.00
Alaska View seat map	9:10 **am**	11:12 **am**	AS355 Total Duration: 2 hr 2 min	0		$97.00	$212.00	$227.00	$257.00
Alaska View seat map	12:40 **pm**	2:45 **pm**	AS341 Total Duration: 2 hr 5 min	0		$97.00	$212.00	$227.00	$257.00
Alaska	3:40 **pm**	5:50 **pm**	AS345	0		$97.00	$212.00	$227.00	$257.00

For Oakland to Seattle, the duration of flight is just over two hours; the flights appear to last about 125 minutes. How could we check that these numbers for flight durations are correct? If we had a random sample of flights between Oakland and Seattle then, using hypothesis tests and confidence intervals, we could answer the questions:

- Is the average duration of the flights between Oakland and Seattle really 125 minutes?
- Can I get an estimate of how long, on average, the flights really are?

The first looks like a question for a hypothesis test, and the second question (since it asks for an estimate) looks like it could be answered by a confidence interval.

The Bureau of Transportation Statistics (BTS) collects detailed information on every commercial flight (see http://www.bts.gov/xml/ontimesummarystatistics/src/dstat/OntimeSummaryDepaturesData.xml), and we do have a random sample of flights from Oakland to Seattle. One of the variables measured by the BTS is the actual duration of each flight, which in our data has the name *ActualDuration*. The variable *ActualDuration* is a ***quantitative*** variable, and therefore the population mean for *ActualDuration* should be denoted with the Greek letter $\boldsymbol{\mu}$. Once again, we do not know the value of the population mean μ, but we *do* need a symbol to refer to what we do not know.

Using the Facts about Sampling Distributions to Get a Confidence Interval

To answer the second question above ("Can I get an estimate…?"), we want a confidence interval, and the formula for the confidence interval should be based on the sampling distribution for sample means $\bar{x}$. (Refer to the box about sampling distributions for sample means $\bar{x}$ on the previous page.) For a confidence interval, the basic idea was to calculate the interval

$(\text{Sample Estimate}) \pm (\text{Margin of Error})$ or, in terms of the sampling distribution:

$(\text{Sample Estimate}) \pm z^*(\text{Standard deviation of the sampling distribution})$. For proportions, we had

$\hat{p} \pm (\text{Margin or Error}) = \hat{p} \pm z^* \sqrt{\dfrac{\hat{p}(1-\hat{p})}{n}}$ because the standard deviation of the sampling distribution

was $\sqrt{\dfrac{p(1-p)}{n}}$. So, it makes sense that the formula for the confidence interval to estimate a population

mean μ from a sample mean $\bar{x}$ to be: $\bar{x} \pm z^*\left(\dfrac{\sigma}{\sqrt{n}}\right)$. This formula makes sense because the ***Facts about Sampling Distributions of Sample Means…*** tells us that the sampling distribution will be approximately Normal (so that is why the z^* is there), and the standard deviation of the sampling distribution is $\dfrac{\sigma}{\sqrt{n}}$.

A sample of flights: To make things more concrete, we have flight data from Oakland to Seattle for March 2009, and from these data we have a simple random sample of just $n = 20$. The first thing we must do is a descriptive analysis of the sample data. Here is a dot plot of the duration of our twenty flights (the variable is *ActualDuration*). Looking at the plot, it appears that most of the flights actually took less time than the hypothesized two hours and five minutes (125 minutes), although there were some flights that were longer.

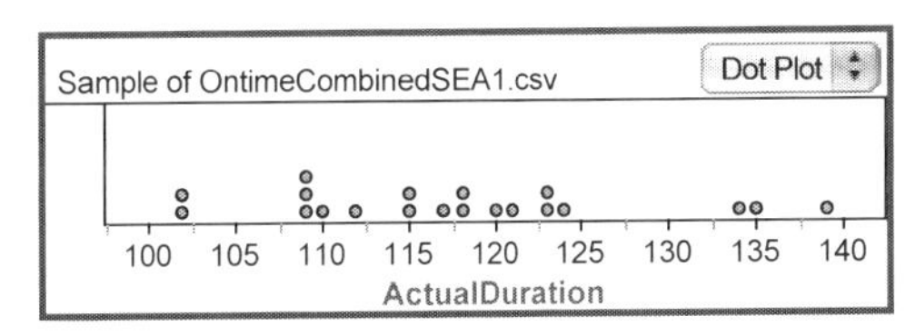

The next thing is to calculate the sample mean $\bar{x}$ and the sample standard deviation s. We find that $\bar{x} = 117.75$ minutes, and that $s \approx 10.151$ minutes. You may begin to think that perhaps the mean duration of 125 minutes is wrong, if we go by this small sample.

Sample of OntimeCombinedSEA1.csv

	ActualDuration
	117.75 10.1508

S1 = mean( )
S2 = s( )

Calculating the Confidence Interval: Trouble! When we go to calculate a 95% confidence interval, we can calculate $\bar{x} \pm z^* \frac{\sigma}{\sqrt{n}} = 117.75 \pm 1.96 \frac{\sigma}{\sqrt{20}}$. However, we realize that we do not know the population standard deviation σ. What we have is the sample standard deviation s. If we use the sample standard deviation $s \approx 10.151$ in place of the unknown population standard deviation σ, we would then calculate:

$$\bar{x} \pm z^* \frac{\sigma}{\sqrt{n}} = 117.75 \pm 1.96 \frac{10.151}{\sqrt{20}} \approx 117.75 \pm 1.96(2.2698) \approx 117.75 \pm 4.45 .$$

This calculation gives us an estimated population mean of $\mu = 117.75$ minutes with a margin of error of 4.45 minutes.

However, can we do this? Are we allowed to simply substitute the sample standard deviation s for the population standard deviation σ? The trouble we have is that there is another source of variability—another source of randomness—in our calculations, and we do not know whether the sampling distribution will still have the shape of a Normal distribution or something different. If we knew the population standard deviation σ then we would also know that the sampling distribution of the sample mean will be approximately Normal. But if we do *not* know the population standard deviation σ then it is possible that the sampling distribution is not Normal but some other shape. Our next job is to understand what happens to the sampling distribution when we do not know σ and want to use the sample standard deviation in its place.

Finding the Sampling Distribution of $\bar{x}$ When Sigma Is Not Known

Once again, you may find it helpful to follow along with the software simulation; we will do a similar "simulation" of the sampling distribution to what we have done before. Recall that we started with a population and had software take many random samples from the population, and for each sample that was taken, Software calculated the sample mean. We then collected together these sample means into a collection. This collection was our approximation of the sampling distribution; it was an approximation because we knew that the complete sampling distribution would require something like 10^{80} samples, which would take too much of our time.

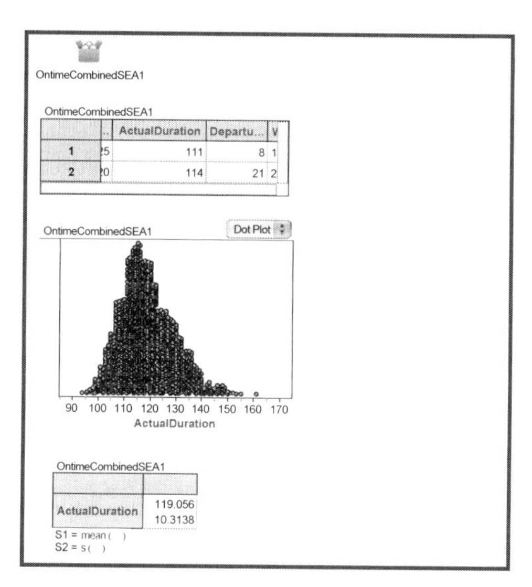

In this simulation, we will have software calculate what is called a ***t-score*** $t = \frac{\bar{x} - \mu}{\frac{s}{\sqrt{n}}}$ instead of calculating just $\bar{x}$ each time a sample is taken. (You may notice that a ***t-score*** has a form like a ***z-score*** using the mean and standard deviation of the sampling distribution.) We have software calculate the *t-score* so that we can see the influence that the sample standard deviation s has on the sampling distribution. For any sample that we have, we want to include the variability that comes from the sample standard deviation s as well as the variability that comes from the sample mean $\bar{x}$. Follow the bullets; what you see on the screen should be similar to the graphics shown here. We start with *all* of the data for a month; we are treating all these data as the population.

- Open the file ***OntimeCombined4SEA***.
- Get a ***Dot Plot*** of the variable *ActualDuration* and summary showing the mean and the standard deviation of the variable *ActualDuration.* The results should be similar to the left side of the graphic shown below; as before, the left side represents the population.

What we see is the population distribution of the variable *ActualDuration.* We see that the shape of the distribution is nearly symmetrical, with perhaps just a hint of right skewness.

- Use software to collect a sample of size $n = 20$ from the collection ***OntimeCombined4SEA*** that includes the variable *ActualDuration.* Arrange the screen so that the population and the sample are both shown, as here.
- Get summary statistics for the variable *ActualDuration* including the mean and the standard deviation.

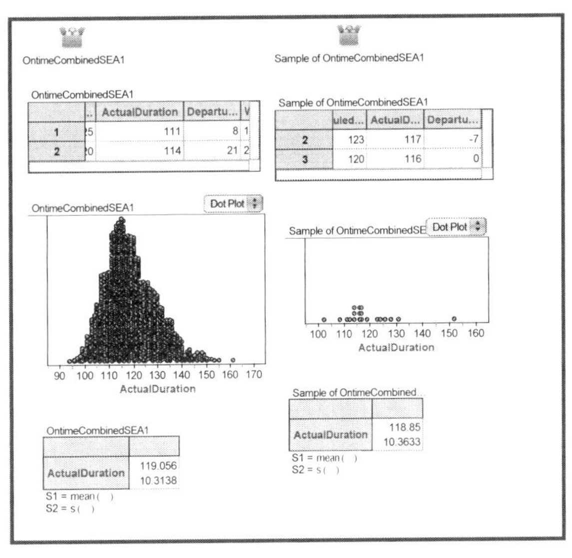

This second set of results is the sample distribution for one random sample of size $n = 20$. Now we are ready to get Software to start building a sampling distribution. We will use the actual population mean $\mu = 119.056$, but because we want to see the effect of using *sample standard deviation* s, we will not use the population standard deviation.

- Have software calculate the t score for this first sample.
- Then, have software do this for a huge number of samples. The way this is done will vary depending on the software. See ***Simulating a Sampling Distribution*** in the ***Software Supplement***.

The software results should show three things: the population, the sample and the sampling distribution. The graphic on the next page comes from the software package called Fathom; different packages have slightly different formats.

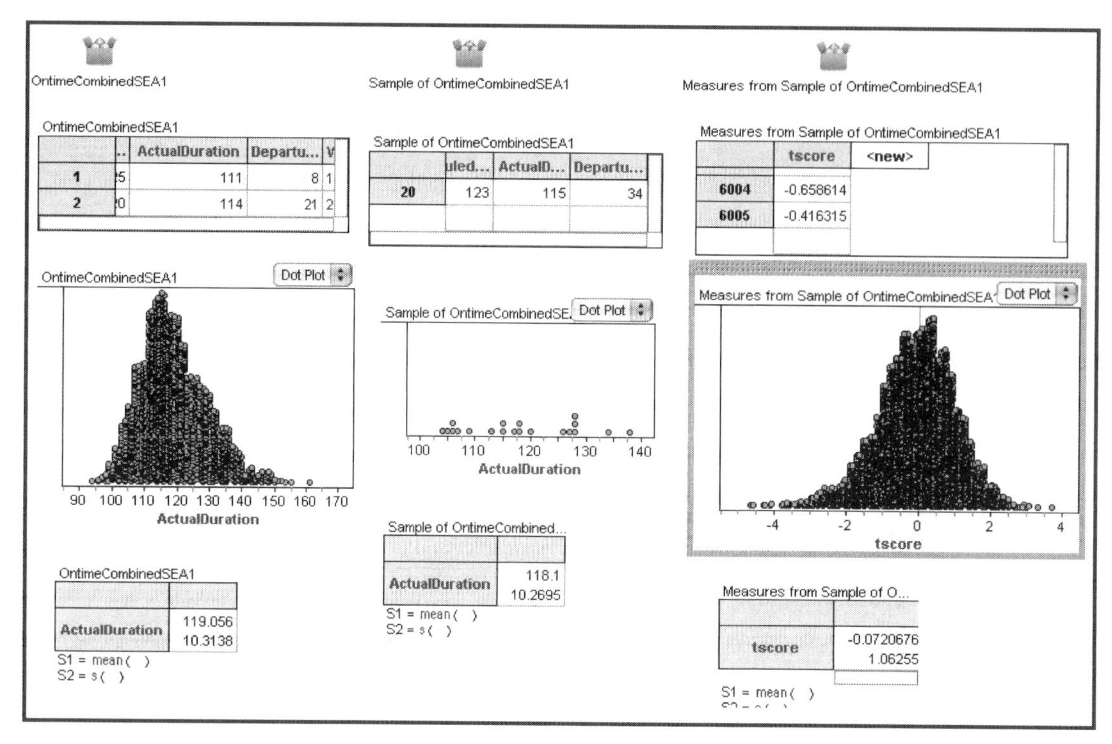

What we see from our simulation. Look at the plot on the right, which is our simulation of the sampling distribution. Notice that it is nearly symmetrical, although there is a hint of left skewness. If this were a Normal distribution then the empirical rule tells us that we expect 99.7% of the distribution (that is nearly 100%) to be within three standard deviations. But here we see that (especially on the left-hand side) the distribution stretches beyond three standard deviations to four standard deviations.

The effect of using the sample standard deviation s instead of the population standard deviation σ gives a distribution that still looks like a Normal distribution but has ***thicker tails*** than a Normal distribution.

This distribution is called a ***t-distribution*** or a ***"Student's t-distribution."*** More importantly for our calculations: the ***t-distribution*** is the sampling distribution that arises when we substitute the standard deviation s instead of the population standard deviation σ in our calculations for a hypothesis test or a confidence interval.

Introduction to the t distributions

Like the chi-square distributions, there are many different t distributions. On the op of the next page are graphics of some t distributions for different degrees of freedom.

Notice that all of the distributions "look Normal" in that all of the t distributions are symmetrical. However, some of the t distributions have markedly thicker tails than the Normal distribution does. They differ by their ***degrees of freedom,*** an idea we encountered when studying the ***chi-square distributions.***

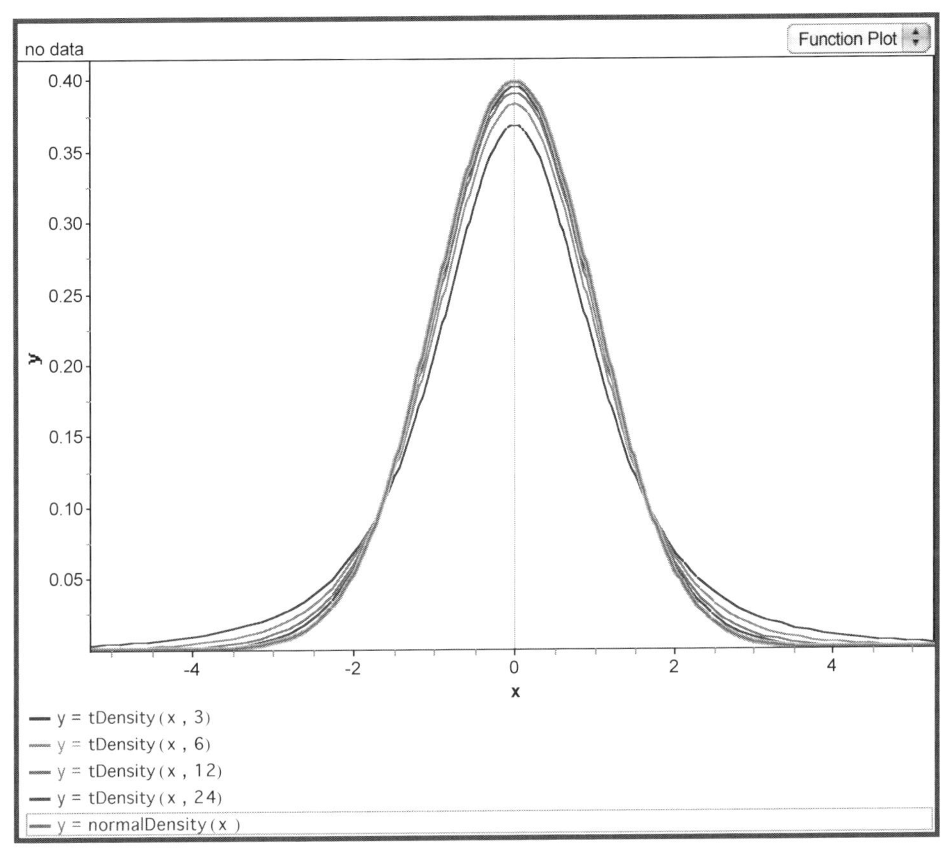

Here are more facts about t distributions that we will use.

Facts about t-distributions

Compared with the standard Normal distribution, Normal with μ =0 and σ =1, the ***t distributions***:

1. Are perfectly symmetrical with $\mu = 0$ (which is also true for the standard Normal distribution)
2. Have "thicker tails" than the standard Normal distribution
3. The "thickness of the tails" is greater the *smaller* the ***degrees of freedom*** where ***df* = *n* - *1***.
4. The *bigger* the *degrees of freedom,* the more a ***t distribution*** approximates the Normal distribution.

With the ***t distributions***, the degrees of freedom depend upon the sample size, $df = n-1$, and so for the example that we have been looking at, we should have $df = n-1 = 20-1 = 19$. Our simulation calculated a t score that acts something like a z score. What a z score does is to translate to the standard Normal distribution, in the Normal Distribution Chart, the Normal distribution with μ =0 and σ =1; the t score does a similar translation when we are using the sample standard deviation s.

Using the t distributions: the plot. First we need a definition. Recall that for the Normal distribution we could say that between $z^* = -1.96$ and $z^* = 1.96$ there would be 95% of the distribution and that in each of the tails beyond the $z^* = -1.96$ and $z^* = 1.96$ there will be 2.5%, adding up to 5%. Because the t distributions have thicker tails than the standard Normal distribution, the corresponding number that includes 95% of the t distribution will always have an absolute value bigger than 1.96. To extend this idea to the t distributions, we use the term ***critical value*** and the symbol ***t**** as defined in the box just below.

Definition of a Critical Value, t*

A ***critical value t**** indicates a value of t that gives a specified percentage (or probability) of a t distribution in the tails—to the right of **t*** or to the left of **−t***—and therefore a specified percentage between **−t*** and **t***.

Notice that the definition of a critical value is very general. The picture illustrates the area between $-t^*$ and t^* for 95. For our example, with $df = n - 1 = 20 - 1 = 19$ instead of $z^* = -1.96$ and $z^* = 1.96$, the values of the t distribution that enclose 95% are the critical values $t^* = -2.093$ and $t^* = 2.093$.

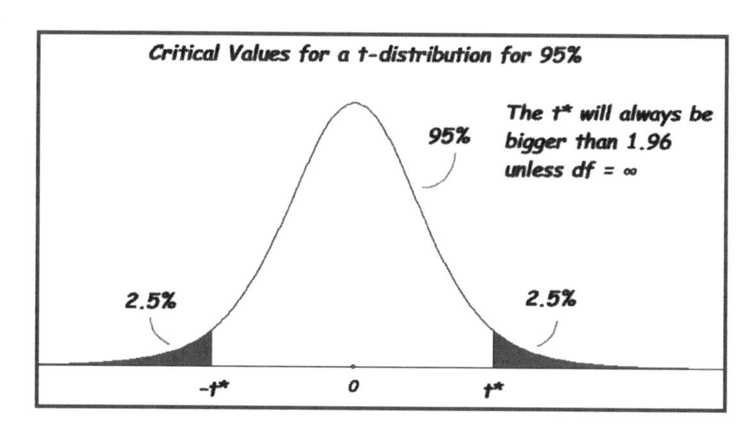

How do we know this number $t^* = 2.093$? As with the chi-square distributions, we have a table that shows these critical values. Part of the table is shown below, where we discuss how to use the table.

What do we do with this number? It is the number that we must use in a confidence interval *instead* of $z^* = 1.96$, so that the confidence interval for the actual duration of flights to Seattle (for *all* flights) will be $\bar{x} \pm t^* \frac{s}{\sqrt{n}} = 117.75 \pm 2.093 \frac{10.151}{\sqrt{20}} \approx 117.75 \pm 2.093(2.2698) \approx 117.75 \pm 4.75$ or the interval 113.0 to 122.50 minutes.

Using the t distributions: the chart. For the infinitely many t distributions, we have a very condensed chart that shows the relationship between the probabilities (or percentages) in the tails of the distributions and the values of the critical values and therefore the probability between **−t*** and **t*** for many different degrees of freedom. Here (on the next page) is a piece of the chart, but to follow the discussion here, you should have the entire chart in your hand or on the screen.

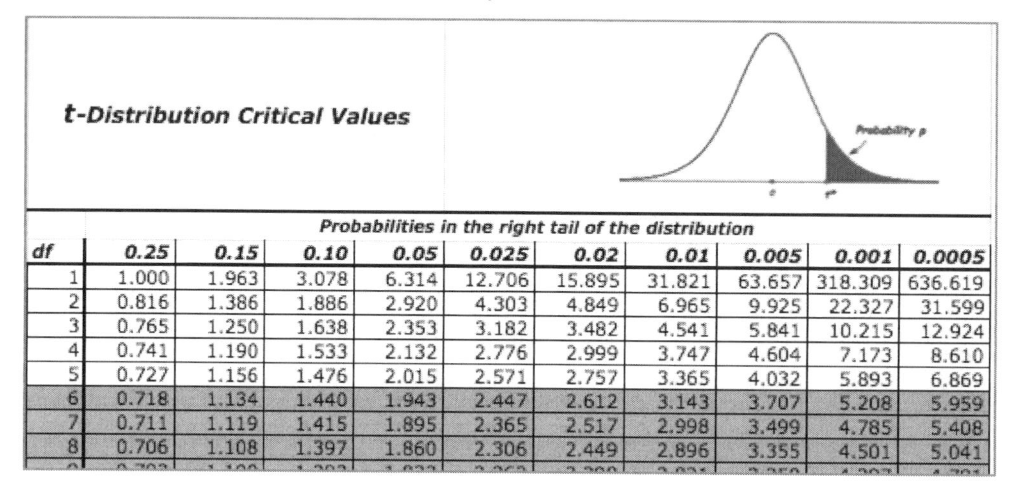

t-Distribution Critical Values

	Probabilities in the right tail of the distribution									
df	0.25	0.15	0.10	0.05	0.025	0.02	0.01	0.005	0.001	0.0005
1	1.000	1.963	3.078	6.314	12.706	15.895	31.821	63.657	318.309	636.619
2	0.816	1.386	1.886	2.920	4.303	4.849	6.965	9.925	22.327	31.599
3	0.765	1.250	1.638	2.353	3.182	3.482	4.541	5.841	10.215	12.924
4	0.741	1.190	1.533	2.132	2.776	2.999	3.747	4.604	7.173	8.610
5	0.727	1.156	1.476	2.015	2.571	2.757	3.365	4.032	5.893	6.869
6	0.718	1.134	1.440	1.943	2.447	2.612	3.143	3.707	5.208	5.959
7	0.711	1.119	1.415	1.895	2.365	2.517	2.998	3.499	4.785	5.408
8	0.706	1.108	1.397	1.860	2.306	2.449	2.896	3.355	4.501	5.041

Notice that on the left-hand side of the chart are the degrees of freedom or ***df***, like the Chi-Square Distribution Chart. (The construction of this chart is basically the same as the Chi-Square Chart, but the distributions are quite different.)

The numbers across the top of the chart give the probabilities (or proportions, or areas) in the right tail of the *t* distribution. Hence, for the probability of 0.025 in the tail of the distribution, we see that for $df = 5$ (which means that the sample size must have been $n = 6$), we have a $t^* = 2.571$. Since the *t* distributions are perfectly symmetrical, we know that the probability to the left of $t^* = -2.571$ —that is, in the left tail—will also be 0.025, and the probability in the two tails together is 0.05 or 5%. Hence the probability between $t^* = -2.571$ and $t^* = 2.571$ is 0.95 or 95%. So if we really did have only $n = 6$ then in the confidence interval formula $\bar{x} \pm t^* \frac{s}{\sqrt{n}}$ we would use $t^* = 2.571$ and *not* $z^* = 1.96$. It is worthwhile to look at the foot of the *t* Distribution Chart and see that the connection between the probability on the right-hand tail and the confidence level is made explicit. The last row shows the values of z^* showing that "at the limit" the *t* distributions get closer and closer (as close as we want for sufficiently large *n*) to the Normal distribution.

250	0.675	1.039	1.285	1.651	1.969	2.065	2.341	2.596	3.123	3.330
400	0.675	1.038	1.284	1.649	1.966	2.060	2.336	2.588	3.111	3.315
1000	0.675	1.037	1.282	1.646	1.962	2.056	2.330	2.581	3.098	3.300
∞	0.674	1.036	1.282	1.645	1.960	2.054	2.326	2.576	3.091	3.291
	50%	70%	80%	90%	95%	96%	98%	99%	99.5%	99.9%
	Confidence Level C									

Things to notice and rounding down. Notice that not all the possible degrees of freedom are shown on the chart and also notice that the values for the t^* are not very different for large degrees of freedom. It is good practice in using the *t* Distribution Chart to "round down," giving a slightly larger t^*. So, if we had a sample size of $n = 140$, so that the $df = 139$, we would use $t^* = 1.657$, which is the value for $df = 125$. Most often we will be using software, so the *t* Distribution Chart and the graphiccs are here primarily to get the idea of the relationships between the *critical values t**, the tail probabilities, and the degrees of freedom.

Applying the t distributions, including conditions for use

Confidence Intervals Here is the Software output for the 95% confidence interval for the sample of $n = 20$ of the flight times from the San Francisco Bay Area to Seattle. One of the big reasons for studying the *t* distributions in some detail is that statistical software generally uses the *t* distributions in their calculations of confidence intervals and the test statistics for hypothesis tests involving means of quantitative variables. The software does this because *in practice* researchers do not know the population standard deviation σ.

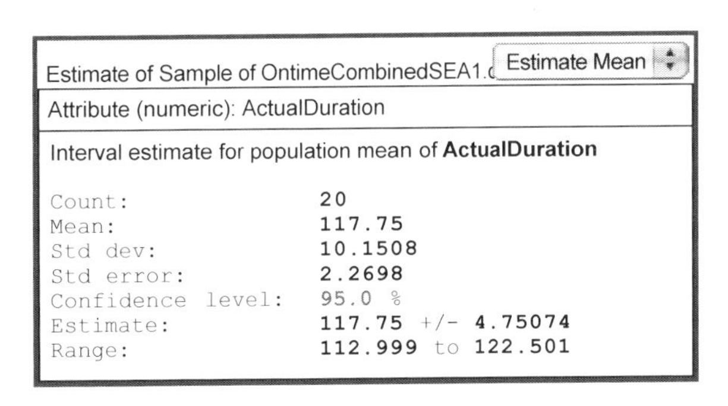

Researchers have samples (or experimental data), so they can always calculate the sample standard deviation *s*. (Even though, with large sample sizes, the *t* distributions are nearly the same as the standard Normal distribution so that it would be feasible to work with the Normal distributions if we were working by hand, statistical software developers tend to use the *t* distributions for all the calculations involving sample means $\bar{x}$.)

Formula for Confidence Intervals for One Population Mean

$$\bar{x} \pm t^* \frac{s}{\sqrt{n}}$$

where t^* depends upon the $df = n-1$ and the confidence level chosen.

This formula can be written $\bar{x} \pm t^*(SE)$, where $SE =$ Standard Error $= \frac{s}{\sqrt{n}}$, so

$ME =$ Margin of Error $= t^*(SE)$.

Be careful not to confuse ***standard error*** with ***margin of error. Standard error*** refers in this context to the sample estimate of the standard deviation of the sampling distribution, and so $SE = \frac{s}{\sqrt{n}}$, whereas ***margin of error*** refers to $t^* \frac{s}{\sqrt{n}}$. Software output often gives both the standard error and also the margin of error, but the margin of error needs to be read as the ± portion of the "estimate" line; in our example shown above, $ME = 4.75074$.

Interpretation of Confidence Intervals for Means The interpretation of a confidence interval for a population mean follows the same pattern as for proportions, so for our estimate of the population mean duration of flight from the San Francisco Bay Area to Seattle, we would say either:

- "With 95% confidence, we can say that the mean duration of all flights from the SF Bay Area to Seattle is 117.75 minutes with a margin of error of 4.75 minutes," or:
- "We are 95% confident that the mean duration of all flights from the SF Bay Area to Seattle is between 113.0 minutes and 122.5 minutes."

Conditions with t distributions

There are two basic conditions for using the *t* distributions for inference.

Conditions for Using the t Distributions for Inference

1. The sample or samples used must be random.
2. The population distribution or distributions must be Normal.

These conditions look rather restrictive; we have encountered many variables whose sample distributions are quite skewed, and we know that random samples are hard to get. It turns out that one—but only one—of these conditions can be "relaxed" or "violated" and the *t* distributions will still yield trustworthy results. The ***random sampling condition*** *cannot* be relaxed; to have trustworthy confidence intervals or results from hypothesis tests, we *must have random sampling.*

Robustness and the 15/40 Rule of Thumb Statisticians call a procedure (such as a formula for a confidence interval) ***robust*** if the procedure gives trustworthy results even when one or more of the conditions for use are relaxed or violated. There are limits to the violation of the conditions, and therefore ***rules of thumb*** have been developed to guide the researcher in using the procedure.

Sometimes there are different rules of thumb followed by different statisticians, and sometimes there are arguments about the robustness of procedures. For the *t* procedures, the Normality condition can be relaxed according to the ***15/40 rule of thumb.*** This rule of thumb relates sample size to the Normality condition of the *t* procedures. The basic idea is that the bigger the sample size, the less important is the Normality condition.

The 15/40 Rule of Thumb

- *Small Samples, $n < 15$* For samples where $n < 15$, the data *must* come from a Normal or nearly Normal distribution. The *t* procedures will not be trustworthy otherwise.
- *Moderate-sized Samples, $15 \leq n \leq 40$* The *t* procedures may cope with some skewness but will not be trustworthy in the face of strong skewness or outliers.
- *Large Samples, $n > 40$* The *t* procedures are *robust* in the face of skewness or outliers where the sample is large and the more so the larger the sample size.

Assessing whether a population is Normal How does one know if the population distribution is Normal or not? One way is by examining the sample distribution to see if there is skewness in the sample distribution. This is a good guide if the sample size n is large enough so that the shape of a distribution can be assessed from a dot plot or stem plot. However, with small samples (such as the n = 20 we were working with), it may not be possible to detect shape. If that is so then what you are left with is thinking carefully about what the population distribution should be; this is actually what statisticians regularly do. For example, for the duration of flight data, we had reason to believe that there would be some flights whose durations would be longer than the scheduled durations because of bad weather and other circumstances, but because the flights were between California and Seattle, we thought that there would be very few such flights.

Meaningful data and nonsensical data It is always wise to look at data carefully and think about the data. We expect the variable *Height* to be Normally distributed, but when looking at the Australian Census at School data, we find some high school students who are apparently one thousand centimeters tall. Not likely! More likely some Australian high-schoolers were not taking the exercise seriously. In the flight data, flights that were cancelled obviously had an *ActualDuration* of zero minutes—completely true, but this is not meaningful for the analysis.

Why we have the t distributions and why they are called "Student's t"

The answers to these questions are related. The *t* test and other procedures were developed in the early part of the twentieth century by William Sealy Gosset. Gosset had degrees in both chemistry and mathematics and worked for the Guinness Brewery in Dublin, Ireland. At the brewery he developed the *t* procedures to aid quality control. His problem was that he only had small samples to work with, and, as we have seen, if we have small samples, we dare not simply use the sample standard deviation s in place of the population standard deviation σ. Gosset's work was original, but when he went to publish his results in scientific journals, the brewery insisted that he not publish under his own name; Gosset chose to publish under the name "Student," and so the *t* procedures have (since that time) often been referred to as "Student's t." For more information, see http://www.gap-system.org/~history/Biographies/Gosset.html.

Summary: Introduction to Inference for Quantitative Variables

- With quantitative variables, we infer from sample means $\bar{x}$ to population means μ. The notation used is shown here.

Notation for means and standard deviations

	Population Distribution	*Sample Distribution*	*Sampling Distribution*
Mean	μ	$\bar{x}$	$\mu_{\bar{x}}$
Standard Deviation	σ	s	$\sigma_{\bar{x}}$

- ***Facts about Sampling Distributions of Sample Means Calculated from Random Samples***
  - ***Shape:*** The shape of the sampling distribution of $\bar{x}$ will be approximately Normal if the sample size n is sufficiently large, even if the population distribution from which the samples were drawn is not Normal.
  - ***Center:*** The mean of the sampling distribution of $\bar{x}$ is equal to the mean of the population distribution. That is, $\mu_{\bar{x}} = \mu$.
  - ***Spread:*** The standard deviation of the sampling distribution of $\bar{x}$ is $\sigma_{\bar{x}} = \frac{\sigma}{\sqrt{n}}$.
- The sampling distribution of sample means $\bar{x}$ presents the researcher with a practical problem in that the population standard deviation σ is not known, so the formulas implied by the ***Facts*** above cannot be applied directly, with the sample standard deviation s substituted for σ. Instead, if s is used then the family of ***t distributions*** must be used.
- ***Facts about t-distributions*** Compared with the standard Normal distribution, Normal with μ =0 and σ =1, the ***t distributions***:
  - Are perfectly symmetrical with $\mu = 0$ (which is also true for the standard Normal distribution)
  - Have "thicker tails" than the standard Normal distribution
  - The "thickness of the tails" is greater the *smaller* the **degrees of freedom** where $df = n - 1$.
  - The *bigger* the *degrees of freedom,* the more a ***t distribution*** approximates the Normal distribution.
- ***A critical value t**** indicates a value of t that gives a specified percentage (or probability) of a t distribution in the tails—to the right of ***t**** or to the left of ***–t**—**and therefore a specified percentage between ***–t**** and ***t****.

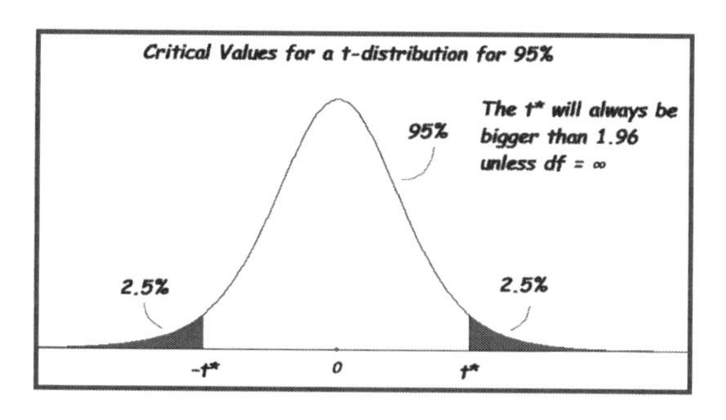

- ***Formula for Confidence Intervals for One Population Mean***

$$\bar{x} \pm t^* \frac{s}{\sqrt{n}}$$

where t^* depends upon the $df = n-1$ and the confidence level chosen. This formula can be written

$\bar{x} \pm t^*(SE)$, where $SE =$ Standard Error $= \frac{s}{\sqrt{n}}$, so $ME =$ Margin of Error $= t^*(SE)$ ***PTO.***

- ***Conditions for Using the t Distributions for Inference***
  - The sample or samples used must be random.
  - The population distribution or distributions must be Normal.
- A ***robust*** procedure in statistics is one that gives trustworthy results even if the conditions for its use are not met. The procedures for inference based upon *t distributions* are robust if the population distributions are not Normal but not robust if randomization is not used. In the situation that the population distributions are not Normal, the ***15/40 rule of thumb*** (below) may be used as a guide.
- ***The 15/40 Rule of Thumb***
  - *Small Samples, n < 15* For samples where $n < 15$, the data *must* come from a Normal or nearly Normal distribution. The *t* procedures will not be trustworthy otherwise.
  - *Moderate-sized Samples, 15 ≤ n ≤ 40* The *t* procedures may cope with some skewness but will not be trustworthy in the face of strong skewness or outliers.
  - *Large Samples, n > 40* The *t* procedures are *robust* in the face of skewness or outliers where the sample is large and the more so the larger the sample size.

§5.2 Hypothesis Testing for One Mean

How long it takes to get to Seattle, continued

In the last section we examined the duration of flights from the San Francisco Bay Area to Seattle. Looking at the advertised duration of flights, we posed two inferential statistical questions:

- Is the average time of all flights 125 minutes or not?
- Can I get an estimate of how long, on average, the flights really are?

Is there? Are there? Or estimate... We found that to do inference for quantitative variables (inferring from sample means $\bar{x}$ to population means μ, we had to use procedures based on the *t distributions* and not the Normal distribution). We were able to calculate an estimate of the population mean duration of flights μ:

$$\bar{x} \pm t^* \frac{s}{\sqrt{n}} = 117.75 \pm 2.093 \frac{10.151}{\sqrt{20}} \approx 117.75 \pm 2.093(2.2698) \approx 117.75 \pm 4.75$$

This gave us an interval estimate for the population mean of $113.00 < \mu_{\text{Actual Duration}} < 122.50$ minutes, 95% confidence interval with a margin of error of 4.75 minutes. To do this calculation, we found that we used the *t* distributions, with the sample standard deviation *s* in the calculation.

However, the first question posed above implies a hypothesis test where $H_0 : \mu = 125$ minutes, and $H_a : \mu \neq 125$ since it is a version of an "Is there..." or "Are there..." type of statistical question.

For proportions, we had ***five steps*** for a hypothesis test. We will follow these five steps, with an additional preliminary step to look at the data.

Step 0: Looking at our sample: First, we do a descriptive analysis of the sample data. We have already seen that one of the ***conditions*** for using the *t* distributions is that the population from which we sample is Normal. One of the ways that we can sometimes judge the Normality of the population is shape of the sample distribution. We also know that the shape of the population distribution is important if the sample size is small—the ***15/40 rule.*** However, with a small sample size, it is sometimes hard to detect the shape of the distribution; all we can say from the sample about the shape of the distribution is that there does not appear to be marked skewness; what we see suggests symmetry.

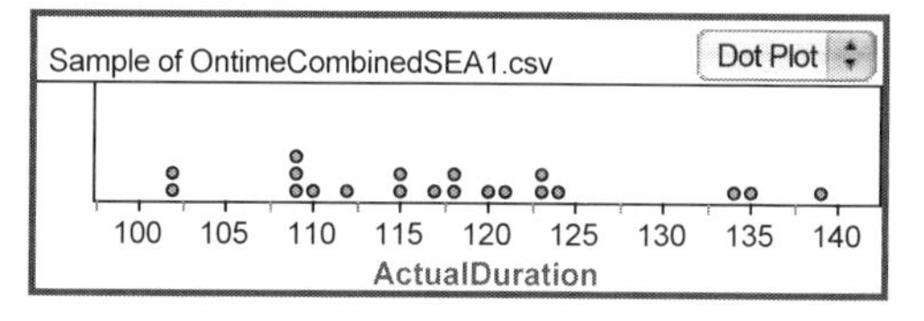

Sample of OntimeCombinedSEA1.csv

	ActualDuration
	117.75
	10.1508

S1 = mean( )
S2 = s( )

Next, we calculate the sample mean $\bar{x}$ and the sample standard deviation *s*. Again, we need to have notation to help distinguish the sample means and standard deviations from the population means and standard deviations. For a hypothesis test, we must have notation to identity our ***hypothesized population mean.*** The notation is shown below.

Step 1: Setting up the hypotheses: The first step in doing a hypothesis test is to set up the hypotheses, and with that we will need notation to keep track of the hypothesized population mean. On the next page, the notation is introduced.

Notation for Hypothesis Testing for Means

	Population	Sample
Mean:	μ	$\bar{x}$
Standard Deviation:	σ	s

- The *null hypothesis* that the population mean μ is equal to μ_0 is written: **H_0: $\mu = \mu_0$**
- The *alternate hypothesis* that the population mean μ is not equal to μ_0, written: **H_a: $\mu \neq \mu_0$**

  where μ_0 is the value of the hypothesized mean.

The μ_0 here functions in the same way that the p_0 did in hypothesis tests for proportions; it represents the value of the population mean μ that we are testing, but now we are dealing with *means* and *not* proportions. For our example of the duration of flights to Seattle, we would write:

$$H_0 : \mu = 125 \text{ minutes}$$
$$H_a : \mu \neq 125 \text{ minutes}$$

Step 2: Checking the conditions: The first condition says that the sample must be drawn randomly from the population. These data are a random sample of the information on the Bureau of Transportation Statistics website, so we have satisfied the randomness condition. For the Normality condition, we would be a bit uncertain if we had only the sample, although (see ***Step 0*** above) the sample does not reveal extreme skewness or outliers. The ***15/40 rule*** says that with $n = 20$ we should be wary of skewness in the population distribution.

Step 3: Calculating the test statistic: When we did hypothesis tests for proportions, we used the z score (to get the p-value) because the sampling distribution for $\hat{p}$ is a Normal distribution. If we use the sample standard deviation s in our calculations then the sampling distribution involves the t distributions, and so the test statistic will be the ***t score*** (introduced in §5.1), which the equivalent of the z score. The t score has the same form as the z score but uses the sample standard deviation instead of the population standard deviation σ.

$$t = \frac{(\text{Sample result}) - (\text{Mean of Sampling Distribution})}{\text{Standard Deviation of Sampling Distribution}} = \frac{\bar{x} - \mu_0}{\frac{s}{\sqrt{n}}}$$

In our example, the sample mean is $\bar{x} = 117.75$ minutes, the sample standard deviation is $s = 10.151$, the sample size is $n = 20$, and our hypothesized mean is $\mu_0 = 125$ minutes, so that we calculate:

$$t = \frac{\bar{x} - \mu_0}{\frac{s}{\sqrt{n}}} = \frac{117.75 - 125}{\frac{10.151}{\sqrt{20}}} \approx -3.194 .$$

Test Statistic for Testing a Single Population Mean μ

$t = \dfrac{\bar{x} - \mu_0}{\frac{s}{\sqrt{n}}}$ where μ_0 is the value used in the null hypothesis H_0: $\mu = \mu_0$.

Step 4: Finding the p-value: A reminder of what a *p*-value is (see *§4.2*): a *p*-value is the probability of getting a *test statistic as extreme as or more extreme* than the one observed when compared with the null hypothesized value for the population. It measures the "rarity" of our observed value if the null hypothesis is true and just random variation is operating.

Our test statistic in our example comes out to be $t = -3.194$; if this were a *z* score for a Normal distribution, we would know that the *p*-value would be very small. But we are dealing with the *t* distributions, and we know that the *t* distributions have thicker tails than the Normal, so it *may* be that the *p*-value is not as small as we think. So how do we determine the *p*-value when we are using a *t* distribution? There are two answers; one is to approximate using the ***t Distribution Chart*** and the second way is to use software.

We use the ***t Distribution Chart*** in the same way we used the ***Chi-Square Distribution Chart***. We first locate our degrees of freedom in the left-hand column of the chart, so that we look at $df = 19$. Then, for our example, ignoring the negative sign (we are using the absolute value of $|t| = |-3.194| = 3.194$), we find the two numbers that surround 3.194 are 2.861 and 3.579. Then, reading the probabilities in the right tail from the top of the chart, we see that our 3.194 will have a probability between 0.001 and 0.005. However, this probability is the probability only in the right tail, and because our alternate hypothesis is two-sided (we hypothesized that the population mean could be either larger or smaller than $\mu_0 = 125$), we need to consider both sides of the *t* distribution. Hence, we need to multiply this interval by two. When we do this, we get $0.002 < p - value < 0.010$, since $2(0.001) = 0.002$, and $2(0.005) = 0.010$. From the ***t Distribution Chart*** we can only approximate that the *p*-value is between 0.002 and 0.01.

	Probabilities in the right tail of the distribution									
df	***0.25***	***0.15***	***0.10***	***0.05***	***0.025***	***0.02***	***0.01***	***0.005***	***0.001***	***0.0005***
1	1.000	1.963	3.078	6.314	12.706	15.895	31.821	63.657	318.309	636.619
2	0.816	1.386	1.886	2.920	4.303	4.849	6.965	9.925	22.327	31.599
3	0.765	1.250	1.638	2.353	3.182	3.482	4.541	5.841	10.215	12.924
4	0.741	1.190	1.533	2.132	2.776	2.999	3.747	4.604	7.173	8.610
5	0.727	1.156	1.476	2.015	2.571	2.757	3.365	4.032	5.893	6.869
6	0.718	1.134	1.440	1.943	2.447	2.612	3.143	3.707	5.208	5.959
	⋮	⋮	⋮	⋮	⋮	⋮	⋮	⋮	⋮	⋮
18	0.688	1.067	1.330	1.734	2.101	2.214	2.552	2.878	3.610	3.922
19	0.688	1.066	1.328	1.729	2.093	2.205	2.539	2.861	3.579	3.883
20	0.687	1.064	1.325	1.725	2.086	2.197	2.528	2.845	3.552	3.850
21	0.686	1.063	1.323	1.721	2.080	2.189	2.518	2.831	3.527	3.819

Actually, the information that the *p* value is in the interval $0.002 < p - value < 0.010$ is sufficient for us to interpret the hypothesis test. If our level of significance is α=0.05, the interval for the *p* value tells us that we have a rare test statistic. The *p*=value indicates that *if* $\mu_0 = 125$, we have observed a *rare* rather than *reasonably likely* event in our random sample.

Why can we ignore the negative sign in $t = -3.194$ and use the absolute value of *t*? We can ignore the negative sign because we know that the *t* distributions are perfectly symmetrical, so that the area (or probability) in the right tail, to the right of 3.194, is the same as the area (or probability) to the left of $t = -3.194$. The ***t Distribution Chart*** has been constructed to take advantage of the symmetry of the *t* distributions; the chart only needs to show one side.

This way of approximating the p-value can be referred to as the "little box" method because we can depict the values of the t score test statistic and the probabilities in the tail, as a little box. Alternately, we can let software do a more accurate calculation, as in the output shown. Some software will also show a picture of the sampling distribution of the test statistic.

	Probability in tail	
	0.005	**0.001**
t-score	2.861	3.579

```
From Summary Statistics                         Test Mean
Attribute (numeric): unassigned

Ho: population mean of ActualDuration equals 125
Ha: population mean of ActualDuration is not equal to 125

Count:        20
Mean:         117.75
Std dev:      10.1508
Std error:    2.26979
Student's t:  -3.194
DF:           19
P-value:      0.0048
```

We see that the p-value = 0.0048 and that agrees with our approximation from the ***t* Distribution Chart**, since 0.0048 is in the interval $0.002 < p - value < 0.010$. The shaded-in portions of the t distribution with $df = 19$ show (together) the small p-value. The shaded areas are barely visible because the p-value = 0.0048 is so small, indicating that our sample *ActualDuration* mean $\bar{x} = 117.75$ would be very rare *if* it were true that $\mu_0 = 125$.

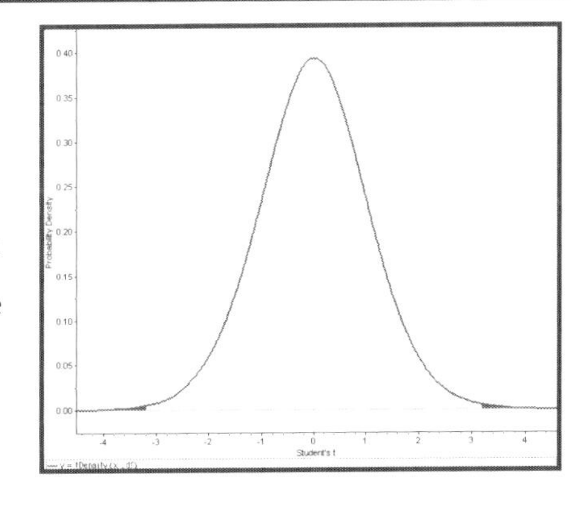

Step 5: Interpret the results in the context of the data: Our hypotheses were that for all flights starting at SF Bay Area and ending at Seattle, the population mean *ActualDuration* $\mu = 125$ minutes; the alternate hypothesis claimed that the mean duration $\mu \neq 125$ minutes. Our random sample of $n = 20$ flights gave us a sample mean of $\bar{x} = 117.75$ minutes for the variable *ActualDuration* and we found that the probability of seeing this sample mean $\bar{x} = 117.75$ or one more extreme than this was only 0.0048—very rare, *if* it were true that $\mu = 125$ minutes. Hence, since the p-value is so small, we can say that the test was *statistically significant* and that we have sufficient evidence contrary to the null hypothesis that the mean actual duration of flights to Seattle is $\mu = 125$ minutes.

In our interpretation, we related the essentials of the evidence (the p-value) and we did this in the context of our test—that is, in the amount of time that flights take to get to Seattle from the SF Bay Area.

How we can read a two-sided hypothesis test from a confidence interval

We met the idea of the connection between a confidence interval and hypothesis test before in the context of tests for the difference of two proportions (*§4.3*). Here is the reasoning: if a hypothesis test is significant then we know that the $\bar{x}$ is in the *rare* region (as opposed to the region of *reasonably likely* $\bar{x}$) for the μ_0 we are testing. Then we also know that the confidence interval that we calculate based on that $\bar{x}$ will *miss* or *not capture* the μ_0. If the $\bar{x}$ is in the "rare" region, the distance between the $\bar{x}$ and the μ_0 is bigger than the margin of error that gets added and subtracted from the $\bar{x}$.

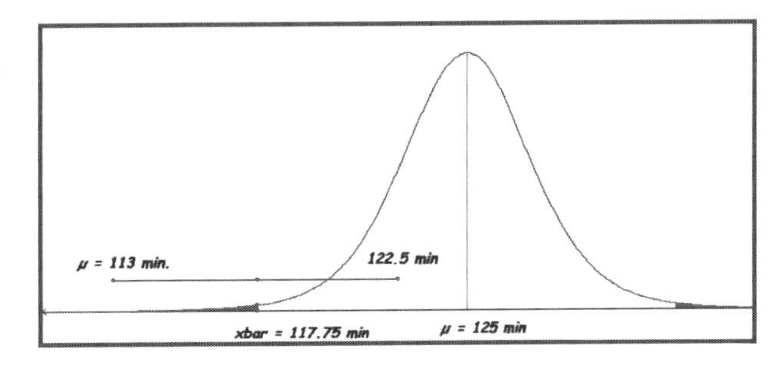

Our analysis of the *ActualDuration* of flights to Seattle illustrates this connection. The confidence interval for the population mean *ActualDuration* came out to be $113.00 < \mu_{\text{ActualDuration}} < 122.5$ minutes; this interval does *not* include the hypothesized mean time of $\mu_0 = 125$ minutes. And we found that our hypothesis test of $H_0 : \mu = 125$ versus $H_a : \mu \neq 125$ was *statistically significant* (the p-value was 0.0048). Now, think of what would have happened if we had got a reasonably likely $\bar{x}$; in that case, the confidence interval would have captured (or crossed over or hit) the $\mu_0 = 125$ minutes, since the distance between the $\bar{x}$ and μ_0 is smaller than the margin of error.

Confidence Intervals and Hypothesis Tests When a confidence interval

- ***excludes*** an hypothesized value then the related **two-sided** hypothesis test will be *statistically significant;*
- ***includes*** an hypothesized value then the related **two-sided** hypothesis test will not be *statistically significant.*

However, the principle only works for **two-sided** hypothesis tests.

This principle connecting confidence intervals and hypothesis tests works in reverse as well; if we have a *statistically significant* two-sided hypothesis test then we know that if we used the same sample data to calculate a confidence interval, that confidence interval would *exclude* the hypothesized value. This principle means that in some ways confidence intervals are more useful inferential techniques than hypothesis tests, since confidence intervals give all the information we would get from a two-sided hypothesis test and an interval of plausible values for the population value.

Another Example: A One-Sided Test

Despite the nice connection between confidence intervals and hypothesis tests for two-sided tests, one-sided hypothesis tests are also important. Here is an example. Let us suppose someone (call her Cindy) has another sample of *ActualDuration* data and wants to use these data to do her own hypothesis test. There will be one difference, however; Cindy has concluded that the proper way to challenge the null hypothesis $H_0 : \mu = 125$ should be a one-sided alternate hypothesis $H_a : \mu < 125$. She thinks that $H_a : \mu < 125$ in her alternate hypothesis because she thinks that airlines give themselves more time on average than they actually need to make the flight. Therefore, the actual flight durations should, on average, be shorter than the announced flight durations. What Cindy is doing is bringing more information to the test, and that is common in the use of hypothesis tests.

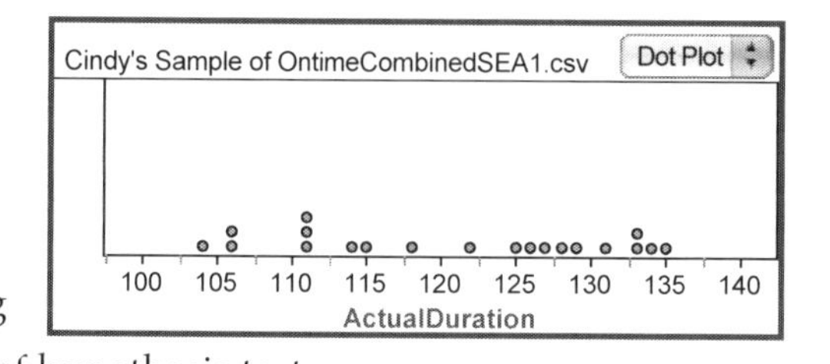

Step 0: Descriptive analysis: Here are the results of the descriptive side of Cindy's analysis. The sample mean $\bar{x} = 120.95$ minutes for Cindy's sample is higher than $\bar{x} = 117.75$ minutes for the sample that we had, but it is still less than the hypothesized $\mu_0 = 125$ minutes. The standard deviation $s = 10.4553$ is similar, and the plot does not show obvious skewness, although it could be there.

Cindy's Sample of OntimeC...

ActualDuration	120.95 10.4553 20

S1 = mean( )
S2 = s( )
S3 = count( )

Step 1: Setting up the hypotheses: Cindy wants a one-sided test, since she has reason to believe that the population mean μ will be less than the $\mu_0 = 125$ minutes, so her hypotheses will be:

$$H_0 : \mu = 125 \text{ minutes}$$
$$H_a : \mu < 125 \text{ minutes}$$

It is in the alternate hypothesis that the one-sided test differs from the two-sided test; the two-sided test would have $\mu \neq 125$ minutes.

Step 2: Checking the conditions: The conditions and their assessment will be the same as with the example we went through. Once again, Cindy is using data that come from a random sample of the population of flight records. Once again, Cindy has a small sample of $n = 20$, so we need to be cautious about outliers or evidence of extreme skewness in the population distribution of the variable *ActualDuration.*

Step 3: Calculating the test statistic: For a one-sided test, the calculation of the test statistic is the same as for a two-sided test. For Cindy's sample where $\bar{x} = 120.95$, $s = 10.4553$, and $n = 20$, she will calculate

$$t = \frac{\bar{x} - \mu_0}{\frac{s}{\sqrt{n}}} = \frac{120.95 - 125}{\frac{10.4553}{\sqrt{20}}} \approx -1.732$$

This does not look nearly as extreme as the test statistic we got earlier. To see whether it is extreme enough to challenge the null hypothesis, we need to calculate the p-value.

Step 4: Finding the p value: Again, we can approximate the p-value using the ***t distribution Chart***, with the "little box" method. We will ignore the negative sign of our test statistic since the right hand tail (which the chart shows is the same as the left-hand tail). Here is the "little box" for $t = -1.732$, and this shows that our approximate p-value is between 0.025 and 0.05. We would express this as $0.025 < p - value < 0.050$. However, here we did *not double* the probabilities shown in the chart; we did not double because we are conducting a one-sided test instead of a two-sided test. As before, we could let software do the work.

	Probability in tail	
	0.05	**0.025**
t-score	1.729	2.093

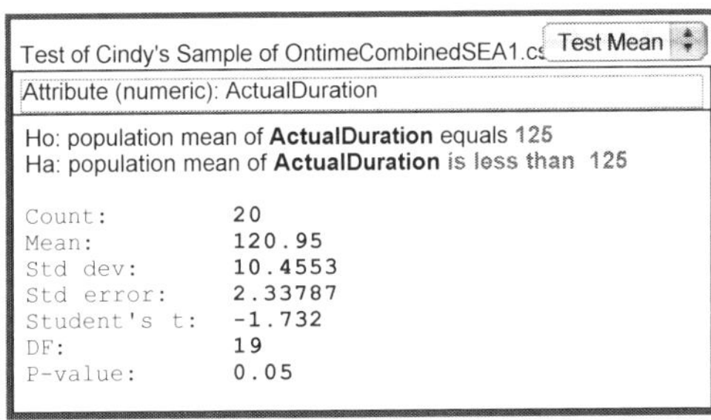
Test of Cindy's Sample of OntimeCombinedSEA1.cs Test Mean
Attribute (numeric): ActualDuration

Ho: population mean of **ActualDuration** equals 125
Ha: population mean of **ActualDuration** is less than 125

Count:	20
Mean:	120.95
Std dev:	10.4553
Std error:	2.33787
Student's t:	-1.732
DF:	19
P-value:	0.05

The plot showing the sampling distribution of the test statistic t-score shows the p-value shading only on the left side. This is because the hypothesis test we are doing is *one-sided* and not *two-sided.*

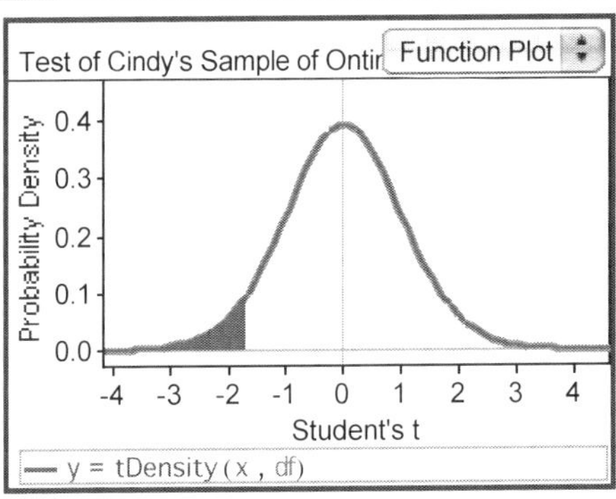

Step 5: Interpret the results in the context of the data: For the Software output, the p-value appears to be "exactly" 0.05, which is the most common value of α, the "line" that divides "significant" from "not significant." However, the p-value shown is rounded, and software can show the p-value to more decimal places. If we do "round" to seven decimal places, then we see that the p-value is 0.0497388. (See the next page.)

So what conclusion do we come to "on the boundary"? Is the test statistically significant or not? One lesson we learn from getting an example "on the boundary" is that there is nothing magical or "cut in stone" about the number $\alpha = 0.05$ (or the number $\alpha = 0.01$ or any other value of α.) What we can say is that we *do* have some evidence for the alternate hypothesis that the mean flight time is less than 125 minutes. The evidence is that *if* the mean flight time were really $\mu_0 = 125$ *then* the probability that we would get sample mean as small as $\bar{x} = 120.95$ or smaller is about 5%.

no data	
Drop an attribute here	
	0.0497388
S1 = tCumulative(−1.732, 19)	

In the conditional probability notation of §1.2, we could write the *p*-value as $P(\bar{x} \le 120.96 \mid \mu_0 = 125) \approx 0.05$. This rather elegant way of writing the *p*-value says: "Given the null hypothesized value $\mu_0 = 125$ minutes, the probability that we will see a sample mean $\bar{x}$ of 120.96 minutes or less is 5%." Another way of putting our conclusion is that the sample mean of $\bar{x} = 120.95$ is unlikely (probability about 5%) to have come about by random sampling variation *and* a population mean of $\mu_0 = 125$.

Lessons from Cindy's sample results One lesson is (as mentioned above) not to look for hard and fast rules that tell you: "When you see this number, say this..." or "When you see that number, say that..." Real data analysis is often not that simple and must take into account many pieces of evidence. The kinds of statistical analysis we are doing here primarily answer the question: "Is what I am seeing consistent or inconsistent with random variability with the null hypotheses I have set up?"

Second lesson: In research that uses these inferential techniques, you may well find that *p*-values rather than declarations of statistical significance or non-significance are reported. Reporting *p*-values rather than stating that a result is statistically significant or not allows the reader (often another researcher) to make up his or her own mind on the basis of what that reader considers *rare.*

Can anything else go wrong? Data analysis in the face of random variation

The answer is yes and not because of bad or wrong calculations. We can do all of the calculations correctly but because of random variation still come to the wrong conclusions. We can do the calculations for confidence intervals correctly and still "miss" the population mean μ in our interval, and we can make the same kinds of "errors stemming from randomness" in hypothesis tests.

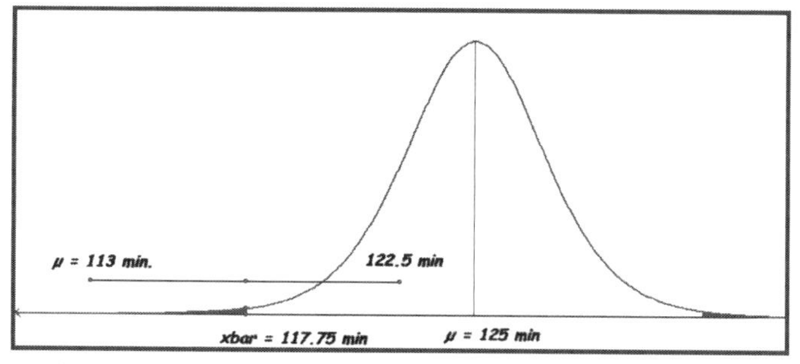

Confidence Intervals. The issues can be understood by considering confidence intervals. The confidence interval for our first sample was $113.00 < \mu_{\text{ActualDuration}} < 122.5$. We would say that we are "95% confident that the mean flight duration from the SF Bay Area is between 113 minutes and 122.5 minutes." Can we be wrong? Yes. We could have, just by random chance, chosen a *rare* sample. If we did, our confidence interval would not capture the true population mean μ duration of flight. The graphic shown here illustrates this. We are 95% confident that the mean duration of flight to Seattle is *not* as long as (for example) 124 minutes, but we are not 100% confident.

Hypothesis Testing Can we "miss the mark" with hypothesis testing in a similar fashion? Yes, it is possible, and it is such a possibility that statisticians have a special terminology for the ways in which we can miss the mark. There are essentially two kinds of "errors" (and they are *not* errors in calculation) that can be made. They are called **Type I** and **Type II errors.**

Type I and Type II Errors

Type I Error: If the null hypothesis is *actually true,* but our test *rejects* the null hypothesis

Type II Error: If the alternate hypothesis is *actually true,* but our test *does not reject* the null hypothesis

We can illustrate these two types of errors using Cindy's hypothesis test.

μ =125 true, then what? Suppose it *actually* is true that the mean duration of flight is $\mu = 125$ minutes, but Cindy decides (based on her sample and calculations) to reject the null hypothesis and go with her alternate hypothesis, which is that $\mu < 125$ minutes. In this situation, Cindy would be making a **Type I error.** She has rejected the null hypothesis when it is actually true.

μ <125 true, then what? Suppose it is *actually* true that that the mean duration of flight is less than 125 minutes, that is $\mu < 125$, but Cindy decides on the basis of her test to *not to reject* the null hypothesis. She did not reject the null hypothesis when she should have. Cindy would be making a **Type II error.** She has failed to reject the null hypothesis when she should have.

For any hypothesis test, you should be able in the *context* of any hypothesis test to identify what a **Type I error** and what a **Type II error** will be.

To keep these two types of errors straight, some people find this diagram useful. Since Cindy is on the boundary, she could be in danger of making either one of these two errors, but we are always in danger of making one or the other of these errors when doing hypothesis testing.

		Reality	
		H_0 True	H_a True
Result of Hypothesis Test	Rejects H_0	Type I Error	Correct Decision
	Does not Reject H_0	Correct Decision	Type II Error

Power and how it works Statisticians are concerned about both types of errors, but in general a **Type II error** is the more problematic of the two. For one thing, it is very easy to see the probability of making a Type I error. That probability is just α; if we have chosen $\alpha = 0.05$ then the probability that we will get a rare result by random sampling variability is just 5%. However, Type II errors are harder to control. The smaller one tries to make the probability of making a Type I error, the bigger will be the probability of making a Type II error. Because of this, statisticians have a more useful term to get at what they are trying to achieve. The technical term is called **power,** and its definition is given below.

Power

The **power** of the hypothesis test is the *probability* that a hypothesis test opts for the alternate hypothesis (or correctly detects the alternate hypothesis) when the alternate hypothesis is *actually* true.

In general, other things being equal, the bigger the sample size *n*, the greater the power of a test.

Calculations of the power of a hypothesis test are very, very complicated, partly because they depend upon the unknown reality behind the alternate hypothesis. The calculations are complicated, but they are also important. Expect to see references to the power of a test in technical reports. Some software has helpful calculations and graphics to show the connections between power, the difference between the true value of the mean and the hypothesized value of the mean, and the sample size. Here is some output from the statistical package program called StatCrunch.

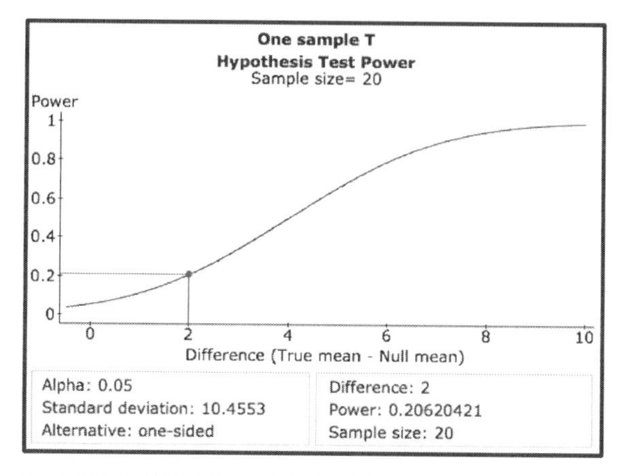

However, one result of these calculations is that by increasing sample size n, the **power** is also increased. Bigger samples are better and give more confidence in our results. In the StatCrunch output, a difference of two minutes between the hypothesized mean of $\mu = 120$ minutes and the true mean has only a 21% chance of being detected if we have only a sample size of $n = 20$. However, if we increase the sample size to $n = 160$, we have a 78% chance of detecting a true mean that is two minutes from the hypothesized mean of $\mu = 120$.

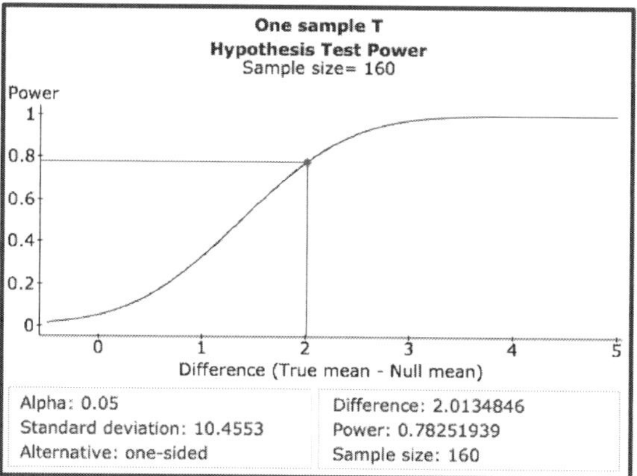

The StatCrunch output is also helpful in that it allows you to see how bigger differences between the true and hypothesized mean are more likely to be detected than small differences.

Summary: Hypothesis tests for one mean

- The logic and terminology for hypothesis tests for means are the same as for proportions, but the formulas are different, being based upon the sampling distribution that arises when the sample mean $\bar{x}$ and sample standard deviation s are used, namely the t distributions. Specifically,
  - The form of the null and alternate hypotheses is the same but refer to population means μ rather than population proportions p.
  - The meaning of the test statistic is the same, but the formula is different (see below).
  - The meaning of a p-value is the same.
  - The interpretation of the results of a hypothesis test follows the same pattern; the term *statistically significant* has the same meaning.

Five steps for hypothesis tests for one mean

Step 1: *Set up the null and alternate hypotheses using an idea for the population mean* μ_0.

$$H_0 : \mu = \mu_0 \qquad H_a : \begin{cases} \mu \neq \mu_0 \text{ or} \\ \mu < \mu_0 \text{ or} \\ \mu > \mu_0 \end{cases}$$

Step 2: *Check the conditions for a trustworthy hypothesis test.*
The sample must be a simple random sample.
The population must be Normal; however, since the t procedures are robust for large sample sizes, judge whether to proceed using the 15/40 rule.

Step 3: *Calculate the test statistic* $t = \dfrac{\bar{x} - \mu_0}{\frac{s}{\sqrt{n}}}$ *using the sample mean* $\bar{x}$ *and sample standard deviation and the hypothesized* μ_0.

Step 4: *Calculate the p-value by getting* $P(\bar{x} \text{ more extreme} | \mu = \mu_0)$ *using the absolute value of the test statistic t you calculated in Step 3.*

Step 5: *Evaluate the evidence that the p-value and the test statistic give you to determine whether your test successfully challenges the null hypothesized population mean* μ_0 *or not. Give an interpretation in the context of the data using the terminology of "statistical significance" and "rejecting the null hypothesis."*

- ***The 15/40 rule of thumb*** for the use of the t distributions introduced in connection with confidence intervals also applies to the use of t distributions for hypothesis tests.
  - *Small Samples, $n < 15$* *For samples where $n < 15$, the data must come from a Normal or nearly Normal distribution. The t procedures will not be trustworthy otherwise.*
  - *Moderate-sized Samples, $15 \leq n \leq 40$* *The t procedures may cope with some skewness but will not be trustworthy in the face of strong skewness or outliers.*
  - *Large Samples, $n > 40$* *The t procedures are robust in the face of skewness or outliers where the sample is large and the more so the larger the sample size.*
- ***Confidence Intervals and Hypothesis Tests*** When a confidence interval
  - *includes* an hypothesized value then the related hypothesis test will *not* be *statistically significant*
  - *excludes* an hypothesized value then the related hypothesis test *will* be *statistically significant*
  - This principle works only for *two-sided* hypothesis tests.
- ***Type I and Type II Errors***
  - *Type I Error:* If the null hypothesis is *actually true,* but our test *rejects* the null hypothesis
  - *Type II Error:* If the alternate hypothesis is *actually true* but our test *does not reject* the null hypothesis
  - The statements above are general and should be translated into the context of the data to be useful.
- ***Power***
  - The *power* of the hypothesis test is the *probability* that a hypothesis test opts for the alternate hypothesis when the alternate hypothesis is *actually* true.
  - Generally, the larger the sample size used, the larger the power, other things being equal.

§5.3 Comparing Two Measures in One Collection

Blood pressure measurements: once or twice?

When you visit your physician, you may have your blood pressure measured more than once. For some people, visiting the doctor is a scary experience, and so the first time the measurement is made, blood pressure may be elevated just because they are nervous. It makes sense to take the measurement a second time when the patient may be somewhat more relaxed. Our statistical questions are:

- *If we measure the blood pressure of a person twice, do we find a difference between the measurements that is bigger than what we would expect by sampling variation?*
- *If there is a difference (beyond just sampling variation) between the first and second measurements of a person's blood pressure, can we estimate the size of this difference?*

The first question looks as though it should be answered by a hypothesis test. If we do *not* find a significant difference between the measurements then we may conclude that, for many people, one measurement should be sufficient. However, finding a significant difference would suggest that there is something beyond sampling variation (being scared? the "white coat" effect?) that is causing the difference in the two measurements, and we would like to have an estimate of the size and direction of the difference. The second question looks like a job for a confidence interval.

A short introduction to blood pressure measurement. There are two numbers that come from the measurement of blood pressure. One number is called ***systolic blood pressure,*** and the second one is called ***diastolic blood pressure,*** and both measurements are in millimeters of mercury, or mm Hg. Normal blood pressure is expressed with both numbers in a "fraction" ***systolic/diastolic*** and the numbers that are ideal for adults are 120/80. For more information (especially on the medical difference between the two) see http://www.new-fitness.com/Blood_Pressure/numbers.html. When we speak of *two* measurements of blood pressure, what we mean is that we will have two values for the *systolic* and two measures for the *diastolic*. In the case table above, case 502 is a twenty-nine-year-old married female, and for her first measurement she had 130/74; for her second measurement she had 128/76. In all, there are four numbers, but the important thing for us is that there are two measures being made. This section is how we analyze such data. First we need a definition for these kinds of data.

NHANES Blood Pressure Data

	Sex	Age	Marital	Systolic1	Diastolic1	Systolic2	Diastolic2	Pulse	DiffSys2Sys1
499	Male	26	Living wi...	106	54	106	58	84	0
500	Male	28	Married	118	80	112	78	80	-6
501	Male	39	Never m...	140	74	132	68	50	-8
502	Female	29	Married	130	74	128	76	86	-2
503	Male	53	Married	142	86	144	82	54	2
504	Male	21	Never m...	114	62	118	70	60	4

> ***Paired comparison data*** *consist of two comparable measures within essentially one collection.*

When it makes sense to get the difference between two measures (to subtract one measure from another) then we say we have ***two comparable measures***. We will often express these two comparable measures as ***two comparable variables*** in a spreadsheet so they will appear as two columns, so that we will end up analyzing another variable that is the *difference* of the two variables' values.

As an example, it makes sense to determine whether there is a difference between the two measurements of systolic blood pressure. For case 502, the difference between the *second* measurement of systolic blood pressure and the *first* measurement is $128 - 130 = -2$. It would *not* make sense to subtract the person's *Age* from that person's *Pulse*; these two variables are *not* comparable variables.

Examples of Paired Comparison Data:

- The cases are students, and each student takes a mathematics placement test twice; it makes sense to see if the score on the second attempt differs from the first.
- The cases are places, and every month a measure of air pollution is taken; for each place, it makes sense to see if there is a change in pollution (a difference) between a later time and an earlier time.
- The cases are gas stations selling both gasoline and diesel fuel; it makes sense to find the difference between the price of regular gasoline and diesel fuel.

All of these examples involve ***comparable measures*** (usually expressed as variables), and in each situation, there is *one* collection of cases: *one* collection of students, *one* collection of places, *one* collection of gas stations. The definition says "essentially ***one collection***"; in §5.4, we will look at comparisons between ***two independent collections.*** We can change the three examples above so that we would have one measure (or one variable) but from *two independent collections*.

Examples of Two Independent Collections: Not Paired Comparison Data:

- The cases are students, and each student takes a mathematics placement test twice; we compare male and female students for one of the tests.
- The cases are places, and every month a measure of air pollution is taken; we compare places in the northern and southern parts of a state or province for the same month.
- The cases are gas stations selling both gasoline and diesel fuel; we compare the prices of gasoline for stations in two different places.

For the first example of *two independent collections*, comparing males and females is considered two independent collections even if they are all in one data set. If random sampling has been used, the fact that a particular male is in the sample is independent of whether a particular female is there. Moreover, it would not make sense to subtract a female score on the placement test from a male score on a placement test. In the second example, the same principle operates: even though the data may have been collected together, the places chosen can be considered independent.

Answering our statistical questions

Descriptive analysis. The first thing that we should always do is to look at our sample data. As an example, we will examine the blood pressure data for the females in our sample and leave the analysis of the data for males as one of the exercises. We will look at three variables initially: *Systolic1*, which is the first measure of systolic blood pressure, *Systolic2*, which is the second measure of systolic blood pressure, and the difference between the second and first measures, calculated as *DiffSys2Sys1 = Systolic2 − Systolic1*.

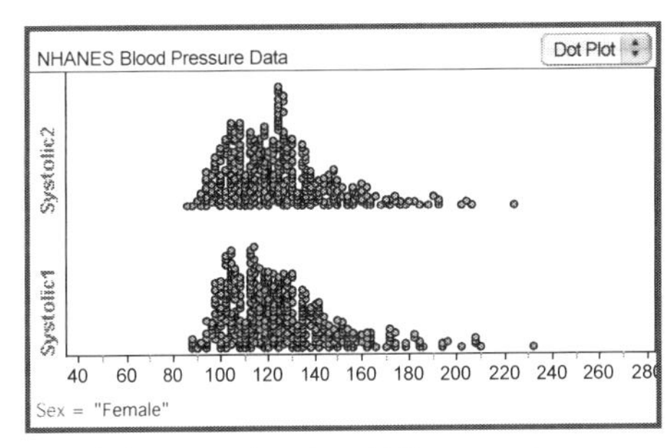

Shown below are the dot plots of each of these for the entire sample. (The one for the difference is on the next page.)

Recall that for case 502, the twenty-nine-year-old married female, the difference between the *second* measurement of systolic blood pressure and the *first* measurement is $128 - 130 = -2$, so that she is one of the dots in the stack of dots in the plot for the variable *DiffSys2Sys1*. To compare two measures in one collection, we will work entirely with the differences, so, in this instance, we will look entirely at the variable *DiffSys2Sys1*. We continue our descriptive analysis by getting the mean and standard deviation of the variable *DiffSys2Sys1*. It appears that, on average, the mean difference between the second measurement and the first measurement is $\bar{x}_{Diff} = \bar{x}_{DiffSys2Sys1} \approx -1.807$ mm Hg, and that the standard deviation is $s_{Diff} = s_{DiffSys2Sys1} \approx 6.484$ mm Hg. On average, the second measure is slightly lower than the first (by –1.81 mm Hg), but there is some variability in the differences, as measured by the standard deviation of the differences of 6.484 mm Hg. The shape of the distribution of differences between the measures shows some left skewness but has a single peak, like a Normal distribution.

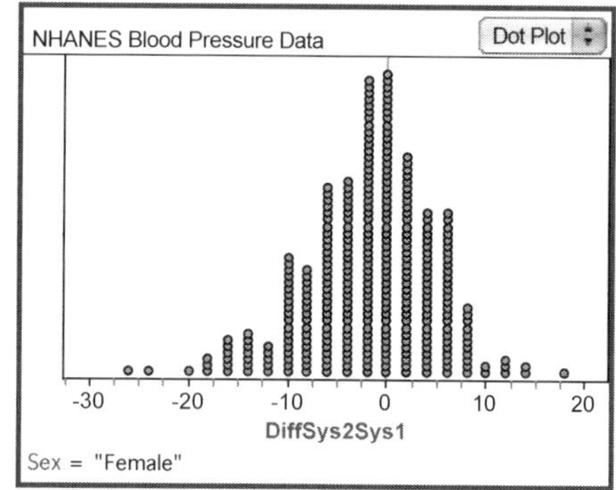

NHANES Blood Pressure Data

	DiffSys2Sys1
	-1.80745
	6.48412
	322

S1 = mean( )
S2 = s( )
S3 = count( )

Sex = "Female"

Inferential analysis. Our first statistical question was whether we find a difference between the measurements that is bigger than what we would expect by sampling variation. We shall see that the hypothesis test follows the form of the hypothesis test for one mean using the *t* procedures, although the mean that we are examining is a mean of differences; that is because we are dealing with essentially one collection. We need some notation to keep things straight for our hypothesis test.

Notation for Paired Comparison Data.

	Sample	Population
Mean	$\bar{x}_{Diff}$	μ_{Diff}
Standard deviation	s_{Diff}	σ_{Diff}

Step 1: Setting up hypotheses: Our basic idea is that there should be *no* difference between the measurements apart from sampling variation. A challenge to that idea is that perhaps the first measure is different—maybe higher but maybe lower—than the second measure. We would therefore set up:

$$H_0 : \mu_{Diff} = 0$$
$$H_a : \mu_{Diff} \neq 0$$

Here we are using a preconceived idea that the mean difference in measurements is *zero*. We would normally indicate the hypothesized value for a hypothesis test with the symbol μ_0, so here we are saying that our preconceived idea is that $\mu_0 = 0$. It is possible that we would have a preconceived idea for the difference that is not zero (no difference), but the most usual test tests the idea of no difference against the idea that the difference is not equal to, or greater than, or less than zero. In symbols, for the alternate hypothesis we will have either $H_a : \mu_{Diff} \neq 0$ or $H_a : \mu_{Diff} > 0$ or $H_a : \mu_{Diff} < 0$ and for the null hypothesis: $H_0 : \mu_{Diff} = 0$.

Step 2: Checking the conditions: Just as the hypotheses followed the pattern of the one mean test, so do the conditions but with respect to the *differences*. Recall that we have two conditions; one is that data be a random sample, and the second is that the population (of differences) be Normal. However, the Normality condition can be relaxed if the sample size is sufficiently large, according to the 15/40 rule. Our data are a random sample, and the sample size ($n = 322$) is large enough that the t procedures are ***robust*** with the amount of skewness we see in the sample data. We can proceed.

Steps 3 and 4: Calculating the test statistic, getting the p-value: Since we are using the t procedures, the test statistic is a t score calculated with the data on the differences between the two measurements. Hence, using $\bar{x}_{Diff} = -1.80745$ and

$s_{Diff} = 6.48412$,

$$t = \frac{\bar{x}_{Diff} - 0}{\frac{s_{Diff}}{\sqrt{n}}} = \frac{-1.80745 - 0}{\frac{6.48412}{\sqrt{322}}} \approx \frac{-1.80745}{0.361346} \approx -5.002$$

Test of NHANES Blood Pressure Data | Test Mean

Attribute (numeric): DiffSys2Sys1

Ho: population mean of **DiffSys2Sys1** equals **0**
Ha: population mean of **DiffSys2Sys1** is not equal to **0**

Count:	322
Mean:	-1.80745
Std dev:	6.48412
Std error:	0.361346
Student's t:	-5.002
DF:	321
P-value:	< 0.0001

Sex = "Female"

The value of $t = -5.002$ looks extreme (the t distribution for $n = 322$ is not too different from the Normal distribution). Checking the ***t Distribution Chart*** for df = 250 shows that $t = -5.002$ is off the chart, indicating a very small p-value.

Step 5: Interpreting the Result It does appear that there is a tendency for the first reading of blood pressure to be higher—making the difference between the second and first measures negative. With our sample size, we are confident that this result is not just sampling variability. On the other hand, the difference is not large, being less than 2 mm Hg on average. We can go on and get an estimate of the mean size of the difference. For that we need a confidence interval.

Calculating and Interpreting a Confidence Interval Once again, the t procedures will follow the pattern of the t procedures for one mean. Using $\bar{x}_{Diff} = -1.80745$ and $s_{Diff} = 6.48412$, and $t^* = 1.969$ for df = 250, we have

$$\bar{x}_{Diff} \pm t^* \frac{s_{Diff}}{\sqrt{n}} = -1.80745 \pm 1.969 \frac{6.48412}{\sqrt{322}} \approx -1.80745 \pm 0.71149$$

and this leads to a confidence interval of $-2.52 < \mu_{Diff} < -1.10$ mm Hg. Here is software output for the confidence interval, which, once again, uses the procedures for a single mean.

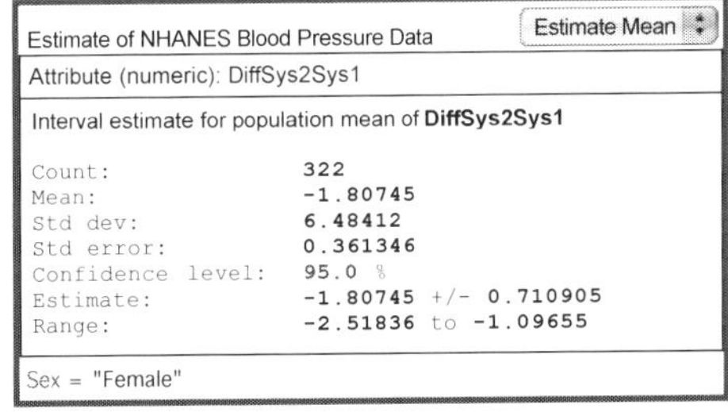
Estimate of NHANES Blood Pressure Data | Estimate Mean

Attribute (numeric): DiffSys2Sys1

Interval estimate for population mean of **DiffSys2Sys1**

Count:	322
Mean:	-1.80745
Std dev:	6.48412
Std error:	0.361346
Confidence level:	95.0 %
Estimate:	-1.80745 +/- 0.710905
Range:	-2.51836 to -1.09655

Sex = "Female"

As expected, the two measures *exclude* the null hypothesized $\mu_0 = 0$ since the associated two-sided hypothesis test was statistically significant. The numbers are negative because the *first* measure *Systolic1* is higher than *Systolic2* for many people.

We can certainly say with 95% confidence that we expect (for females) the second measure of systolic blood pressure to be about 1.8 mm Hg lower than the first measure, with a margin of error of about 0.72 mm Hg. However, the numbers are quite small; our hypothesis test shows that the difference does not stem from random sampling variation—from chance variation—but the difference is small nonetheless. The lesson is that statistically significant differences are not necessarily practical differences.

Statistical significance and practical significance Hypothesis testing tests whether what we actually see in our data could have been "merely" the product of the random variation. Hypothesis testing does not indicate the importance of what we have seen in the context of our data. Is a difference of 2 mm Hg an important difference? It may not be an important difference in the context of measuring blood pressure. The result may lead us to look for the kinds of people or the kinds of situations that produce bigger differences. For example, it may be that the "routine" testing of the NHANES survey is different from the situation that people face when they "go to see their doctor." Or it may be that we could examine the NHANES data to find the characteristics of people for whom there is a big difference between the second and first measurements.

Creating Pairs: Matched Pair Designs

Often, data come with comparable measures "built in," as in the examples shown at the beginning of this section. But not always. When it is impossible to have a paired comparison, it is sometimes possible to approximate a paired comparison by making what are almost pairs. Matched pair designs are quite common in medical studies, where the idea can be called a ***case control study.*** Here are two examples of a matched pair design, one a medical study and the other involving houses for sale.

Does using a cell phone cause cancer? If you think that there is a possibility that using cell phones leads to certain kinds of cancer, it is obviously impractical (and unethical) to create an experiment where people are randomly assigned to a group in which each person is given a cell phone and told to use the cell phone as much possible or to a second group that would be prohibited from ever using a cell phone. Then wait (years, perhaps?) to see whether the incidence of cancer is higher in the cell phone group. Instead, what is more likely to be done is to observe the cell phone usage of those who already have cancer and compare it to the cell phone usage to those who do not have cancer. However, in medical studies of this kind, the cancer cases are *matched up* by sex, age, and other variables to people who have cancer. Here is how the matched pair or case controls is described in one such study[7] that was carried out in two parts of the UK on the relationship between risk of glioma (a type of brain cancer) and cell phone usage:

> ...one control per case was individually matched on age, sex and general practice after the patient with glioma had been interviewed. Nonparticipating controls were replaced.

This means that a patient without glioma (a control) was chosen so that the control patient was the same age and sex as the glioma patient and had the same physician—and that if the "control" decided not to participate in the study, another was found. That the cases and controls had the same general practitioner meant that they probably lived in the same area.

---

[7] Hepworth, Sarah J, *et al.* Mobile phone use and risk of glioma in adults: case control study *British Medical Journal,* 2006, 332:883–887

The idea was to control at least three variables that might have some effect on the development of glioma. The matching process can be quite complicated or it can be quite simple.

Example of Matched Pairs: Location, Location, Location The average sale price of houses for sale in two different places may be different for any number of reasons. One way of assessing the effect of location on the price of a house is to compare houses that are alike (or as alike as we can get them) in all respects *except* location. Here is a very crude attempt to do just that. Houses in Menlo Park and houses in Palo Alto that were part of a sample of the houses for sale in 2007–2008 were arranged in rank order by the measure of the size of the house, *SqFt*. Then the biggest house—the one with biggest *SqFt*—in Palo Alto was matched (or paired) with the biggest house in Menlo Park, the second-largest house in Palo Alto paired with the second-largest house in Menlo Park, and so on, until the smallest house in Palo Alto was matched with the smallest house in Menlo Park. Our matching procedure is extremely crude; it would be better to consider other variables, such as the number of bedrooms, the number of bathrooms, the age of the house, in making the matched pairs. However, even with this crude matching, what we have done is to calculate, for each matched pair of houses—one in Palo Alto, one in Menlo Park—the difference in sale prices. The plot here shows the distributions of the areas of the houses (measured by the variable *SqFt*) for Palo Alto and Menlo Park. (The top distribution is for Palo Alto, and the second distribution is for Menlo Park.)

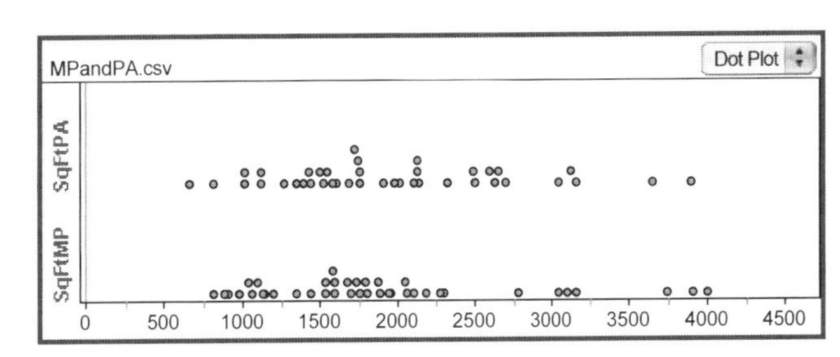

The second plot compares the sale prices of the houses in the two places; again, the first distribution is for Palo Alto, and the lower one is for Menlo Park.

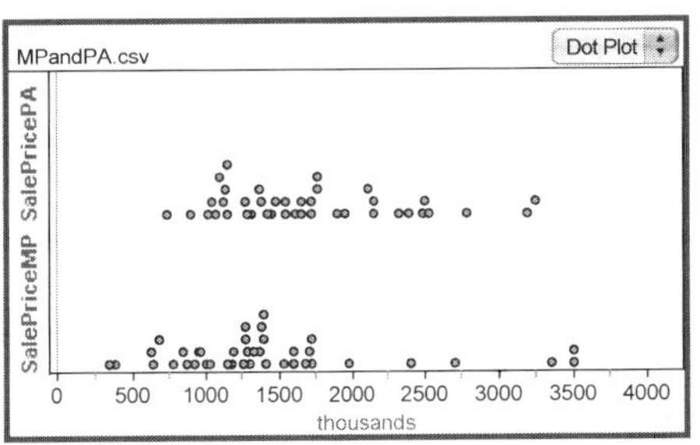

Now that we have the data matched, the variable that we will calculate for each matched pair is *DiffSalePrice = SalePricePA − SalePriceMP*. That is, we will calculate the difference in price between two houses with very similar sizes. It is for this variable (the difference between two measures) that we will calculate a confidence interval. The confidence interval for the variable: *DiffSalePrice = SalePricePA − SalePriceMP* with the results shown here. Using the formula for paired comparisons, and t^* for $df = 39$

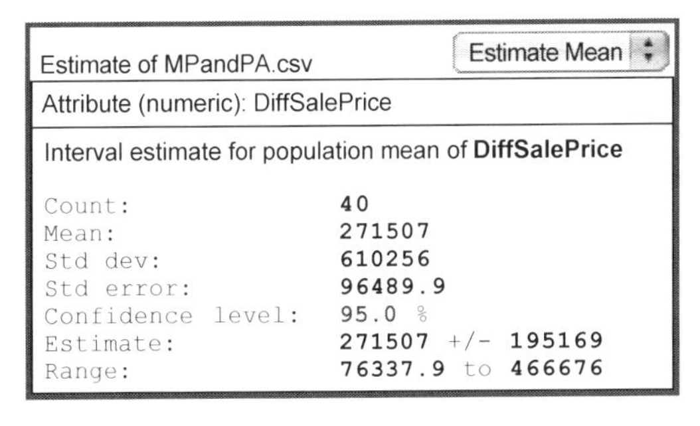

$$\bar{x}_{Diff} \pm t^* \frac{s_{Diff}}{\sqrt{n}} = 271507 \pm 2.023 \frac{610256}{\sqrt{40}}$$

$$\approx 271507 \pm 195169$$

The interval does not include zero; if we were doing an hypothesis test where the null hypothesis was that there was no difference in prices for the same size houses then our $\mu_0 = 0$ would be rejected. Hence, we see that there *is* a significant difference in price between Palo Alto and Menlo Park.

Summary: Paired Comparison Data

- The formulas for inference for paired comparison data are essentially the same as the formulas for one mean, but the variable that is being analyzed is the difference of *two comparable measures* for *one collection* of data.
- The ***notation for paired comparison data*** reflects the fact that the variable being measured is the difference between two measures (or variables).

	Sample	***Population***
Mean	$\bar{x}_{Diff}$	μ_{Diff}
Standard deviation	s_{Diff}	σ_{Diff}

- ***Paired comparison*** data may either come from:
  - One collection in which there are comparable measures whose difference may be calculated, or
  - Two collections in which the cases have been matched, usually because the matched cases have similar or the same values for one or more variables. *Paired comparisons* that have been created in this fashion are called ***matched pairs.***

Summary: Hypothesis Test for Paired Comparison Data

Step 0: *Do a descriptive analysis of the two quantitative variables in the pair and also the variable that represents the difference between the variables for each pair.*

Step 1: *Set up the null and alternate hypotheses for the mean difference μ_{Diff} using an idea for the population mean difference. It is very common to have $\mu_0 = 0$, indicating no mean difference between the measures, but it is possible to test the hypothesis using a null hypothesis with a mean that is something other than zero. (Here, on the left, $\mu_0 = 0$ is shown—on the right, something other than zero; μ_{Diff_0} can be some other number.)*

$$H_0 : \mu_{Diff} = 0 \qquad H_a : \begin{cases} \mu_{Diff} \neq 0 \\ \mu_{Diff} < 0 \\ \mu_{Diff} > 0 \end{cases} \qquad H_0 : \mu_{Diff} = \mu_{Diff_0} \qquad H_a : \begin{cases} \mu_{Diff} \neq \mu_{Diff_0} \\ \mu_{Diff} < \mu_{Diff_0} \\ \mu_{Diff} > \mu_{Diff_0} \end{cases}$$

Step 2: Check the conditions for a trustworthy hypothesis test.

3. *The sample must be a simple random sample.*
4. *The population must be Normal; however, since the t procedures are robust for large sample sizes, judge whether to proceed using the 15/40 rule.*

Step 3: *Calculate the test statistic $t = \dfrac{\bar{x}_{Diff} - \mu_{Diff_0}}{s_{Diff}/\sqrt{n}}$ using the sample mean difference $\bar{x}_{Diff}$, sample standard deviation for the differences s_{Diff}, and the hypothesized μ_{Diff_0}.*

Step 4: *Calculate the p-value by getting $P\left(\bar{x}_{Diff} \text{ more extreme} \mid \mu_0 = \mu_{Diff_0}\right)$ using the absolute value of the test statistic t calculated in Step 3.*

Step 5: *Evaluate the evidence that the p-value and the test statistic give you to determine whether your test successfully challenges the null hypothesized population mean μ_{Diff_0} or not. Give an interpretation in the context of the data using the terminology of "statistical significance" and "rejecting the null hypothesis."*

Summary: Confidence Interval for Paired Difference Data

$$\bar{x}_{Diff} \pm t^* \frac{s_{Diff}}{\sqrt{n}}$$

where t depends on the df = n – 1 and the level of confidence. The conditions for trustworthy confidence interval are the same as for a hypothesis test.*

§5.4 Comparing Means: One Measure, Two Collections

Blood pressures for two independent collections

To provide some continuity with the last section, but also to *contrast* this section with the last one, our examples will continue to be about blood pressure. Our initial statistical question is:

Is there a difference between men and women in average (or mean) systolic blood pressure?

We have **one quantitative variable—**the variable *systolic* blood pressure— but now we are comparing two groups or **two collections:** males and females. In the last section (***Paired Comparison***), we analyzed *two measures* but in *one collection*. The two collections we are comparing must be ***independent samples.*** If the groups that we compare are *parts* of a random sample, such as the males and females in our sample here (where we are analyzing blood pressure), then the groups *are* independent. Parts of random samples are independent. This situation of independent groups—or collections—is quite different from what we analyzed in §5.3. In §5.3 we had two measures or variables for just one group. Here we have *two* independent groups but just *one* measure; we have *one* variable that is measured in each of the groups. The situation here is the same that we encountered for the difference of proportions for two independent groups or collections.

Our statistical question is the kind (an "Is there…?" type of question) that should be handled by a hypothesis test, and so we have a situation comparable to the situation when we compared two proportions. As in that situation, we need notation for all of the numerical quantities we will encounter.

Notation for Comparing a Quantitative Variable between Two Collections:

	Sample Size	*Population Mean*	*Sample Mean*	*Population Standard Deviation*	*Sample Standard Deviation*
Collection 1:	n_1	μ_1	$\bar{x}_1$	σ_1	s_1
Collection 2:	n_2	μ_2	$\bar{x}_2$	σ_2	s_2

As before, we will replace the numbered subscripts with more informative letters that relate to the context, but we need the numbered subscripts to write formulas in general. So we will write null and alternate hypotheses using μ_F and μ_M rather than μ_1 and μ_2; the important thing is that we need to have distinct names for all the quantities we use.

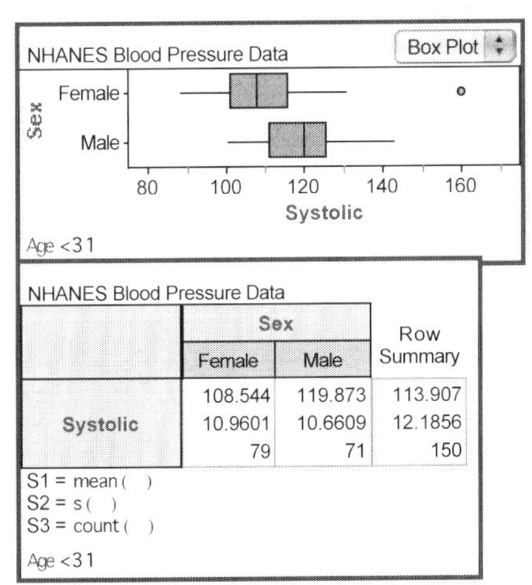

NHANES Blood Pressure Data

	Sex: Female	Sex: Male	Row Summary
Systolic	108.544	119.873	113.907
	10.9601	10.6609	12.1856
	79	71	150

S1 = mean( )
S2 = s( )
S3 = count( )
Age <31

Step 0: Some Descriptive Analyses We will restrict our statistical question to men and women between the ages of twenty and thirty. Our question now becomes:

Is there a difference between men and women in mean systolic blood pressure (for people between the ages of twenty and thirty)?

From the box plot of the distributions of *Systolic* by *Sex*, the distributions look nearly symmetrical and from the summary table showing means and standard deviations, it does appear that women in this age group have lower levels of blood pressure than men. We see that, using our notation: $\bar{x}_F = 108.544$, $s_F = 10.9601$, $n_F = 79$ $\quad$ $\bar{x}_M = 119.873$, $s_M = 10.6609$, $n_M = 71$

Step 1: Null and Alternate Hypotheses The question we originally asked implies a two-sided alternate hypothesis, since we made no prediction about the direction of the difference between men and women. (It would be bad practice to base the test on what we happen to see in the data at hand.) So we have:

$$H_0 : \mu_F - \mu_M = 0$$
$$H_a : \mu_F - \mu_M \neq 0$$

These hypotheses can also be expressed as $H_0 : \mu_F = \mu_M$, $H_a : \mu_F \neq \mu_M$, but it is a good idea to get accustomed to thinking of the difference as being zero (or not zero) since that is how we will calculate the test statistic. In words, our test is based upon the null hypothesis that the mean *systolic* blood pressure does *not* differ between men and women in the age group twenty to thirty years.

Step 2: Checking the Conditions We will again be using the *t* procedures, and the conditions for the *t* procedures for *two* collections are similar to the conditions we have seen before:

- The samples must be random samples independently drawn from two populations.
- The population distributions must both be Normal.

The words "independently drawn" mean that we cannot have data that are paired up in some way. The samples of men and women cannot be from a matched pair design, for example. The men and women cannot be (for example) husbands and wives or brothers and sisters. Our NHANES data are a random sample, and so within that sample, the men and the women can be regarded as independently drawn random samples.

The second condition we have seen before, and we have seen that the *t* procedures are *robust* if the sample sizes are large enough; to decide what is "large enough," we have the ***15/40 rule***. For comparing means, we can use the *15/40 rule* with the sum of sample sizes $n_1 + n_2$. So here, although both $n_F = 79$ and $n_M = 71$ are each over 40, what is important is $n_F + n_M = 79 + 71 = 150$. In any case, in our example, the sample distributions of the variable *Systolic* are both nearly symmetric, with one outlier. We can safely proceed.

Conditions for one measure, two independent collections comparisons

- The samples must be simple random samples independently drawn from two populations.
- The populations must have Normal distributions, however, since the t procedures are robust for large sample sizes. Use the 15/40 rule but with the sum of sample sizes $n_1 + n_2$.

Step 3: Calculating the test statistic The *form* of our test statistic should be:

$$\frac{(\textit{Difference of Sample Means}) - (\textit{Hypothesized Difference of Population Means})}{\textit{Standard deviation of Sampling Distribution for Difference of means}}$$

If we knew the population standard deviations, this general formula would be a *z* score:

$$z = \frac{(\bar{x}_1 - \bar{x}_2) - (\mu_1 - \mu_2)}{\sqrt{\frac{\sigma_1^2}{n_1} + \frac{\sigma_2^2}{n_2}}}$$

However, researchers generally do not know the values of the population standard deviations σ_1 and σ_2 while researchers *do* know the values of the sample standard deviations s_1 and s_2.

Hence, substituting the sample values for the population values, and therefore using the t distributions, the ***test statistic*** will be:

$$t = \frac{(\bar{x}_1 - \bar{x}_2) - (\mu_1 - \mu_2)}{\sqrt{\frac{s_1^2}{n_1} + \frac{s_2^2}{n_2}}}$$

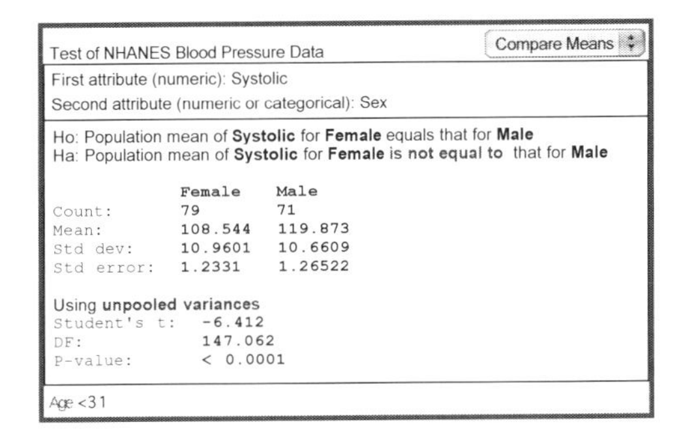

Test of NHANES Blood Pressure Data — Compare Means
First attribute (numeric): Systolic
Second attribute (numeric or categorical): Sex

Ho: Population mean of **Systolic** for **Female** equals that for **Male**
Ha: Population mean of **Systolic** for **Female** is **not equal to** that for **Male**

	Female	Male
Count:	79	71
Mean:	108.544	119.873
Std dev:	10.9601	10.6609
Std error:	1.2331	1.26522

Using **unpooled variances**

Student's t:	-6.412
DF:	147.062
P-value:	< 0.0001

Age <31

So for our example we will have:

$$\begin{aligned} t &= \frac{(\bar{x}_F - \bar{x}_M) - 0}{\sqrt{\frac{s_F^2}{n_F} + \frac{s_M^2}{n_M}}} \\ &= \frac{(108.544 - 119.873) - 0}{\sqrt{\frac{(10.9601)^2}{79} + \frac{(10.6609)^2}{71}}} \\ &\approx \frac{-11.329}{\sqrt{1.52055 + 1.60077}} \\ &\approx \frac{-11.329}{1.76673} \\ &\approx -6.412 \end{aligned}$$

In the formula, the zero after the $(\bar{x}_F - \bar{x}_M)$ indicates that our null hypothesis of *no* difference in the population means between men and women; that is $H_0: \mu_F - \mu_M = 0$.

A value of t = -6.412 looks fairly extreme, and we think that the value of the test statistic shows that *if* the null hypothesis were true and there were really no difference between the mean *Systolic* blood pressure for men and women then what we have seen in our sample is very rare just by sampling variation—so rare as to prompt us to doubt the null hypothesis of no difference.

Step 4: Getting a p-value. However, it turns out that the sampling distribution of the *test statistic* shown above does *not* follow a t distribution. Fortunately, the t distributions can be used with this test statistic with the appropriate choice of ***degrees of freedom.*** We have two choices of degrees of freedom. One of these choices is appropriate for working by hand. This is ***Choice 1:*** the degrees of freedom will be calculated by taking the smaller of the two $df = n - 1$ for our two samples. For our example we will have $df = n - 1 = 71 - 1 = 70$. ***Choice 2***, used by software calculates the degrees of freedom by the formula: $df = \dfrac{\left(\frac{s_1^2}{n_1} + \frac{s_2^2}{n_2}\right)^2}{\frac{1}{n_1 - 1}\left(\frac{s_1^2}{n_1}\right)^2 + \frac{1}{n_2 - 1}\left(\frac{s_2^2}{n_2}\right)^2}$. When this formula is used, the result is often not an integer. This formula looks formidable to calculate, but it is not at all a problem for software and is actually the better choice. The disadvantage of *Choice 1* is that for small sample sizes, the power of the hypothesis test can be low. On the next page is software output for this hypothesis test. Notice that the degrees of freedom is $df = 147.062$ and notice that the p value is very small, as we predicted it would be.

Step 5: Interpretation Even if we had gone with Choice 1 for the calculation of the degrees of freedom, the test statistic $t = -6.412$ is "off the chart," and so the p value is clearly less than 0.0005. Hence, our test is *statistically significant* and we have evidence that there *is* a difference in the mean *Systolic* blood pressure for men and women. From the evidence that we have, it appears that women between the ages of twenty and thirty have, on average, lower *systolic* blood pressure. How much lower is the mean for the women than for the men? We need an estimate, and so a confidence interval.

Confidence Intervals: One measure with two collections

The *form* of a confidence interval should be predictable. In general, we have been calculating

$$(\textit{Sample Estimate}) \pm t^*(\textit{Standard Deviation of the sampling Distribution})$$

and so we should be able to predict, given what we have seen for the hypothesis test, that the formula will be

$$(\bar{x}_1 - \bar{x}_2) \pm t^* \sqrt{\frac{s_1^2}{n_1} + \frac{s_2^2}{n_2}}$$

For our example, if we have Choice 1, where the $df = n - 1 = 71 - 1 = 70$, we see that for a 95% confidence interval, we should have $t^* = 1.994$. So we will calculate the confidence interval for the population mean difference in *systolic* blood pressure between men and women—that is, our estimate of $\mu_F - \mu_M$ as shown below.

$$\begin{aligned}(\bar{x}_F - \bar{x}_M) \pm t^* \sqrt{\frac{s_F^2}{n_F} + \frac{s_M^2}{n_M}} &= (108.544 - 119.873) \pm 1.994 \sqrt{\frac{(10.9601)^2}{79} + \frac{(10.6609)^2}{71}} \\ &\approx -11.329 \pm 1.994\left(\sqrt{1.52055 + 1.60077}\right) \\ &\approx -11.329 \pm 1.994(1.76673) \\ &\approx -11.329 \pm 3.523\end{aligned}$$

We have a ***margin of error*** of 3.523 mm Hg, and this can be written as an interval as $-14.852 < \mu_F - \mu_M < -7.806$. We can say, with 95% confidence, that women in the age group twenty to thirty have mean *systolic* blood pressure that is between 7.8 mm Hg and 14.9 mm Hg lower than the mean *systolic* blood pressure for men in that age group. The signs are negative because we calculated $\bar{x}_F - \bar{x}_M$, and the sample mean for men was higher than the sample mean for women.

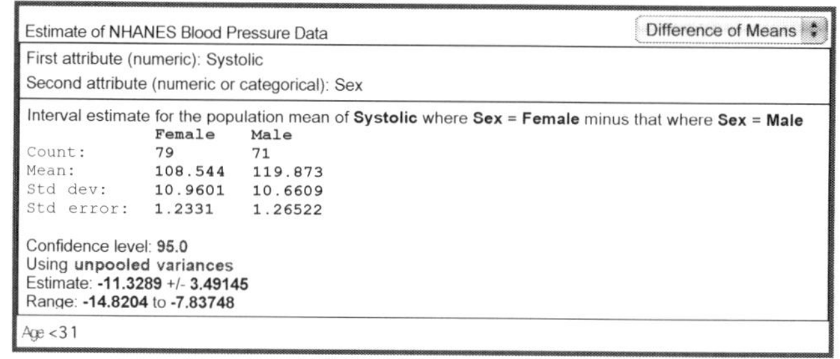

The confidence interval calculated by software is just slightly narrower than the one we calculated "by hand.". The reason for that is that Software has used the t^* based upon the degrees of freedom of $df = 147.062$, rather than $df = 70$.

Comparing means from two *independent* collections, either with confidence intervals or with hypothesis tests, is an extremely common form of analysis.

Summary: One measure, two independent collections

- It is quite important to distinguish ***one measure, two independent collections*** (the subject of this section) from ***two measures, essentially one collection*** (the subject of the preceding section, §5.3).
  - Another name for ***one measure, two independent collections*** is ***a two independent sample comparison*** (this section: §5.4).
  - Another name for ***two measures, essentially one collection*** is ***paired comparisons*** (§5.3).

Summary: Hypothesis Test for Comparing One Measure for Two Collections

Step 0: *Do a descriptive analysis of the quantitative variable in the two samples. Get graphics and summary statistics.*

Step 1: *Set up the null and alternate Hypotheses for the population mean difference* $\mu_1 - \mu_2$.

$$H_0 : \mu_1 - \mu_2 = 0 \qquad H_a : \begin{cases} \mu_1 - \mu_2 \neq 0 \\ \mu_1 - \mu_2 < 0 \\ \mu_1 - \mu_2 > 0 \end{cases}$$

Step 2: Check the conditions for a trustworthy hypothesis test.

5. *The samples must be simple random samples independently drawn from two populations.*
6. *The populations must be Normal; however, since the t procedures are robust for large sample sizes, judge whether to proceed using the 15/40 rule, but with the sum of sample sizes* $n_1 + n_2$.

Step 3: *Calculate the test statistic* $t = \dfrac{(\bar{x}_1 - \bar{x}_2) - 0}{\sqrt{\dfrac{s_1^2}{n_1} + \dfrac{s_2^2}{n_2}}}$ *using the sample means and sample standard deviations.*

Step 4: *Calculate the p-value by getting* $P(\bar{x}_1 - \bar{x}_2 \text{ or more extreme} \mid \mu_1 - \mu_2 = 0)$ *using the absolute value of the test statistic t calculated in Step 3.*

Step 5: *Evaluate the evidence that the p-value and the test statistic gives in order to determine whether the test successfully challenges the null hypothesized population mean* $\mu_1 - \mu_2 = 0$. *Give an interpretation in the context of the data using the terminology of "statistical significance" and "rejecting the null hypothesis."*

Summary: Confidence Interval for Comparing One Measure for Two Collections

$$(\bar{x}_1 - \bar{x}_2) \pm t^* \sqrt{\frac{s_1^2}{n_1} + \frac{s_2^2}{n_2}}$$

where t depends on the level of confidence and degrees of freedom calculated either from:*

- *the smaller of* $df = n_1 - 1$ *and* $df = n_2 - 1$ *or*
- $df = \dfrac{\left(\dfrac{s_1^2}{n_1} + \dfrac{s_2^2}{n_2}\right)^2}{\dfrac{1}{n_1 - 1}\left(\dfrac{s_1^2}{n_1}\right)^2 + \dfrac{1}{n_2 - 1}\left(\dfrac{s_2^2}{n_2}\right)^2}$

The conditions for trustworthy confidence interval are the same as for a hypothesis test.

§5.5 Analysis of Variance

Introduction: which types of cars are more economical?

Cars and trucks come in various shapes and sizes, and they differ in their fuel economy. Every year the US Department of Energy and the Environmental Protection Agency publish their Fuel Economy Guide for all of the cars sold in the United States (see www.fueleconomy.gov).

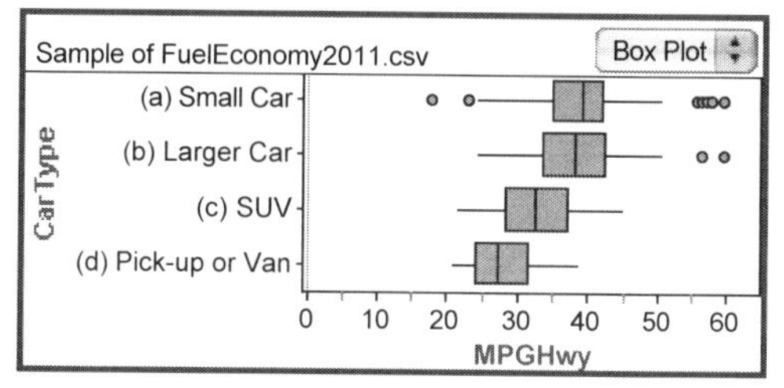

The Fuel Economy Guide reports fuel consumption as city miles per gallon (MPG), highway MPG, and a combined measure. Here are box plots showing the distributions of various types of vehicles for highway MPG and a summary table showing the means, standard deviations, and counts for each of the type of vehicles for highway MPG.

Sample of FuelEconomy2011.csv

	CarType				
	(a) Small Car	(b) Larger Car	(c) SUV	(d) Pick-up or Van	Row Summary
MPGHwy	39.1168 8.38481 88	38.3084 6.74679 86	32.9032 5.72931 87	28.1216 4.89911 63	35.0958 7.88651 324

S1 = mean ( )
S2 = s ( )
S3 = count ( )

It appears (as we would expect) that small cars are more economical (greater miles per each gallon) than SUVs, pick-up trucks, or vans. Are the differences statistically significant? We would be tempted to compare the types of vehicles pair-wise—small cars compared with larger cars, small cars compared with SUVs—but there is a hypothesis test, called ***analysis of variance***, or ***ANOVA,*** that looks at all of the categories at once.

ANOVA is a hypothesis test that answers the question: "Is there a significant difference in three or more means?" The null and alternate hypotheses for this test are always the same, no matter how many groups we have, as shown below.

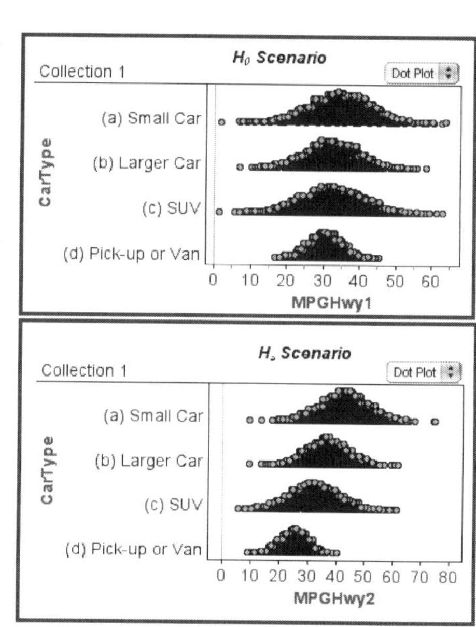

Null and alternate hypotheses for ANOVA

$H_0 : \mu_1 = \mu_2 = \cdots = \mu_k$

H_a : *At least one of the k means differs from the others*

where k indicates the number of different groups in the test.

It is useful to think graphically what the populations would look like if the null hypothesis is true (the "H_0 scenario") and what the populations would look like if the alternate hypothesis is true (the "H_a scenario"). In the "H_0 scenario" the distributions all have the same center, but in the "H_a scenario" (shown below) at least one of the distributions has different center. It is the job of ANOVA to detect which of the scenarios our data most resembles, given that there will be sampling variation that may make data from populations that are the "H_0 scenario" look like an "H_a scenario" and vice versa.

Ideas behind the ANOVA test and the test statistic. Instead of comparing means as we did with the "comparing independent means test," we will compare two different *variances* (hence, "analysis of variance"). The two variances are ***mean square groups (MSG)*** and ***mean square error (MSE). MSG*** measures how much the group means vary around the mean of all the groups taken together (the grand mean of all of the data). It is sometimes known as variance *between* the groups. ***MSE*** measures the average variance *within* the groups.

Then, the idea is if the MSG (variance *between* the group means) is large compared with the MSE (variance *within* the groups), we have evidence that the means are more varied—more different—than the data as a whole and, hence, evidence in accord with the H_a, the alternate hypothesis, that there are differences between the means. This is what is pictured in the left-hand graphic below. On the other hand, if the MSG and the MSE come out to be about the same then we have evidence that the means are no more varied than the data as a whole and evidence consistent with the null hypothesis that there is no difference between the group means. This is picture in the right-hand graphic below.

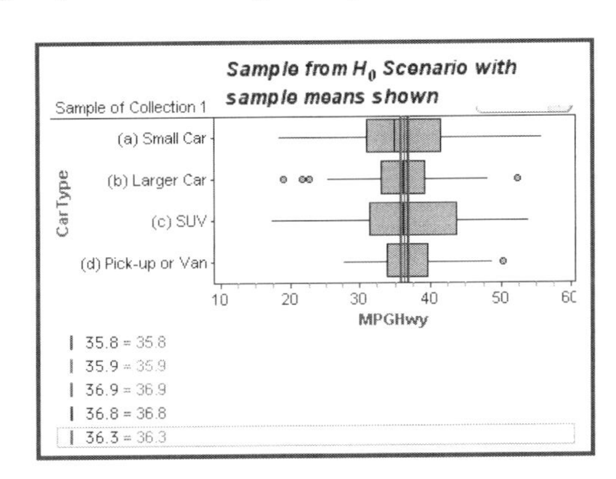

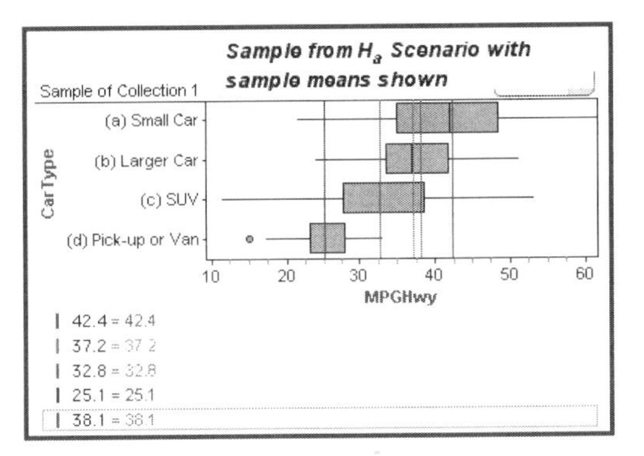

The way we compare the MSG to the MSE is by means of a ratio, called the ***F statistic***, as defined in the box below. This becomes the test statistic for the ANOVA test (instead of a *t* or *z*).

Definition of the F test statistic

$$F = \frac{MSG}{MSE} = \frac{\text{Variation betweeen the means of the groups}}{\text{Variation within the groups}}$$

From what was said above, it follows that if ***F*** is large (so that the MSG is much bigger than the MSE), we have evidence for the alternate hypothesis, and if ***F*** is near to one (or even less than one) then we have evidence consistent with the null hypothesis. The two examples above are deliberately extreme; the calculations show that $F = \frac{MSG}{MSE} = \frac{26.84}{56.64} \approx 0.474$ for the left-hand graphic and $F = \frac{MSG}{MSE} = \frac{3436.61}{61.87} \approx 55.546$. The formulas for the MSG and the MSE are given in the box just below.

Formulas for MSG and MSE

$$MSG = \frac{n_1(\bar{x}_1 - \bar{x})^2 + n_2(\bar{x}_2 - \bar{x})^2 + \cdots + n_k(\bar{x}_k - \bar{x})^2}{k-1} \qquad MSE = \frac{(n_1 - 1)s_1^2 + (n_2 - 1)s_2^2 + \cdots + (n_k - 1)s_k^2}{N-k}$$

where: $\bar{x}$ is the mean for all of the data taken together;

$\bar{x}_1, \bar{x}_2, \cdots, \bar{x}_k$ are the means of the 1, 2,...k groups;

$n_1, n_2, \cdots, n_k$ are the sample sizes in each of the groups, and

$N = n_1 + n_2 + \cdots + n_k$ is the total sample size for all the groups together.

Of course, we need a sampling distribution to tell us how big an F statistic must be to detect a difference in the means, but let us work on some real data to see how the formulas work.

An example of the calculations: a sample from the Fuel Economy Guide

Since ANOVA is a hypothesis test, we will follow the five steps of the hypothesis test. For ANOVA, there is a standard way of displaying the calculations and the p-value that we must also discuss.

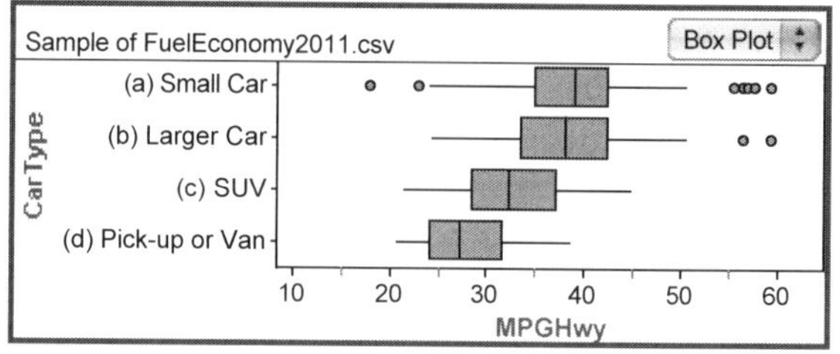

Here are the box plots for the variable *MPGHwy* (highway MPG). The plots suggest that there are differences in the fuel economy of the different types of vehicles. SUVs and truck-based vehicles do poorly (have lower MPG) compared with cars, especially small cars, although the variability with the small car group appears to be great.

Step 1: Setting up the hypotheses. As mentioned above, the null and alternate hypotheses are standard, although it is a good idea to label the groups for reference. Hence:

$$H_0: \mu_{Small} = \mu_{Larger} = \mu_{SUV} = \mu_{Pick-up}$$
$$H_a: \textit{At least one of the four means differs from the others}$$

Step 2: Checking the Conditions. We will return to this step.

Step 3: Calculating the F test statistic. To calculate F, we obviously must first calculate the MSG and the MSE, and for that we need the means and the standard deviations for each of the groups, shown in the table. Below are the calculations:

Sample of FuelEconomy2011.csv

	CarType				Row Summary
	(a) Small Car	(b) Larger Car	(c) SUV	(d) Pick-up or Van	
MPGHwy	39.1168	38.3084	32.9032	28.1216	35.0958
	8.38481	6.74679	5.72931	4.89911	7.88651
	88	86	87	63	324

S1 = mean ( )
S2 = s ( )
S3 = count ( )

$$MSG = \frac{n_{Small}\left(\bar{x}_{Small} - \bar{x}\right)^2 + n_{Larger}\left(\bar{x}_{Larger} - \bar{x}\right)^2 + n_{SUV}\left(\bar{x}_{SUV} - \bar{x}\right)^2 + n_{Pick-up}\left(\bar{x}_{Pick-up} - \bar{x}\right)^2}{k-1}$$

$$= \frac{88*(39.117-35.096)^2 + 86*(38.308-35.096)^2 + 87*(32.903-35.096)^2 + 63*(28.122-35.096)^2}{4-1}$$

$$= \frac{88*(4.021)^2 + 86*(3.212)^2 + 87*(-2.193)^2 + 63*(-6.974)^2}{3}$$

$$= \frac{1422.823 + 887.257 + 418.405 + 3064.111}{3} = \frac{5792.596}{3} = 1930.87$$

$$MSE = \frac{\left(n_{Small}-1\right)s^2_{Small} + \left(n_{Larger}-1\right)s^2_{Larger} + \left(n_{SUV}-1\right)s^2_{ISUV} + \left(n_{Pick-up}-1\right)s^2_{Pickup}}{N-k}$$

$$= \frac{(88-1)*(8.385)^2 + (86-1)*(6.747)^2 + (87-1)*(5.729)^2 + (63-1)*(4.899)^2}{324-4}$$

$$= \frac{6116.816 + 3869.371 + 2822.644 + 1488.012}{320} = \frac{14296.843}{320} = 44.678$$

From these calculations, we calculate the test statistic: $F = \frac{MSG}{MSE} = \frac{1930.865}{44.678} \approx 43.22$.

The Display Table for ANOVA. For the ANOVA test there is a standard way of laying out all of these numbers, once they have been calculated. For our example above, here is an ***ANOVA Table.*** To become familiar with this ANOVA table, it is good to relate the numbers to the calculations.

Test of Sample of FuelEconomy2011.csv — Analysis of Variance

Response attribute (numeric): MPGHwy
Grouping attribute (categorical): CarType

Source	Degrees of Freedom	Sum of Squares	Mean Square	F Statistic	P Value
Groups	3	5792.96	1930.99	43.221	0.0000
Error	320	14296.7	44.68		
Total	323	20089.7			

Alternative hypothesis: The population means of **MPGHwy** grouped by **CarType** are not equal

If it were true that the population means of **MPGHwy** were equal (the null hypothesis) and the sampling process were performed repeatedly, the probability of getting a value for F greater than or equal to the observed value of **43.2209** would be **< 0.0001**.

You should be able to recognize the value of the ***F statistic*** by comparing the numbers in the table with the calculations laid out above. In the column labeled ***Mean Square***, you should see the values of the MSG and MSE. The value of MSG is at the intersection of the ***Mean Square*** column and ***Groups*** row. Likewise, the value of the MSE is at the intersection of the ***Mean Square*** column and ***Error*** row. The column labeled ***Degrees of Freedom*** shows the denominators of the calculations of MSG and MSE. For our example, the MSG denominator is $k-1=4-1=3$ and the MSE denominator is $N-k=324-4=320$. You see both these numbers in the ***Degrees of Freedom*** column. The numbers in the column labeled ***Sum of Squares*** are the numerators of the calculations of MSG and MSE. The last row (***Total***) represents the calculation for the overall variance in the collection. Below there is a graphic showing how the entries in the table relate to the formulas. The graphic is based on softwre output and also shows an interpretation of the results.

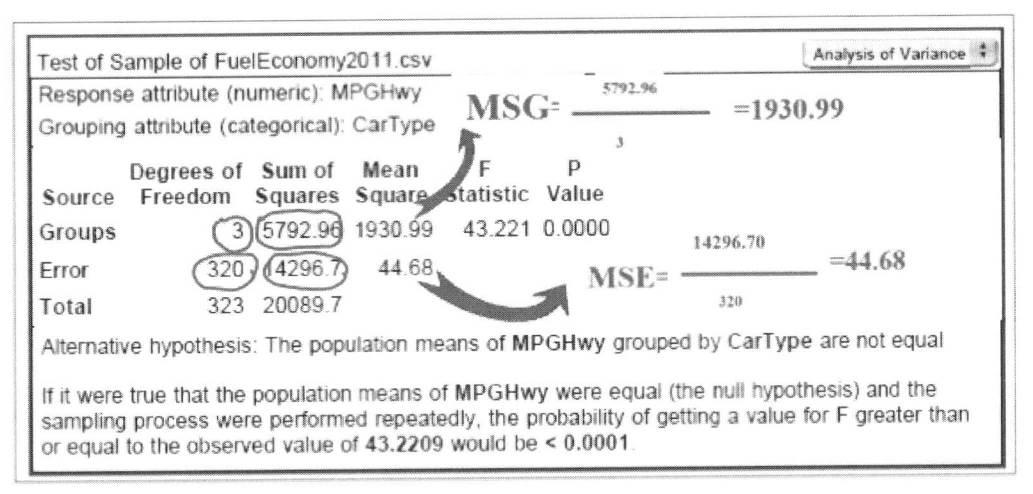

Test of Sample of FuelEconomy2011.csv — Analysis of Variance

Response attribute (numeric): MPGHwy
Grouping attribute (categorical): CarType

$$MSG = \frac{5792.96}{3} = 1930.99$$

Source	Degrees of Freedom	Sum of Squares	Mean Square	F Statistic	P Value
Groups	3	5792.96	1930.99	43.221	0.0000
Error	320	4296.7	44.68		
Total	323	20089.7			

$$MSE = \frac{14296.70}{320} = 44.68$$

Alternative hypothesis: The population means of **MPGHwy** grouped by **CarType** are not equal

If it were true that the population means of **MPGHwy** were equal (the null hypothesis) and the sampling process were performed repeatedly, the probability of getting a value for F greater than or equal to the observed value of **43.2209** would be **< 0.0001**.

Step 4: Getting the p-value. The ***p-value*** is also shown in the ANOVA table. In this instance, the *p*-value is shown as 0.0000, which indicates $p < 0.0001$, and this is the information we need. The test is significant, and the data show evidence that there are differences among the means of MPG for the different types of cars.

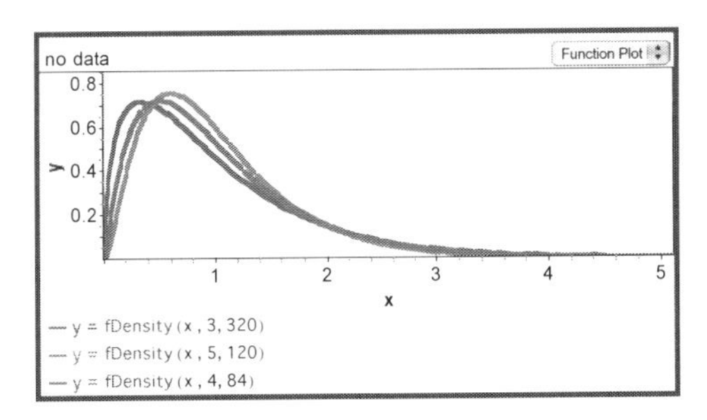

From what sampling distribution was this *p*-value calculated? ANOVA uses a family of distributions called the ***F distributions*** whose shape depends upon the degrees of freedom. Here are some examples. Instead of relying upon a chart, we will depend upon software to give the *p*-value for a given ***F*** test statistic value.

Our graphic includes the F distribution for numerator $DF = 3$ and denominator $DF = 320$, and we can see quite clearly that the F test statistic of $F = 43.22$ is "way off the chart" to the right; the reported $p < 0.0001$ is consistent with what we see. The shaded portion showing the p-value would be so small as to be invisible. However, if we had a test with a big p-value, we would be able to see the p-value as a shaded-in area.

Another Example: automatic or manual?

Sample of FuelEconomy2011.csv

	TransType			Row Summary
	(a) Manual	(b) Semi-Automatic	(c) Automatic	
MPGComb	30.1096 6.8485 40	29.1302 8.70352 30	32.5844 9.23502 18	30.2819 8.03171 88

S1 = mean( )
S2 = s( )
S3 = count( )

CarType = "(a) Small car"

It is an old idea that cars with manual transmissions rather than automatic transmission get better gas mileage—their MPH should be higher. We can test this with the category of small cars. The Summary Table (above) shows the means and standard deviations for *MPGComb* (the combination of city and highway MPG). There appears not to be much difference between the MPG for the transmission types, and in fact the automatic cars do slightly better. We should expect that the ANOVA test will be consistent with the null hypothesis, and that the p-value will be relatively large.

Test of Sample of FuelEconomy2011.csv — Analysis of Variance

Response attribute (numeric): MPGComb
Grouping attribute (categorical): TransType

Source	Degrees of Freedom	Sum of Squares	Mean Square	F Statistic	P Value
Groups	2	136.41	68.2060	1.059	0.3514
Error	85	5475.82	64.4214		
Total	87	5612.23			

Alternative hypothesis: The population means of **MPGComb** grouped by **TransType** are not equal

If it were true that the population means of **MPGComb** were equal (the null hypothesis) and the sampling process were performed repeatedly, the probability of getting a value for F greater than or equal to the observed value of **1.05875** would be **0.35**.

CarType = "(a) Small car"

The test statistic F is near to the value "one," which is what we should see for the equal means situation, and the p-value is large: $p = 0.35$. The shaded-in portion of the test statistic distribution shows the size of the p-value.

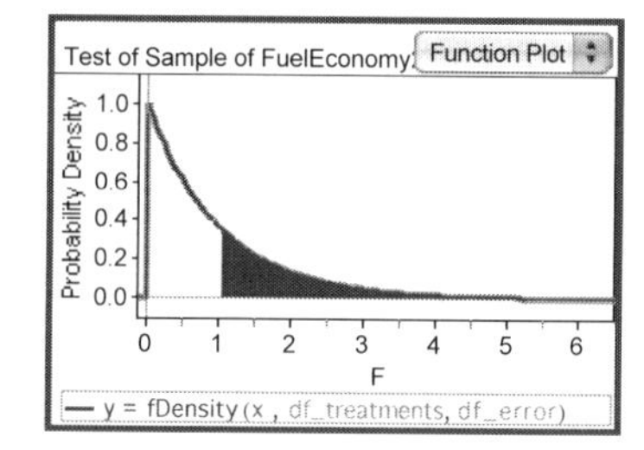

Conditions for ANOVA: Robustness again. Here are the conditions that must be met for the ANOVA procedure to work well. You will see that two of the conditions can be "relaxed" under certain circumstances. In other words, ANOVA procedures are to some extent robust.

Conditions for ANOVA

- The samples must be k ***independent simple random samples***.
- The populations from which each of the k samples is drawn must be ***normally distributed***.
- The populations from which each of the k samples is drawn must have the ***same standard deviation***.

The first condition regarding random sampling is "not negotiable"; samples that are convenience samples or other types of non-random collections make the ANOVA procedures suspect. There are ANOVA procedures for other types of random sampling that can be used, but those are beyond the scope of a first course. The second condition should be investigated in the same way it was for the t procedures; outliers and extreme skewness can cause problems if the sample sizes are very small. There is a rule of thumb regarding the third condition, which is expressed in the box on the next page.

Rule of thumb for equal standard deviations

ANOVA has been found to give reliable results if the largest of the *k* standard deviations is no greater than twice the smallest of the *k* standard deviations.

Step 2: Checking the conditions. We see that since the sample that includes the four types of vehicles was randomly taken from the Fuel Economy Guide, the first condition is met. The box plots show outliers on both sides of some of the vehicle groups' distributions, but the distributions appear symmetrical, and the sample sizes are large. The largest standard deviation is for the small cars $s_{Small} = 8.385$ but it is less than *twice* the size of the smallest standard deviation for the pick-ups: $2 * s_{Pick-up} = 2 * 4.899 = 9.798$. For the transmission types example, since the largest standard deviation $9.235 < 2*(6.849) = 13.698$, the rule of thumb shows that we can proceed safely.

Step 5: Interpretation: Car Type Example We have chosen a rather uncontroversial example; of course, small cars are more fuel-efficient than larger cars, and these in turn more economical than SUVs or pick-up trucks and vans. Our data bear out by the small *p*-value what we knew. By the appearance of the box plots, it seems that there are differences among all the types of cars; however, ANOVA (like the chi-square test) can only tell us that there is some significant difference somewhere and cannot pinpoint where that difference is. That must be done with further analysis.

Step 5: Interpretation: Transmission Type Example. For the transmission types, the ANOVA test shows that our original idea that manual transmission cars get better mileage was not correct. Now it may be that we are looking at different kinds of cars. A better way to analyze this question is to choose just those cars that are offered with either a manual or an automatic transmission and compare them.

Summary: Analysis of Variance

- ***ANOVA*** is a hypothesis test to detect whether data show a difference among group means for more than two independent means in a population. If there are just two means, use a t test.
- ***Null and Alternate Hypotheses for ANOVA***

$H_0 : \mu_1 = \mu_2 = \cdots = \mu_k$

H_a : *At least one of the k means differs from the others*

- ***Test Statistic for ANOVA***

$$F = \frac{MSG}{MSE} = \frac{\textit{Variation betweeen the means of the groups}}{\textit{Variation within the groups}}$$

where $$MSG = \frac{n_1(\bar{x}_1 - \bar{x})^2 + n_2(\bar{x}_2 - \bar{x})^2 + \cdots + n_k(\bar{x}_k - \bar{x})^2}{k-1}$$ and

$$MSE = \frac{(n_1 - 1)s_1^2 + (n_2 - 1)s_2^2 + \cdots + (n_k - 1)s_k^2}{N - k}$$

- ***Sampling distributions for the ANOVA test are the family of F distributions***. There is a different distribution for each combination of numerator and denominator degrees of freedom.
- ***ANOVA Display Table***

Source	*Degrees of Freedom*	*Sum of Squares*	*Mean Square*	*F Statistic*	*p Value*
Groups	*Denominator of MSG*	*Numerator of MSG*	*MSG*	*MSG/MSE*	
Error	*Denominator of MSE*	*Numerator of MSE*	*MSE*		
Total	*Denominator of Variance*	*Numerator of Variance*			

- ***Conditions for ANOVA***

- The samples must be k ***independent simple random samples***.
- The populations from which each of the k samples is drawn must be ***Normally distributed***.
- The populations from which each of the k samples is drawn must have the ***same standard deviation.***

- ***Rule of thumb for equal standard deviations***

  ANOVA has been found to give reliable results if the largest of the k standard deviations is no greater than twice the smallest of the k standard deviations.

§5.6 Reading the Mental Map

Getting the Connections Right

For a complex subject like statistics, it is helpful to have a mental map that will organize the material so that it "hangs together." The map can be schematic, like a transit map. Even without having ridden the train, you know that the tracks are not likely to be the straight lines shown in the diagram; however, the diagram still works to show the relationships of the various lines and where they go. As a review, we explore a "transit map" for the introductory statistics course. The map is shown in reduced form below.

Stat Land

We can think of the map as a map of a place called "Stat Land," whose geography is the content of our introductory statistics course. The course has two related "regions"— ***Descriptive Statistics*** and ***Inferential Statistics***—separated by a "river," which is the theory of sampling distributions.

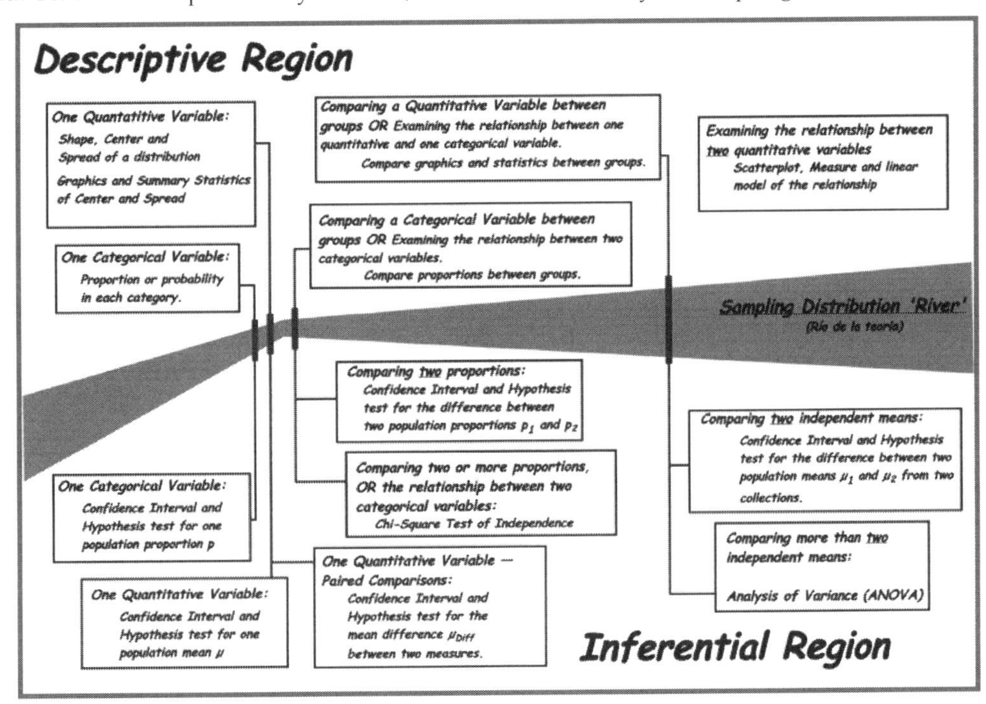

Descriptive and Inferential Statistics The goal of ***descriptive statistics*** is to understand data and be able to communicate its form by using graphics and summary statistics. The goal of formal ***inferential statistics*** is to generalize beyond the data at hand; however, this can only be done with the theory of sampling distributions, with the procedures of ***hypothesis testing*** and ***confidence intervals***, but only under specified ***conditions***. You can think of the "bridges" in the diagram as representing the conditions, but the bridges are there also because proper inferential statistical analysis *starts* with descriptive statistical analysis. After description, then, if the samples are random and sufficiently large, we can "cross the bridge" to inferential analysis.

Categorical and Quantitative Variables The other important distinction seen in the map is the difference between ***categorical*** and ***quantitative*** variables. The summary statistic that we use with categorical variables is a proportion, whereas, with quantitative variables, it makes sense to depict their distributions with graphics such as dot plots, box plots, or histograms and then to calculate measures of the location (center) such as ***mean*** and ***median*** and of spread, such as the ***IQR*** and ***standard deviation***.

Comparing Groups or Examining Relationships between Variables When we are comparing groups or collections, we compare the distributions and the summary statistics for a quantitative variable in the two (or three or four or more) groups; or if we have a categorical variable, we compare the proportions between the two (or three or four or more) groups. However, there are two ways of looking at what we are doing; both ways are legitimate and often end up with the same calculations, but they do involve a slightly different stance.

- We can regard the comparisons of either quantitative or categorical variables as being between groups. If we compare the hours that male students work with the hours that female students work then we are regarding our comparison as between groups.
- We can regard the comparison as indicating the relationship between variables; in our example, we would speak of the relationship between the variable *Students Work Hours* and the variable *Gender.*

Four Kinds of Statistical Questions If you look at the map carefully, you will see that there are four kinds of comparisons that we typically ask, and for the first three of these we have developed formal inferential techniques.

- Comparing data from either a *categorical* or *quantitative* variable with a standard. Are 10% of people left-handed or is the mean flight time to Seattle 125 minutes?
- Comparing a *quantitative* variable by the categories (or groups) of a *categorical* variable. Is there a difference in the mean age of day community college students and evening community college students?
- Comparing a *categorical* variable by the categories (or groups) of a *categorical* variable. Are female college students more likely or less likely to a have a tattoo?
- Examining the relationship between two *quantitative* variables. At what rate does the price of a house increase with the size of the house?

To analyze the relationship between two quantitative variables (the last question above), we use graphic called a ***scatterplot*** and then (if appropriate) applied a *linear model* to the data. Using a linear model, we were able to find a best-fitting line—the *least squares regression* line—and also get both an assessment of the direction and strength of the linear model (using the *correlation coefficient r)* and also get a measure of how well the linear model fits the data using the *coefficient of determination* R^2. There are important formal inferential techniques for linear models not covered in this course.

Mathematical models We have used mathematical models several times. One use was mentioned just above; the simplest model for the relationship between two variables is a linear—simply a straight line—model, which we expressed as $\hat{y} = a + bx$. The other model that we used extensively is the *Normal distribution* model. We use the Normal model to characterize some distributions but more in the "inferential region" part of the course. We used the Normal model there because sampling distributions often (under certain conditions) follow a Normal distribution. The *t distributions* and the *chi-square distributions* are also formal mathematical models.

The Language of Stat Land: "Probability spoken here." Throughout Stat Land, the language and symbols of probability are used. (If Stat Land really had roads, the road signs would be in "probability.") The symbols that are used may differ; for example, if we are describing the proportion of female students who speak two or more languages, we would express this as $P(X \geq 2 \mid F)$, but if we were conducting a hypothesis test we would express the sample proportion as $\hat{p}_F$ where it is understood that the symbol refers to the proportion who speak two or more languages. However, wherever we use the language of probability, it can always refer to the *likelihood* or the *chance* of an event occurring. Wherever we use the language, it must be true that a probability is between zero and one since "p-values" are probabilities, and it makes no sense to write, "The p-value is 2.13."

Reading the Stat Land Map in the Face of a Statistical Question

From www.cars.com we have data on used cars that are advertised there. Data were collected from Boston, Chicago, Dallas, and the San Francisco Bay Area. Data were collected on Audi A4, BMW 3 Series, Mercedes C-class, Infiniti G-35, and Lexus IS. The data were collected in July 2009. Before we answer any statistical question, we need to see what kind of sample we have.

Summer 09 Used Cars

	Make1	Place1	Price	Miles	Age	Convert...	Body	Seller	Distance
1131	BMW 3 S...	Chicago	34995	5654	0.416667	Not Conv...	Sedan	Dealer	21
1132	BMW 3 S...	Chicago	34991	10637	0.416667	Not Conv...	Sedan	Dealer	29
1133	BMW 3 S...	Chicago	34991		0.416667	Not Conv...	Sedan	Dealer	29
1134	BMW 3 S...	Chicago	33995	8252	0.416667	Not Conv...	Sedan	Dealer	30
1135	BMW 3 S...	Chicago	33995	8590	0.416667	Not Conv...	Sedan	Dealer	30
1136	BMW 3 S...	Chicago	33991	10882	1.33333	Not Conv...	Sedan	Dealer	18
1137	BMW 3 S...	Chicago	33900	14000	1.33333	Not Conv...	Coupe	Private S...	15

What kind of sample do we have? Do we have simple random sample? What can we regard as our population? The answer to the first question is no because the cars were not chosen by a random process from a complete list of used cars for sale—a list that does not exist in any case. Since *all* of the cars (that is, all Audi A4, all BMW 3 Series, etc.) advertised on www.cars.com were selected on the dates and places that were chosen, we have something like a cluster sample. Is it random? We need to think about what the population could be. The population is not *all* cars (that is, all Audi A4, all BMW 3 Series, etc.) since we have selected only cars that were offered for sale. And the population is not all cars that were offered for sale because cars are sold in other ways than through www.cars.com—there are other Internet sites that have used cars, and there are many other ways of selling used cars. Cars that are sold by dealers that are not put on Internet sites may be different; they may be older, or less expensive, or more expensive. We do not know since we have no data on those cars. We appear to have a cluster sample (over four places) of cars that are sold through one specific Internet site, www.cars.com. It may even be that the cars that are offered in the summer differ in some systematic way (age, body style, etc.) from those that are offered in the winter. We may be safe in saying that what we have can be regarded as a random cluster sample of cars (again, that is all Audi A4, all BMW 3 Series, etc.) offered for sale in the summer— perhaps just in the summer of 2009. When we interpret our results, we have to take into account what our sample is.

A statistical question and how we answer it One of the variables measured is whether the seller is a dealer or a private seller; we can ask:

If a private seller is selling a car, will the price be lower (or possibly higher) than if a dealer offers the car?

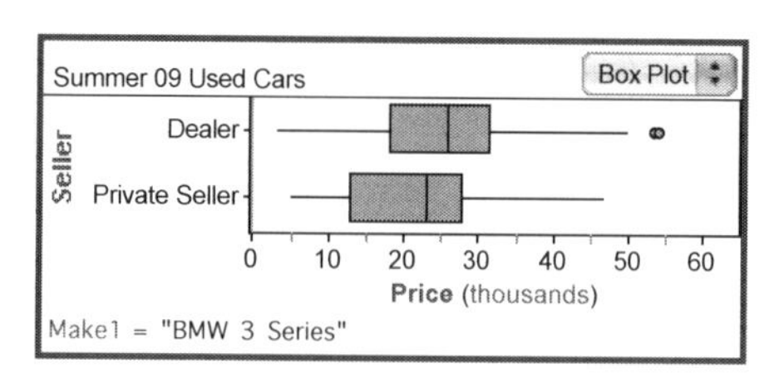

What kind of variables do we have? *Price* is a ***quantitative*** variable, and the variable *Seller* is ***categorical.*** This means that we are comparing a quantitative variable between groups, which is the middle top box in the *Descriptive Region* of the ***Stat Land Map.*** What are the appropriate analyses? For the descriptive analysis, we need a graphic and summary statistics that compare two distributions. Here they are for the BMW 3 Series.

Summer 09 Used Cars

	Seller: Dealer	Seller: Private Seller	Row Summary
Price	25340.30	21177.00	24899.15
	25998.00	22976.50	25991.00
	9729.05	8961.27	9729.83
	13790.00	14994.00	13948.00
	675.00	80.00	755.00

S1 = mean( )
S2 = median( )
S3 = s( )
S4 = iqr( )
S5 = count( )

Make1 = "BMW 3 Series"

It does appear that the cars being sold by private sellers are generally cheaper than the cars being sold by the dealers on www.cars.com. Is this a big enough difference to be beyond sampling variation? If we had random samples, we could use a hypothesis test for ***comparing two independent means***. Since we have a large sample total size ($n = 755$), and since the *Price* distributions are not extremely skewed, the normality condition is not a problem although we must remember that we are not certain that we have a random sample. Here is the output; the *p*-value is very low; what we see is unlikely to be just sampling variability.

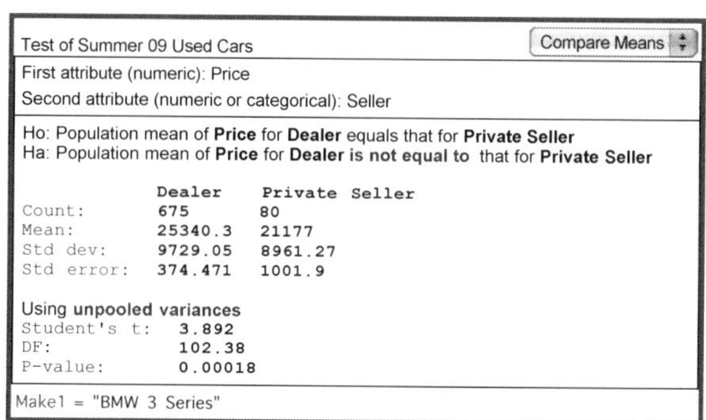

Test of Summer 09 Used Cars — Compare Means

First attribute (numeric): Price
Second attribute (numeric or categorical): Seller

Ho: Population mean of **Price** for **Dealer** equals that for **Private Seller**
Ha: Population mean of **Price** for **Dealer is not equal to** that for **Private Seller**

	Dealer	Private Seller
Count:	675	80
Mean:	25340.3	21177
Std dev:	9729.05	8961.27
Std error:	374.471	1001.9

Using **unpooled variances**
Student's t: 3.892
DF: 102.38
P-value: 0.00018

Make1 = "BMW 3 Series"

However, although we have shown that on average the cars sold by private sellers cost less than those sold by dealers, we could ask: are the cars comparable? Perhaps the private sellers are selling older cars, and for that reason the cars have lower prices.

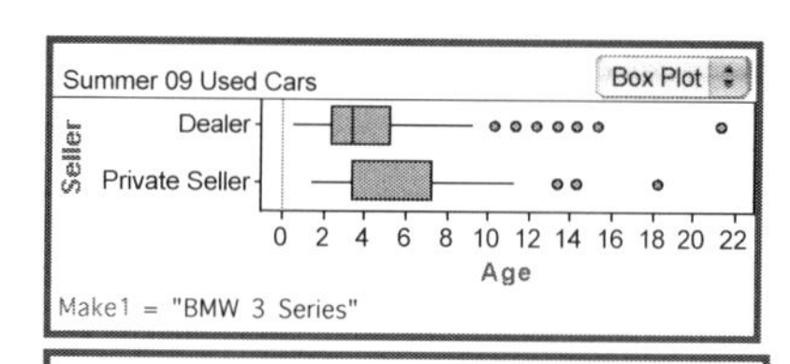

A second statistical question:

If a private seller is selling a car, will the age be older than if a dealer offers the car?

Summer 09 Used Cars

	Seller: Dealer	Seller: Private Seller	Row Summary
Age	3.79	5.12	3.92
	3.33	3.33	3.33
	2.87	3.28	2.94
	3.00	4.00	3.00
	724.00	80.00	804.00

S1 = mean( )
S2 = median( )
S3 = s( )
S4 = iqr( )
S5 = count( )

Make1 = "BMW 3 Series"

Again, we are comparing one quantitative variable (*Age*) between two groups, so we will do the same kind of analysis but with *Age* instead of *Price*. The descriptive analysis here appears to show that it is true that the cars being sold by private sellers are older on average than the cars being sold by dealers, and this may account for the lower prices of the private seller used BMW 3 Series. It may also be that there is a different mix of body styles sold.

A statistical question with two categorical variables

Do the proportions of coupes, sedans, and convertibles differ by the variable Seller?

We should immediately recognize that the variables *Body* (indicating the body style of the car) and *Seller* are both *categorical*. (On the **Stat Land Map** we look at the box nearest the "river" in the *Descriptive Region*.) For descriptive analysis, we need a table showing the number of the various body styles being offered for sale by dealers and by private sellers.

Summer 09 Used Cars

		Body					Row Summary
		Convertible	Coupe	Hatchback	Sedan	Wagon	
Seller	Dealer	55	91	2	554	12	714
	Private Seller	19	11	0	22	2	54
Column Summary		74	102	2	576	14	768

S1 = count( )
make1 = "BMW 3 Series"

Since we are regarding *Seller* as the explanatory variable, we will want to compare the proportion of convertibles (for example) according to whether they are being sold by a dealer or a private seller. So we will calculate $P(Conv \mid Dealer) = \frac{55}{714} \approx 0.077$ compared with $P(Conv \mid Private\ Seller) = \frac{19}{54} \approx 0.352$ as well as $P(Sedan \mid Dealer) = \frac{554}{713} \approx 0.776$ compared with $P(Sedan \mid Private\ Seller) = \frac{22}{143} \approx 0.407$, and we are surprised! Our results tell us that the private sellers are much more likely to be selling convertibles than the dealers and less likely to be selling sedans. The difference is big enough that we suspect that the inferential analysis will be statistically significant. The relevant test here will be a ***chi-square test of independence*** since the categorical variable *Body* has more than two categories. Here is the Software output. Notice that we have done an *Add Filter* to do the analysis only for the coupes, sedans, and convertibles since there were too few wagons and hatchbacks for the conditions for the inferential procedure to be met. Notice also that the chi-square test of independence does not tell us the details of the differences between the dealers and the private sellers; the test just tells us that there *is* a difference and that, whatever difference it is, it cannot be explained by sampling variability. The direction of the differences is surprising because we think that convertibles are generally *more expensive* than sedans, other things being equal. This time, we want to include the variable Age as well as the variable Price in our analysis, so we will look at the *relationship* between price and age for the convertibles and non-convertibles (coupes, sedans, wagons, and hatchbacks).

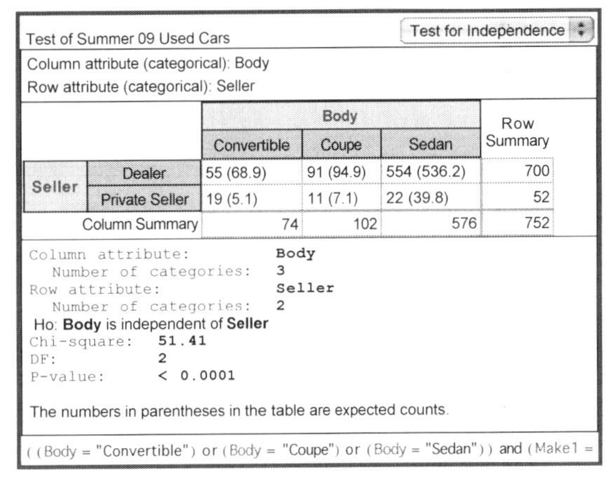

Test of Summer 09 Used Cars — Test for Independence

Column attribute (categorical): Body
Row attribute (categorical): Seller

		Body			Row Summary
		Convertible	Coupe	Sedan	
Seller	Dealer	55 (68.9)	91 (94.9)	554 (536.2)	700
	Private Seller	19 (5.1)	11 (7.1)	22 (39.8)	52
Column Summary		74	102	576	752

Column attribute: Body
Number of categories: 3
Row attribute: Seller
Number of categories: 2
Ho: **Body** is independent of **Seller**
Chi-square: 51.41
DF: 2
P-value: < 0.0001

The numbers in parentheses in the table are expected counts.

((Body = "Convertible") or (Body = "Coupe") or (Body = "Sedan")) and (Make1 =

A statistical question about the relationship between two quantitative variables

Is the relationship between the variables Age and Price different for convertible and non-convertible BMWs?

Since we have *two quantitative variables*, and we are examining the relationship between them, we will look at a scatterplot of the data and see if it is appropriate to fit a linear model to the data. On the **Stat Land Map** we are still in the Descriptive Region (above the "river") and on the right-hand side. The Software output is shown below with two least squares regression lines, one for the convertible BMWs and the other one for all the other body styles—the coupes, sedans, etc.

What we see is that the slopes of the least squares lines for the convertibles and non-convertibles are nearly the same; for each year a BMW ages, the value goes down about 2,900 dollars by our calculations. However, the least squares regression line for the convertibles is always about five thousand dollars above the least squares regression line for the non-convertibles. We also see some scatter in the data (not much, however) and some curviness to the plot, both of which make the linear model not fit well. Still, the *Coefficients of Determination* R^2 (= r^2) show that for the convertibles, 65% of the variability in *Price* is accounted for by the linear model on *Age*, and for the non-convertibles, 78% of the variability in *Price* is accounted for by the linear model on *Age*.

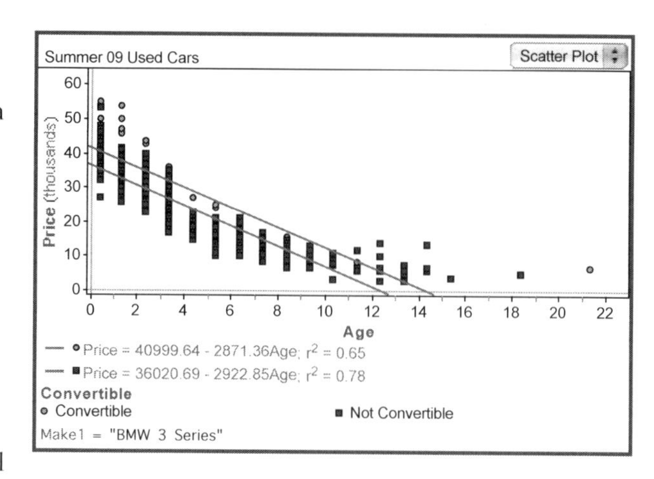

The End of This Story: What is the best analysis? We started out asking whether BMWs sold by private sellers are more (or less) expensive than BMWs sold by dealers. The answer overall, on average, is that the BMWs sold by private sellers are less expensive. We then tried to find out why this may be. It turns out that the private sellers have a tendency to sell cars that are older (and therefore cheaper) than the dealers. However, it also turns out that the private sellers are far more likely to sell convertibles, and convertibles are generally more expensive! However, the best way to answer our question would be to do an analysis of *matched pairs* of cars. We were able to match each of the private seller car with a dealer car of the same place (Boston, Chicago, Dallas, or SF), age, and whether a convertible or not, and chose the car with the nearest miles driven (there were, of course, far more dealer-sold cars than private seller cars). Here is the dot plot showing the difference of price. The variable *DiffPrice* = *Price2 – Price1*, where *Price 2* is the price for the dealer car, and *Price1* is the price for the private seller car. If there were no difference, we would expect a distribution to have a mean of zero. (In the ***Stat Land Map*** we are on the Descriptive Region on the top left-hand box.) Here are the summary statistics. From a descriptive analysis, it does appear that the dealer cars sell for more. The relevant inferential procedure is a paired-comparison test. Shown here is the software output for this test, giving us evidence that a car offered by a dealer is more expensive than a private seller car.

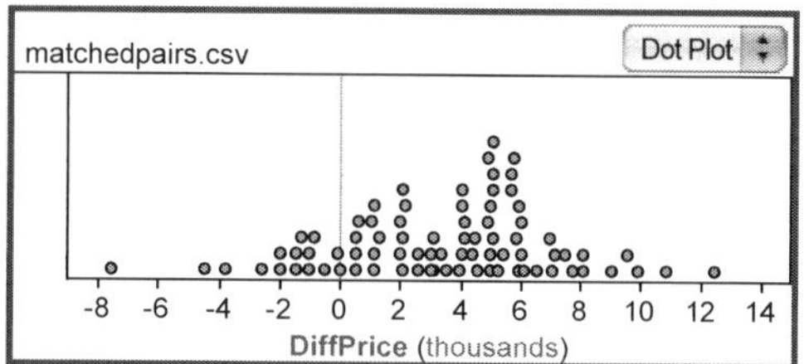

matchedpairs.csv

	Price2	Price1	DiffPrice
	24560.1	21177	3383.14
	10825.5	8961.27	3624.07
	80	80	80

S1 = mean ( )
S2 = s ( )
S3 = count ( )

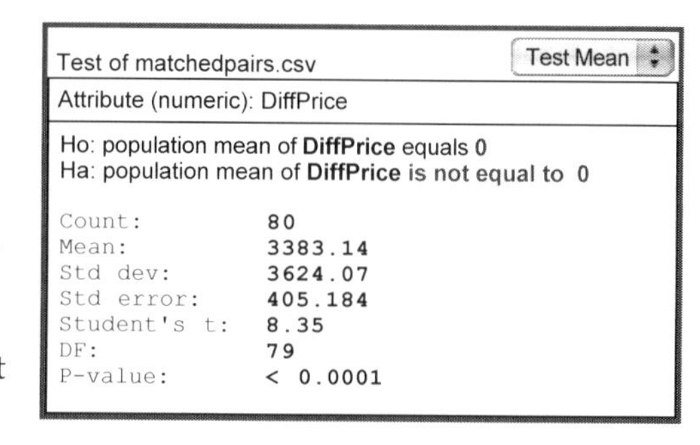
Test of matchedpairs.csv — Test Mean

Attribute (numeric): DiffPrice

Ho: population mean of **DiffPrice** equals 0
Ha: population mean of **DiffPrice** is not equal to 0

Count:	80
Mean:	3383.14
Std dev:	3624.07
Std error:	405.184
Student's t:	8.35
DF:	79
P-value:	< 0.0001

One Collection or Paired Comparison on One Collection

Type	Confidence Interval	Hypothesis Tests	Comments (Mostly Technical Conditions)
One Sample Problem	$CI = \bar{x} \pm t^* \frac{s}{\sqrt{n}}$ Get the t^* using the *t distribution* Table for the row corresponding to the df and the column for the level of significance. $df = n - 1$	1. Hypotheses: $H_0: \mu = \mu_0$ $H_a: \mu \begin{cases} < \mu_0 \\ > \mu_0 \\ \neq \mu_0 \end{cases}$ 3. Test Statistic $t = \frac{\bar{x} - \mu_0}{\frac{s}{\sqrt{n}}}$ 3. Get an estimate of the p value using the t distribution Table with $df = n - 1$	2. The sample must be an SRS (= Sample random sample) from the population. In the population from which the sample is drawn, the variable x must be normal. But see the comments on **robustness** below: Our judgment of the normality of the population can often be made from the sample distribution.
Paired Comparisons	$CI = \bar{x}_{Diff} \pm t^* \frac{s_{Diff}}{\sqrt{n}}$ where $\bar{x}_{Diff} = \bar{x}_1 - \bar{x}_2$, but s_d must be calculated directly from the differences. $s_d \neq s_1 - s_2$ $df = n - 1$	$H_0: \mu_d = 0$ $H_a: \mu_d \begin{cases} < 0 \\ > 0 \\ \neq 0 \end{cases}$ $t = \frac{\bar{x}_{Diff} - \mu_d}{\frac{s_{Diff}}{\sqrt{n}}}$ Get the p value using $df = n - 1$	Distribution of differences is normal in the population; but see comments on robustness below. Our judgment of the normality of the population of *differences* can often be made from the sample distribution of differences.

Robustness:

- By "robust" we mean that the procedures will give reliable results *even if* the one of the technical conditions fails to be met. If the population distribution is not Normal, as it should be under the technical conditions, the t procedures are "robust" if the sample size is sufficiently large. Use the 15/40 Rule to determine whether the sample size is large enough after you look at the shape of the distribution.

  ***n* under 15:** Sample Distribution should be nearly Normal; Otherwise, abandon the procedure.

  ***n* between 15 and 40:** Some skewness allowed, but not extreme skewness or extreme outliers. Consider a transformation for right skewness.

  ***n* over 40:** The Robustness of the *t* procedures can handle skewness for sample sizes greater than 40.

- The t procedures are <u>not</u> robust if we do not have a random sample. Robustness only applies to the Normality Condition.

Two Independent Collections

Type	Confidence Interval	Hypothesis Tests	Comments (Mostly Conditions)
Independent Collections:	$CI = (\bar{x}_1 - \bar{x}_2) \pm t^* \sqrt{\frac{s_1^2}{n_1} + \frac{s_2^2}{n_2}}$ $df = \frac{\left(\frac{s_1^2}{n_1} + \frac{s_2^2}{n_2}\right)^2}{\frac{1}{n_1 - 1}\left(\frac{s_1^2}{n_1}\right)^2 + \frac{1}{n_2 - 1}\left(\frac{s_2^2}{n_2}\right)^2}$	$H_0: \mu_1 - \mu_2 = 0$ $H_a: \mu_1 - \mu_2 \begin{cases} \neq 0 \\ < 0 \\ > 0 \end{cases}$ $t = \frac{(\bar{x}_1 - \bar{x}_2) - 0}{\sqrt{\frac{s_1^2}{n_1} + \frac{s_2^2}{n_2}}}$ df as in CI	The samples must be *independently* drawn and they must be SRS samples from the populations. The population distribution of populations 1 and 2 must each be normally distributed; but see comments on robustness above. Use the guideline for the sum of the sample sizes: $n_1 + n_2$

Two or More Collections:

ANOVA:	**_Hypotheses:_** $H_0 : \mu_1 = \mu_2 = \cdots = \mu_k$ H_a : *At least one of the k means differs from the others* where k indicates the number of different groups in the test.	**_Test Statistic:_** $F = \frac{MSG}{MSE} = \frac{\textit{Variation betweeen the means of the groups}}{\textit{Variation within the groups}}$ where: $MSG = \frac{n_1(\bar{x}_1 - \bar{x})^2 + n_2(\bar{x}_2 - \bar{x})^2 + \cdots + n_k(\bar{x}_k - \bar{x})^2}{k-1}$ and $MSE = \frac{(n_1 - 1)s_1^2 + (n_2 - 1)s_2^2 + \cdots + (n_k - 1)s_k^2}{N-k}$ and where: $\bar{x}$ is the mean for all of the data taken together; $\bar{x}_1, \bar{x}_2, \cdots, \bar{x}_k$ are the means of the 1, 2, . . . k groups; $n_1, n_2, \cdots, n_k$ are the sample sizes in each of the groups, and $N = n_1 + n_2 + \cdots + n_k$ is the total sample size for all the groups.	**_Conditions:_** • The samples must be k ***independent simple random samples***. • The populations from which each of the k samples is drawn must be ***normally distributed***. • The populations from which each of the k samples is drawn must have the ***same standard deviation***. **_Rule-of-Thumb for equal standard deviations:_** • ANOVA has been found to give reliable results if the largest of the k standard deviations is no greater than twice the smallest of the k standard deviations.

Made in the USA
San Bernardino, CA
27 January 2015